WITHDRAWN

ROADSIDE KANSAS

ROADSIDE KANSAS

A Traveler's Guide to Its Geology and Landmarks

REX C. BUCHANAN AND JAMES R. McCAULEY

PHOTOGRAPHS BY JOHN R. CHARLTON

Published for the Kansas Geological Survey by

University Press of Kansas

© 1987 by the University Press of Kansas

Published by the University Press of Kansas (Lawrence, Kansas 66045), which was organized by the Kansas Board of Regents and is operated and funded by Emporia State University, Fort Hays State University, Kansas State University, Pittsburg State University, the University of Kansas, and Wichita State University

Library of Congress Cataloging-in-Publication Data
Buchanan, Rex, 1953–
Roadside Kansas.
Bibliography: p.
Includes index.
1. Geology—Kansas—Guide-books. 2. Natural history—
Kansas—Guide-books. 3. Kansas—Description and travel—
1981–   —Guide-books. I. McCauley, J. R. II. Kansas
Geological Survey. III. Title.
QE113.B83   1987   557.81   87-2013
ISBN 0-7006-0323-9
ISBN 0-7006-0322-0 (pbk.)

Printed in the United States of America
10   9   8   7   6   5   4   3   2   1

TO

Susan Schuette Buchanan
and Vickie McCauley,
both of whom grew up
along U.S. Highway 36 and
have tolerated us for
longer than we deserve

CONTENTS

LIST OF ILLUSTRATIONS

PREFACE

One afternoon, on the way to some field work in central Kansas, we used a guide to follow the geology along I-70. Written shortly after I-70 had opened, the log was twenty years old and was difficult to follow. Landmarks such as roads and bridges had changed. Grass covered some of the rock outcrops; others had simply eroded away, nearly disappearing. Even the geology had changed, though it seemed permanent. Eventually we gave up. Several years later, we published a current guide to the geology of I-70 in the book *Kansas Geology,* and the response convinced us that lots of people were interested in learning more about the rocks, rivers, and features that lined the road.

Thus began *Roadside Kansas.* We picked nine representative highways, and when the opportunity presented itself, we looked for old logs and went out to study the geology ourselves. On the way to meetings or field work—and even on vacations—we drove the highways to check and recheck our results. Over the past three years we have driven through the tabletop flat lands of southwestern Kansas during an ice storm, wandered through the chalk badlands of Logan County in the best heat and wind a Kansas summer had to offer, and dodged traffic and tromped through trash-littered ditches while studying the rocks along the interstate freeways of Kansas City. We have also enjoyed nearly every minute. Though we knew Kansas well at the outset, we learned a great deal.

The result is this book. Along with identification of rock formations, it includes a variety of information about Kansas geology. Some of that information has been known for years. Other information is based on current geological research. Though we tried to include as much information as possible, the book began to get the best of us, requiring that we make difficult choices about the material we included. We generally wrote about things that we thought were important for an understanding of Kansas geology. We included information that might be helpful in understanding the relationship between geology and history. Sometimes we wrote about things that we found interesting; not trivia, really, but the sorts of thing that might liven up a trip. We wrote about things that were little known, things that surprised us. At the same time, we tried to avoid topics that we considered common knowledge. Throughout it all, we stuck to the things that interested us. And we tried to keep a sense of humor.

In the course of writing this book, we had help from a number of sources, many of which are listed in the References section. We also had the aid and support of a number of individuals who read portions of the manuscript and provided other help. In particular, we would like to acknowledge the help of staff members at the Kansas Geological Survey. John Charlton took most of the photographs in this book and printed nearly all of them. He accompanied us on several of the trips and probably learned more about Kansas than he ever wanted to know. Catherine Evans provided several other photos. Don Steeples, Howard O'Connor, and Marla Adkins–Heljeson read portions of the manuscript. Patricia M. Acker drafted many of the line drawings. Finally, the Survey's director, William W. Hambleton, and its associate director, Dean Lebestky, provided the patient support that allowed us to devote the considerable amount of time necessary to complete the book. We also had help from outside the Survey, particularly from the staff at the Kansas Historical Society, Joe Collins at the University of Kansas Museum of Natural History, Dan Merriam at Wichita State University, and many others. In particular, we would like to acknowledge the staff at the University Press of Kansas.

We appreciate the help. As is always the case, however, any errors in the book are the responsibility of the authors. We think the book represents a considerable improvement over the log we used along I-70 five years ago. Whether you're a native Kansan or a traveler, we hope you'll like it.

"There's nothing there." It is not unusual to hear travelers say this in describing some areas of the country, including Kansas. Certainly, parts of Kansas do not possess grand spectacles of nature, bustling cities, or sprawling suburbs. But there is something there. Geology, for one thing. Often it is subtle, something that only the trained eye of a geologist can spot and interpret. But that geology reveals a world that is difficult to imagine, a prehistoric world shaped by powerful, relentless forces. Natural resources are there: water, soil, rocks, minerals, and other treasures that once lured settlers and were later exploited to meet human needs. There are springs, rivers, and hills with little-known names, names that tell interesting stories or reveal something intrinsic to the surrounding countryside. Towns are also there: some thriving, some dying, some evoking only ghostly memories. Each town has a history filled with human dramas worthy of the Broadway stage.

Clearly, there is a lot there; but travelers sometimes must make an effort to find it. This book attempts to explain what's along the state's highways—the geology, natural resources, and landscape of Kansas. However, the history of the state and other aspects of its natural environment are not ignored. All these topics have been researched and keyed to mileposts along nine major Kansas highways, covering more than 2,600 miles. In short, this book tries to fill in the white spaces on the highway map, to show that, at least in Kansas, there is something there.

This book is hardly the first of its kind. Travel books have been a literary mainstay for centuries, and in Kansas they have been popular since settlers moved into the state in the 1800s. Some, such as Zebulon Pike's record of his trek across Kansas in 1806, had a noticeable impact on the state's history and image. Later, more favorable travel books provided information about routes to the Rocky Mountain gold fields or encouraged emigration to Kansas during its territorial days. Most of these books concentrated on the obvious climatic, scenic, and economic features of the state, paying little attention to the geology, which was generally relegated to scientific reports that were either dry as the summer dust or technically incomprehensible to the average reader.

Descriptions of roadside geology began to appear in the twentieth century as people became increasingly mobile. One of the first detailed point-by-point guides to Kansas geology was published by the United States Geological Survey in 1916. Written by the famed geologist Nelson H. Darton, it described the geology along the Santa Fe Railroad route that cut diagonally across Kansas. Later, as travel by automobile became more practical, geologists prepared detailed logs of the geology along highways. Roads provided convenient access to geology, because they often cut through hills, exposing geology that had never before been seen by humans; new, clean road cuts were often easier to study and examine than were any natural outcrops. For geologists, road construction was, and still is, an occasion akin to Christmas.

Many of those road guides, however, were highly technical, describing the geology in terms familiar only to geologists. Usually they concentrated on small, circumscribed areas where the rocks exhibited features that the geologists were interested in. Field guides to the limestones of southeastern Kansas or to the chalks of western Kansas were classics, but they were never particularly helpful to the general public. One collection of road logs appeared in 1963 in Daniel Merriam's *Geologic History of Kansas*. Those logs provided the names of rock formations, streams, and other features for several major routes in Kansas.

More accessible road guides had appeared in the meantime. *Kansas: A Guide to the Sunflower State* (1939) for example, brought together the historic, social, and scenic elements along the state's highways in a detailed guide that has never been equaled. Recently it has been reprinted as *The WPA Guide to 1930s Kansas,* and its popularity continues. That guide, however, did not provide detailed geologic information. Moreover, much of our knowledge about geology has changed since the 1930s. Even the highways have changed. New roads, especially four-lane interstate highways, have been built; old roads have sometimes been rerouted around towns.

In some ways, then, this book aims to fill a gap. It provides a highly detailed guide to the geology and natural features of the entire state, something that has never been attempted before. This should make it valuable to geologists. At the same time, it is written in a nontechnical manner; it includes historic and cultural information that should make it accessible to nearly any traveler, geologist and nongeologist alike. Similar road guides are available for many states, particularly in the West, where mountainous terrain attracts flocks of tourists. Kansas probably lacks such a

The nine Kansas highways covered by road logs in this book

guide because it lacks mountains and oceans and other dramatic natural features. But given a chance, Kansas amply demonstrates a variety of natural features that make the trip worthwhile.

To cover the state, we chose nine highways. We picked some because they carry heavy traffic. Interstate 70 and the Kansas Turnpike, for example, don't necessarily traverse the most interesting geology or scenery in Kansas, but they are used by huge numbers of travelers. We included these and all other four-lane roads, along with five two-lane highways. U.S. Highway 36 in northern Kansas was once a major thoroughfare to Denver; and though traffic on it has decreased, it still remains a major highway in the northern part of the state. We picked other roads because they cut through some of the state's most interesting geology. U.S. Highway 160, in southern Kansas, is the only highway that crosses through both the Red Hills and the Flint Hills, two of the most rugged and colorful parts of the state. U.S. Highway 69 runs along the eastern edge of Kansas, through layer after layer of limestone and past the coal fields and the lead and zinc mines that dominated the area's economy during the early 1900s. Most of U.S. Highway 83, in far western Kansas, crosses the featureless High Plains, but for a few miles it cuts through the chalks of west-central Kansas, making it as geologically interesting as any road in the state. Finally, we picked U.S. Highway 56 partly because of its history—much of this road follows the route of the Santa Fe Trail, constantly passing the sites of old forts and Indian battles—and partly for sentimental reasons. Both of us grew

Santa Fe Trail
Oregon Trail
Smoky Hill Trail-
Butterfield Overland
Dispatch

Chisholm Trail
Parallel Road
Fort Leavenworth-
Fort Scott Military
Road

Coronado's Route
Pony Express Trail
last Indian raid
in Kansas

Trails and historic routes that are followed by today's highways

up along U.S. 56 (though in different parts of the state), so it seemed only appropriate to include it.

In compiling the logs, it was clear that these highways have a great deal in common. Nearly all started out as trails—for Indians, the military, wagon trains, or cattle. They are located where they are for historic reasons, not by chance. The first people to cross the plains generally traveled the path of least resistance, often sticking to hilltop ridges or river bottoms where the ground was less rugged and the travel a little faster. That decision has ramifications today. I-70 seems to go out of its way to avoid interesting geology, some of which can be found only a few miles off the highway. Pioneers looked for even, easy surfaces to cross, and so do highways. A geologically interesting route was probably a pioneer's nightmare, just as a geologically interesting highway creates problems for engineers.

As a result, today's travelers are following roads that have been traveled, in one way or another, for hundreds of years. U.S. 69 in eastern Kansas was an old military road. The Kansas Turnpike, I-35, and I-135 cover various parts of the old Chisholm Trail. I-70 generally follows the old Smoky Hill Trail, which carried pioneers to the Denver gold fields in 1859. In eastern Kansas, U.S. 36 follows the old Pony Express Route and the

Oregon Trail, and in the western end of the state it follows a lesser-known route called the Parallel Trail. And, of course, U.S. 56 follows the old Santa Fe trail.

The roads share other features, mostly in the names of the towns and counties they pass through. Drive Kansas highways for any amount of time, and the similarity of names quickly becomes apparent. Ethnicity seems to be the most obvious source for Kansas place-names, no matter which highway you drive. U.S. 36, for example, passes near settlements such as Scandia and Hanover. I-135 goes near German towns such as Hesston and the Swedish communities of Lindsborg, New Gottland, and Falun. U.S. 69 runs through an array of nationalities, from Frontenac to Capaldo; and I-70 runs past English towns such as Victoria and German–Russian settlements such as Catharine. In southern Kansas, U.S. 160 passes near Runnymede, Oxford, and Cambridge.

The first ethnic group to influence the state's history and place-names was the native American Indian, and in some respects it is a little surprising that Kansas place-names don't reflect more about the state's original inhabitants. Several Kansas counties, such as Comanche, Cheyenne, and Shawnee, are named after Indian tribes, but the Indian names have been largely replaced by European labels, in much the same fashion that settlers removed evidence of previous Indian occupation as the West was settled.

Another source of place-names was the Civil War. For most present-day Kansans, the Civil War is a history book's philosophical discussion about States' rights and slavery. But the reality of the war to contemporary Kansans—even though few battles were actually fought on Kansas soil—is reflected in the names of counties and towns throughout the state. More Kansas counties are named after Civil War soldiers than after any other group, event, or location, providing a constant reminder that the Civil War was the formative event when Kansas became a state in 1861.

In many ways, the roads also share a common geology. All of these highways, at one point or another, run through cross sections of Kansas geology. They all display information about the state's surface and subsurface geology, through dips and hills and streams and draws. Along some routes—through the Flint Hills, say—evidence about the geology is always on display. Long, dramatic road cuts often line the highways, or rim-rock forms natural terraces on the sides of pastures. In western Kansas, where the landscape is just plain flat for long distances, there is evidence of recent erosion and deposition, of a landscape that looks much as it has for the past million years.

While most of these roads were built during the past fifty years, the Kansas landscape has been millions of years in the making. The rocks that form the landscape fall into separate categories, according to the time they were deposited, like chapters in a book. The first chapter of that story isn't even visible at the surface. It is written in rocks formed during the Precambrian Era, mostly igneous and metamorphic rocks that underlie the state in the way that a basement underlies a house. These basement rocks are more than a billion years old, and in Kansas, the Precambrian is covered by layer after layer of sediments. Only the last 350 million year's worth are exposed at the surface, making up the state's landscape. The second chapter, then, is written in the oldest of those sedimentary rocks found at the surface, which appear in southeastern Kansas. These were deposited during the Mississippian Period of geologic history, about 350 million years ago, when a shallow sea covered most of the state. This sea left behind shales and limestones that are buried under more recent sediments across most of the state but that crop out in a tiny portion of Cherokee County. These rocks form the Kansas part of the Ozark Plateau, a region usually associated with Missouri and Arkansas.

More recent rock formations, at the surface in much of the eastern third of the state, make up the third chapter in this story. These are limestones, shales, coals, and sandstones, left behind after the Pennsylvanian Period of geologic history, about 300 million years ago. Again, virtually all of Kansas was covered by a shallow, oscillating sea. As it grew deeper and then shallower, it left behind alternating layers of shale and limestone—generally shale in the shallow parts and limestone in the deeper portions—along with occasional deposits of coal when huge coastal swamps had time to grow, die, decompose, and compact.

The fourth chapter came during the Permian Period of geologic history, about 250 million years ago. For part of the Permian the same pattern of shallow, limestone-depositing seas held true. But this time the oceans left behind different things. In the Flint Hills of east-central Kansas the seas deposited layers of chert along with the limestone, providing an erosion-resistant base for the Flint Hills. A little later in the Permian a huge inland sea covered central Kansas, leaving behind thick layers of salt, gypsum, and other remains of evaporation. Most of these deposits are buried under the center of the state, but in Barber and Clark counties,

Permian-age sandstones, shales, siltstones, and gypsum appear at the surface in an array of geologic formations. Because of the iron oxide in some of the rocks, the Permian has left the area colored a fantastic rust red.

The next chapter was written 135 to 65 million years ago, during the Cretaceous Period. Another series of oceans covered Kansas, but this time, ocean remains were deeper and more widespread. In central Kansas the beaches and deltas of that ocean left behind long stretches of sandstone that today crop out at the surface in reds and browns and yellows. Further west, in north-central Kansas, there is more limestone; even further west are the chalks that were deposited in the deepest part of that ocean. By this point in the geologic history of the state, life had evolved into more varied, more complex forms, and the fossils that it left behind were more exciting. Particularly in the chalks, where leftovers from the oceans include swimming reptiles, fish, sharks, swimming birds, and even flying reptiles.

After the Cretaceous the seas disappeared, and the swimming reptiles and sharks went with it. But the geologic history of Kansas continued. The next chapter was mostly one of deposition—not by seas, but by rivers and wind. By now the Rocky Mountains had appeared on the landscape west of Kansas, and the streams that came out of those mountains were often chock-full of the debris of erosion. The rivers wandered onto the Kansas plains and then dried up, in about the same way they do today. But with millions and millions of years in which to operate, they built up an alluvial plain hundreds of feet deep in western Kansas. Today we call it the High Plains.

Even now the book isn't finished. The last chapter was written only a million years ago—yesterday in geologic terms—when glaciers pushed down from the north and chewed up the northeastern corner of Kansas, leaving clay, silt, and large red quartzite boulders, carried in from Minnesota or South Dakota. Life during this period was teeming and fantastic, producing fossils of huge mammals and birds. When the period ended and the glaciers retreated, winds whipped the recently arrived sediments of dust and sand into huge clouds, which settled over the state, forming vast blankets of loess and dunes of sand. Since then the state has returned to a period of erosion and occasional river-source deposition that we see around us today.

Every highway encounters a different facet of that geologic history. Some pass through rocks of only one period. Others, such as U.S. 36 in northern Kansas, display nearly the entire sequence of Kansas'

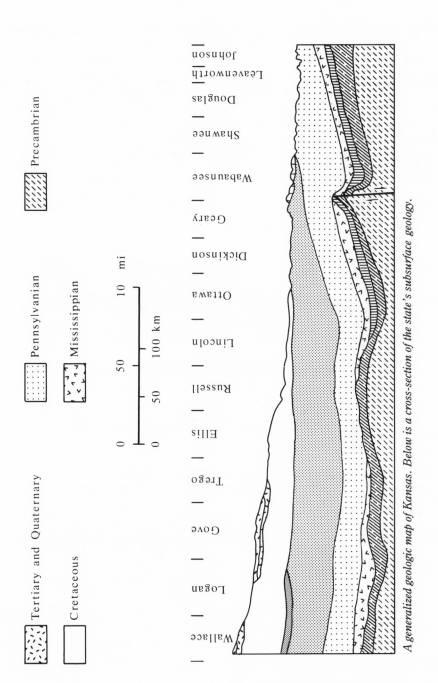

A generalized geologic map of Kansas. Below is a cross-section of the state's subsurface geology.

Tertiary and Quaternary

Cretaceous

Pennsylvanian

Mississippian

Precambrian

| 0 | | 50 | | 10 mi |
| 0 | 50 | | 100 km | |

GEOLOGIC TIMETABLE AND KANSAS ROCK CHART
(Not scaled for geologic time or thickness of deposits)

ERAS	PERIODS	EPOCHS	EST. LENGTH IN YEARS*	TYPE OF ROCK IN KANSAS	MILLION YEARS PAST
CENOZOIC	QUATERNARY	HOLOCENE	10,000 +	Glacial drift; river silt, sand, and gravel; dune sand; wind-blown silt (loess); volcanic ash.	0.010
		PLEISTOCENE	1,990,000		2
	TERTIARY	PLIOCENE	3,000,000	River silt, sand, gravel, fresh-water limestone; volcanic ash; bentonite; diatomaceous marl; opaline sandstone.	5
		MIOCENE	19,000,000		24
		OLIGOCENE	14,000,000		38
		EOCENE	17,000,000		55
		PALEOCENE	8,000,000		63
MESOZOIC	CRETACEOUS		75,000,000	Limestone, chalk, chalky shale, dark shale, varicolored clay, sandstone, conglomerate. Outcropping igneous rock.	138
	JURASSIC		67,000,000	Sandstones and shales, chiefly subsurface. Siltstone, chert, and gypsum.	205
	TRIASSIC		35,000,000		240
PALEOZOIC	PERMIAN		50,000,000	Limestone, shale, evaporites (salt, gypsum, anhydrite), red sandstone; chert, siltstone, dolomite, and red beds.	290
	PENNSYLVANIAN		40,000,000	Alternating marine and nonmarine shale, limestone, sandstone, coal; chert and conglomerate.	~330
	MISSISSIPPIAN		30,000,000	Limestone, shale, dolomite, chert, oölites, sandstone, and siltstone.	360
	DEVONIAN		~50,000,000	Subsurface only. Limestone, predominantly black shale; sandstone.	410
	SILURIAN		25,000,000	Subsurface only. Limestone.	435
	ORDOVICIAN		65,000,000	Subsurface only. Dolomite, sandstone.	500
	CAMBRIAN		70,000,000	Subsurface only. Dolomite, sandstone, limestone, and shale.	~570
	PRECAMBRIAN		1,930,000,000	Subsurface only. Granite, other igneous rocks, and metamorphic rocks.	3,500
			1,100,000,000 +		

KANSAS GEOLOGICAL SURVEY   Eons not shown   *U.S. Geological Survey, Geologic Names Committee, 1980

A geologic timetable showing the periods that make up Kansas geologic history geologic history. In Kansas, U.S. 36 begins and ends in loess hills. It starts in western Kansas, passing young sediments, and then moves east, passing through the sandstones and limestones of the Cretaceous Period, and then into the limestones of the Permian. After a brief encounter with the limestones of the Pennsylvanian, it travels through the glaciated region of northeastern Kansas, providing scenic views of the glacial-deposited loess hills. I-70 follows much the same route, although it takes in more of the Cretaceous sandstones of central Kansas and the Flint Hills to the east.

Another east-west highway, U.S. 160, passes through geology of approximately the same age but provides an entirely different perspective. Again this highway begins out west, in the young sands and silts, passes briefly through rocks of the Cretaceous Period, and moves quickly into the Permian-age Red Hills, where time and erosion have sculpted landscapes unique to Kansas. It then moves back into more recent deposition along the Arkansas River before encountering the finery of the southern extent of the Permian-age Flint Hills. From there it goes into rocks of the Pennsylvanian Period; in southern Kansas, U.S. 160 encounters the sandstone and limestone Chautauqua Hills before moving back into the flatter Pennsylvanian limestones and shales in southeastern Kansas. In some ways, U.S. 160 is the most interesting of these highways, because it is the only one that passes through the Red Hills and the Chautauqua Hills, certainly two of the most underappreciated parts of the state.

East-west highways such as U.S. 160 are bound to take in more geological variety, because the geology of the state has been deposited in layers that are exposed in north-south lines that pass across Kansas like stripes on a zebra. Highways that run north and south, for example, tend to encounter only one or two stripes, while the east-west roads cut across several. U.S. 83, in western Kansas, is a good example. It begins in the young geology of southwestern Kansas; it only briefly passes through the older Cretaceous chalks in the west-central part of the state before again returning to the more recent loess, sands, and Ogallala Formation of northwestern Kansas. That one encounter with the chalk, however, qualifies U.S. 83 as a geologically important route. It is the only highway of this group to cut through the heart of chalk country, although several others come close. It makes that encounter because the Smoky Hill River has eroded away the more recent formations in Gove and Logan counties, exposing the chalk for travelers to view along the highway and on backroads.

A generalized physiographic map of Kansas, showing the major regions of the state

HIGH PLAINS

SMOKY HILLS

ARKANSAS RIVER LOWLANDS

WELLINGTON-McPHERSON LOWLANDS

RED HILLS

GLACIATED REGION

OSAGE CUESTAS

FLINT HILLS UPLANDS

OZARK PLATEAU

CHEROKEE LOWLANDS

CHAUTAUQUA HILLS

0   50   100 miles

0   75   150 kilometers

CHEYENNE   RAWLINS   DECATUR   NORTON   PHILLIPS   SMITH   JEWELL   REPUBLIC   WASHINGTON   MARSHALL   NEMAHA   BROWN   DONIPHAN

SHERMAN   THOMAS   SHERIDAN   GRAHAM   ROOKS   OSBORNE   MITCHELL   CLOUD   CLAY   RILEY   POTTAWATOMIE   JACKSON   ATCHISON   JEFFERSON   LEAVEN-   WYAN-
                                                                                                                                                WORTH   DOTTE

WALLACE   LOGAN   GOVE   TREGO   ELLIS   RUSSELL   LINCOLN   OTTAWA   DICKINSON   GEARY   WABAUNSEE   SHAWNEE   DOUGLAS   JOHNSON

GREELEY   WICHITA   SCOTT   LANE   NESS   RUSH   BARTON   ELLSWORTH   SALINE   MORRIS   OSAGE   FRANKLIN   MIAMI

HAMILTON   KEARNY   FINNEY   HODGEMAN   PAWNEE   STAFFORD   RICE   McPHERSON   MARION   CHASE   LYON   COFFEY   ANDERSON   LINN

                     GRAY   FORD   EDWARDS   PRATT   RENO   HARVEY   GREENWOOD   WOODSON   ALLEN   BOURBON

STANTON   GRANT   HASKELL   KIOWA   SEDGWICK   BUTLER   ELK   WILSON   NEOSHO   CRAWFORD

MORTON   STEVENS   SEWARD   MEADE   CLARK   COMANCHE   BARBER   HARPER   SUMNER   COWLEY   CHAU-   MONT-   LABETTE   CHEROKEE
                                                                                              TAUQUA   GOMERY

Other north-south highways encounter about the same amount of variety. I-135, which is combined with U.S. 81 in this book, begins in Wichita in the young sediments from the Arkansas River and quickly passes Cretaceous-age sandstones and limestones before once again reaching into young, unconsolidated sediments. That route, however, takes the highway tantalizingly close to some really interesting geology. Visible from the road south of Salina are the Smoky Hill Buttes, Coronado Heights, and other remnants of Dakota Formation sandstone. The road passes through only a few outcrops, near Minneapolis, but the sandstone is visible in the hills along much of the route.

U.S. 69 in eastern Kansas is another north-south road that encounters relatively little variation in geology. U.S. 69 crosses over Pennsylvanian limestones and shales of eastern Kansas, with one layer of limestone seemingly being replaced by another that looks much the same. But human interaction with the geology makes this an interesting path: the road is lined by pits, where coal was stripped away; by sinkholes, where old underground lead and zinc mines have collapsed; by huge piles of chat, hauled up from deep underground mines; and by reminders of the ethnic influence of the people who came to work in those mines.

Some of these roads are neither north-south or east-west, but a combination of the two. The turnpike and adjoining I-35 begin as a north-south road in Oklahoma, take a right turn at Wichita, and angle up across the state. These two roads part company at Emporia but join up again, in effect, in Kansas City. Along that route, the roads begin in the young geology of the Arkansas River; later they provide long, leisurely views of the Flint Hills in east-central Kansas. Then they both enter the Pennsylvanian. About the only difference is that the turnpike encounters several locations where glacial debris is visible, whereas I-35 cuts through sandstone hills to the south of Ottawa.

Finally there is U.S. 56, which cuts a giant angled swath across the center of Kansas. It begins out west on the High Plains; then it joins the Arkansas River, passing through geologically young river-deposited material. Along much of the path, sandstone hills of the Cretaceous are visible along the horizon, cropping out in a place or two, such as Pawnee Rock. In central Kansas, U.S. 56 passes through the Flint Hills and then into the Osage Cuestas and sandstones of eastern Kansas before terminating in Kansas City.

Thus, each highway displays its own personality and has its own

Major geologic structures that underlie Kansas

characteristics in terms of the geology it encounters. Some, such as I–135 and U.S. 36, pass through only a few road cuts, and the geology is more evident in features that lie a few miles from the road. Others, such as U.S. 160, take in a progression of constantly changing landscapes, from one geological type to the next. Some roads, such as U.S. 69 and I–35, are identified with a particular type of geology. In the case of I–35, it's the Flint Hills. Along U.S. 69, it is the unending progression of limestones and shales from the Pennsylvanian.

Still, all of these roads encounter geology in one form or another. To understand that geology, it is necessary to understand several geological concepts. To begin with, geologists think in terms of long stretches of time. Millions of years are but a blink. Because of that long-term perspective, geologists imagine things that seem impossible. A few years of erosion may disturb the landscape only slightly, but millions of years can wear down mountains.

To help them talk about such expanses of time, geologists break geologic history into several eras or periods, each having a different length and including rocks deposited during a specific time. In Kansas, those periods represented at the surface include the Mississippian, Pennsylvanian, Permian, Cretaceous, Tertiary, and Quaternary. Those are units of geologic time, but the rocks deposited during those times are generally referred to according to that age. Rocks deposited during the Pennsylvanian Period, for example, are referred to as Pennsylvanian rocks. That distinction, unfamiliar as it may be to some readers, is hardly specific enough,

River drainage basins of Kansas

because a single period can include many different layers of rocks. The Pennsylvanian Period, for example, runs the span from 330 to 290 million years ago, and a rock deposited at the beginning of that period may be entirely different from a rock deposited later on.

To identify different layers of rock, geologists have created several different categories. The most common of these categories are called groups, formations, and members; geologists use these categories to classify rocks in much the way that biologists use families, species, and genera to describe animals. The largest of these categories of rock layers is called a group. Four or five groups, for example, may include all of the rocks deposited during a specific period. The Chase Group, which appears at the surface in the Flint Hills, includes several different rock layers deposited during part of the Permian Period.

Generally, geologists speak of individual rock layers, not as groups, but as formations, an even more specific designation. The Oread Formation, common in eastern Kansas, includes several rock layers that are 50 to 100 feet thick. A familiar formation in central Kansas is the Dakota. A formation is not limited solely to one type of rock. The Dakota is probably most closely identified with sandstone, but the formation also includes layers of clay and shale, rocks that actually make up a far larger percentage of the Dakota than sandstone does. Thus, a formation may encompass a variety of rock types. That is true even when the name of the formation makes it sound untrue. The Oread Limestone Formation includes shales as well as limestone, even though the name indicates otherwise. Suffice it to say that a rock formation simply keeps its same name no matter where it is found. The Lawrence Formation crops out throughout eastern Kansas, but it retains its name regardless of where it is exposed.

While formation names are the most common parlance, geologists break rock layers down even further into members. Members make up formations in the same way that several species of birds, say, make up a family. Thus, the nomenclature takes on the greatest level of specificity with members. Members make up a formation. Formations make up a group. And the rocks of different groups are included under one system, according to the age of deposition. For example, the Oskaloosa Shale Member is part of the Deer Creek Limestone Formation, which is part of the Shawnee Group, which was deposited during the Pennsylvanian Period. Exceptions and variations creep into this theme, so that it becomes even more difficult for the uninitiated to follow the discussion.

However, it isn't necessary to understand the intricacies of stratigraphic nomenclature to understand that each rock layer has its own name. The following guides provide that information, sometimes according to formation, sometimes according to member, and sometimes according to group. In general the logs provide the most specific information available, although at times it can be difficult to determine the exact identification of a rock layer.

Using these guides, then, is not difficult, although a little practice helps. First, begin by realizing that the numbers on the left-hand side of the page correspond with milepost markers along the highways. All federal highways in Kansas are marked with small green signs that occur every mile, primarily as points of reference for construction workers and drivers. Along Interstate highways, these signs are on the right-hand side of the road. Along federal highways, the side varies according to the direction in which you are traveling. At any rate, the features described in this book are located according to these mileposts.

In Kansas the mileposts are numbered along highways from south to north and from west to east. That is, on an east-west highway, such as U.S. 36, the numbers begin at the west, with 0 at the Colorado border, and grow progressively higher to the east. Similarly on the north-south roads, such as U.S. 83, the numbers begin with 0 at the Oklahoma border and then grow larger to the north. That presents no problems as long as you travel from west to east or from south to north. The road guides read in the same direction as you are traveling. If, however, you are going in the other direction, the guide is usable, but you must start at the back (or wherever you join the highway) and then read up the page. While this may seem inconvenient, we assume that people who travel in one direction must eventually return, and at that point the guide is probably easier to use.

Other things should be kept in mind. First, much of our information is presented in fractions of a mile, for the sake of precision. These fractional locations are easy to find by using a car odometer. The feature at 358.3 on I-70, for example, is the Kansas River at Topeka. If you are driving from west to east, clock 0.3 of a mile east of milepost 358 to find the feature. Headed west, you must subtract 358.3 from 359.0 (the last milepost you will see before the river), and then clock 0.7 of a mile from 359. In most cases, the features at the fractional points are obvious even without using an odometer, although an odometer is often necessary in order to pinpoint a specific rock formation. Remember that all the fractional mileage numbers are approximate and that every car's odometer measures the miles a little differently. Also keep in mind that even the most permanent geologic features change over time. Some rock layers may be visible during the winter, when there is less covering vegetation, but not in the summer. Other features simply erode away and are no longer as visible as when this book was written, itself a lesson in the nature of change.

Once you become accustomed to the logs, you'll see that they have a variety of uses. For instance, you need not follow a highway an entire distance to make use of them. If you join I-70 at Salina, for example, you can find that entry in the log and begin following the features as they appear. The log also provides information on the distance from one feature to another. For instance, if you are eastbound at Abilene and want to find how far it is to the next rest area (which we have included with the geologic and historical features), you need only subtract 275 (the milepost at Abilene) from 294 (the milepost at the rest area near Goose Creek) to see that the rest area is 19 miles away.

Finally, a couple of warnings. We do not recommend that you stop to collect rocks or fossils at any outcrop along these highways. It is illegal to stop along Interstate highways for anything but an emergency. Some of the two-lane highways carry nearly as much traffic and have even less shoulder space, making them equally dangerous to stop along as an Interstate. We recommend that if you want to examine an outcrop, exit from the highway at a county road and follow the outcrop to the nearest road cut where you can stop and look without endangering yourself or traffic. Also keep in mind that many geologic features are on private property and that you must have the landowner's permission before entering. Most landowners are happy to allow you onto their property, and they deserve to be treated with courtesy.

Used properly, these logs should illuminate the landscape. That, at least in part, is the purpose of this book: to provide the information about the Kansas landscape as it appears along several major highways; to name and describe the creeks and hills and rock layers; to provide a few details about their formation; to help readers learn the kind of rock they're looking at—be it limestone or shale or sandstone—and to tell them the name; to understand how the rock was deposited and when, and the kinds of fossils or minerals that might be included; to describe the subsurface geology that drivers pass above and never see. At the same time, we hope to provide a little information about the state's history, its inhabitants, and their interaction with the environment. And to show that there is plenty there.

160·U.S. HIGHWAY·160

From the High Plains to the Coal Fields

0.0   *The Kansas/Colorado state line.* * Westbound U.S. 160 runs through southern Colorado and into the Four Corners area, where Arizona, Utah, Colorado, and New Mexico come together. To the east it extends 480 miles along the southern edge of Kansas. Kansas is 411 miles from west to east, but the distance across the state on U.S. 160 is significantly longer because the highway takes a meandering route, particularly in some of the more scenic areas such as the Red Hills in southwestern Kansas and the Flint Hills in south-central Kansas.

In extreme southwestern Kansas, however, the land is tabletop flat, and topographic relief is rare. The High Plains of western Kansas were deposited in the past few million years by streams that worked their way out of the Rocky Mountains, carrying loads of sand, gravel, silt, and other debris. As those streams dried up, they deposited a vast alluvial sheet that covers much of the western third of the state. In addition, this region was uplifted and tilted toward the east when the Rocky Mountains were raised up, accounting for much of the additional elevation in southwestern Kansas. The elevation here is 3,665 feet, the highest point along U.S. 160 in Kansas, nearly 3,000 feet above its low point, near Independence in eastern Kansas. Between here and Johnson, the land slopes to the east at about 20 feet per mile.

*The most prominent or noticeable feature at each entry in this guide is in italics.

The Coolidge sinkhole, about 27 miles north of U.S. 160 in Hamilton County, opened up in 1929, probably the result of a collapsing cavern in underlying rocks.

0.1 *Saunders.* The Kansas/Colorado border, west of here, marks the division between the mountain and central time zones. The two Kansas counties immediately to the north, Hamilton and Kearny, are in the mountain time zone.

1.5 About 27 miles north is the *Coolidge sinkhole,* named for a small town in Hamilton County. This sinkhole appeared suddenly along a county road on 18 December 1929, and by the following summer it was 104 feet in diameter and 68 feet deep. Sinkholes are generally associated with limestone terrains in humid areas. While Cretaceous shales and chalky limestones crop out near the Coolidge sinkhole, the climate is hardly humid. Other limestones exist just below the surface, and geologists have suggested that a collapsing cavern in the Greenhorn Limestone may have caused the sinkhole.

The Coolidge sink is one of many topographic depressions and ancient sinkholes that line up along Bear Creek and the North Fork of Bear Creek. The Bear Creek fault also occurs along this trend, cutting a bow-shaped path through southwestern Hamilton County, dipping into northern Stanton and Grant counties, then swinging northeast into Kearny County. This fault marks the southern edge of the Syracuse Uplift, a broad east-west area where the land has been raised up.

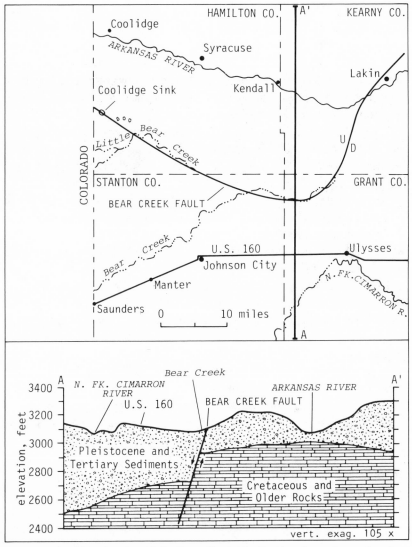

The map shows the location of the Bear Creek Fault. Also shown is a subsurface depiction of the fault.

The Bear Creek fault may also provide an avenue for ground water and occasional flows in Bear Creek to reach highly soluble rocks underground. The subsurface here includes evaporites—Permian gypsum and salt that underlie much of southwestern Kansas—that can be easily dissolved by ground water. Geologists theorize that water could move down and dissolve out cavities in those evaporites, which in turn could collapse and create sinkholes at the surface.

2.5 Along Bear Creek, 2.5 miles north of here, and an equal distance to the south, there are outcrops of Cretaceous *sandstone from the Dakota Formation.** These isolated outcrops, as well as a few others along the Bear Creek fault in southern Hamilton County, are more than 100 miles from the nearest large exposure of the Dakota, which is in Hodgeman County north of Dodge City. Dakota sandstone, usually associated with outcrops in central Kansas, was probably brought to the surface here on the eastern flank of a broad uplift called the Las Animas Uplift, which is centered in southeastern Colorado.

4.3 The road passes through a small area of *sand hills*. Nearly all of southwestern Kansas is covered by a blanket of loess, a wind-deposited silt of Pleistocene age. Silt and sand are blown into numerous tracts of sand dunes, which make up the only recognizable ''hills'' along U.S. 160 in this part of Kansas.

Three miles south is Sand Arroyo Creek. *Arroyo* is the Spanish word for an intermittent stream, which accurately describes many of the rivers and streams in southwestern Kansas. The average annual precipitation here is less than 16 inches, and most of that soaks into the ground rather than running off. In addition, heavy irrigation during the past few decades has dramatically lowered water tables in the area, so that aquifers no longer contribute water to many rivers and creeks during times of low stream flow. As a result, many streams here have all but dried up, carrying water only during heavy thunderstorms.

8.9 The elevation is 3,500 feet. Though the landscape is flat,

---

*This guide follows, insofar as possible, the conventions of capitalization regarding stratigraphic nomenclature. That is, the first letters of formation names are shown in upper case. Zeandale Limestone, for example, is in upper case because it is a formation name. Member names are shown in upper case only when the complete name is given; otherwise, member names are shown in lower case. The Wamego Shale Member of the Zeandale Limestone is shown in upper case because it is a complete proper name. However, when only part of the member's name is shown, the second part of the name is in lower case, as in Wamego shale.

relief can be found three miles north, where Bear Creek parallels the highway and cuts into the High Plains.

9.5   *Manter*. This town was named after an official of the Santa Fe Railroad, which runs through the town and along the south side of the highway. Although Manter's population was only 205 in 1980, it was the second-largest town in Stanton County, behind only Johnson. The total population of the county is 2,339, which makes it the third-least-populated county in Kansas.

15.3   Small area of *dune sand*.

16.6   Junction with *K–27* and the west city limits of *Johnson*. K–27 runs 22 miles south to Richfield, then on to Elkhart and the Oklahoma border.

17.7   Junction with *U.S. 270* and *K–27*. K–27 runs 28 miles north to Syracuse, then on through the western row of Kansas counties before terminating south of Haigler, Nebraska.

18.4   *Johnson's* business district to the south. Johnson County, in northeastern Kansas, is named after the Reverend Thomas Johnson, who established a mission for Shawnee Indians in present-day Kansas City. Johnson City is named after Alexander Johnson, one of Thomas Johnson's twelve children. Alexander was a lieutenant in the Kansas militia during the Civil War, but the town is probably named after him because he was later the land commissioner for the Santa Fe Railroad.

18.7   East city limits of *Johnson*, the seat of Stanton County. The county courthouse, completed in 1926, is a brick building at the north end of the business district.

21.7   A *feed lot* is visible 0.5 miles to the south. With the coming of irrigation to the High Plains, farmers can produce crops, such as corn and alfalfa, that support extensive cattle-feeding operations. This is one of many feed lots that have sprung up in the area. Packing plants, in turn, have moved to southwestern Kansas to handle the finished beef.

This is also the approximate western edge of the *Hugoton Gas Area*. In terms of proven reserves, this is the largest natural-gas field in the United States; it underlies the eastern half of Stanton County and all or parts of Finney, Grant, Hamilton, Haskell, Kearny, Morton, Scott, Seward, and Stevens counties in southwestern Kansas, as well as parts of Oklahoma and Texas. The field was discovered in Seward County in 1922, but the richest part of the Hugoton is under Stevens County, near the town that supplied the field's name. Since the field was discovered, Stevens County alone has produced over 5 trillion cubic feet of gas from the Hugoton; it continues to lead the state in natural gas production, pumping over 92 billion cubic feet in 1984. Stanton County, on the western edge of the Hugoton, was the eighth-largest gas-producing county in Kansas during 1984, pumping over 22 billion cubic feet. Almost two-thirds of that production came from the Hugoton, and wells into the field are scattered along both sides of the road from here to northern Seward County.

The Hugoton gas field is located in a broad geologic basin called the Hugoton Embayment, a bowl-shaped depression that was buried under layers of younger sediments. The Hugoton Embayment is the northern extension of a much deeper basin in Oklahoma and the Texas Panhandle, which is called the Anadarko Basin. The Hugoton Embayment is bounded by the Las Animas Arch on the west and the Central Kansas Uplift and Pratt Anticline on the east. On the south, this embayment plunges into the Anadarko Basin. In southwestern Kansas, Precambrian igneous and metamorphic formations are buried 9,500 feet underground, deeper than anywhere else in the state. As a result, some of the deepest oil and gas wells in Kansas history have been drilled in the sediments that cover the Precambrian in this corner of the state.

24.2 The highway rises slightly here as it passes over a small area of *dune sand.*

28.7 The town of *Big Bow,* a mile south, is named after a Kiowa chief. The Kiowas were one of several nomadic Plains Indian tribes that roamed southwestern Kansas during the 1700s and 1800s. They originally lived west of the Black Hills, but they drifted south over the years. In some ways the Kiowas fit the stereotype of hard-riding, buffalo-hunting Plains Indians. The Kiowas roamed as far east as the Medicine Lodge area, where in 1867 they signed a treaty agreeing to move out of Kansas into a 3-million-acre reservation between the Red and Washita rivers in western Oklahoma.

30.7 *Stanton/Grant county line.* Stanton County is named after Edwin M. Stanton, secretary of war under Abraham Lincoln. After Lincoln's assassination, Stanton was fired by Andrew Johnson, who replaced him with Ulysses S. Grant, who later became president and supplied the name for Grant County.

31.5–33.0 During this 1.5 miles, the highway passes through a *large elliptical surface depression,* thousands of which dot the High Plains from Nebraska south into Texas and New Mexico. Geologists call them High Plains depressions, although they are locally referred to as lagoons or buffalo wallows. In fact, buffalo may have caused these depressions. Buffalo like to wallow in mud, and areas of standing water may have attracted large herds of bison when they roamed freely over the plains. The congregation of hundreds of these hooved animals could have killed the short grass sod, leaving the silty soil unprotected in dry weather. The unrelenting wind could have finished the excavation of a depression by blowing away the loose soil. However, buffalo are just one hypothetical source of these depressions, which may also be created by the dissolution of underlying rocks or the downward sifting of the silty soil.

32.7   A mile south there is a railroad siding, all that remains of a town called *Stano*. Nearby is a plant that produces gasoline from natural gas, rather than refining it from petroleum. By chilling the natural gas as it comes out of the ground, certain chemical components are given off, including propane, butane, gasoline, and other liquids. Natural gas, which is composed primarily of methane, may contain smaller amounts of ethane, propane, isobutane, nitrogen, hydrogen, helium, and other gases. Straight from the ground, natural gas is generally odorless; the odor is added later to aid in detecting leaks.

33.0   A *compressor station* is on the north side of the highway. Natural gas from the Hugoton field is transported away from southwestern Kansas via pipelines; such stations are used to develop pressure so that the gas can be forced through the pipeline.

34.7   The town of *Sullivans Track* is 0.8 miles south.

39.0–40.4   *Ulysses,* the seat of Grant County, has a population of 4,653, making it the largest town in the four southwesternmost counties in Kansas. Much of southwestern Kansas was not settled until the late 1800s. Some farmers who moved onto the High Plains believed that cultivating the land would increase rainfall—"the plow will bring the rain," they said— making it possible to farm areas that were too dry before their arrival. But in the 1890s, bad weather and harsh economic conditions forced many farmers to abandon their claims and leave western Kansas. The population of Grant County is now nearly 7,000. Between 1887 and 1900 the county's population dropped from 2,716 to 422, and the population of Ulysses fell from 1,500 to 40. At the same time, the city's bonded indebtedness grew to $84,000. To escape financial disaster, the entire city, except for the schoolhouse and the courthouse, was moved two miles northwest, and the city of New Ulysses was formed at the site of present-day Ulysses.

39.6   Junction with *U.S. 270* and *K-25*. K-25 runs south to Hugoton and north to Lakin. About 12 miles south, along the Cimarron River, is Wagon Bed Springs, a noted stopping point along the Santa Fe Trail. Wagons on the trail often followed the Arkansas River until a point in

The bed of the Cimarron River near Wagon Bed Springs, south of Ulysses, Grant County

Gray County to the northeast. Then they cut away from the Arkansas, southwest toward the Cimarron. The 60-mile stretch between these two rivers was called the Dry Route, or La Jornada, because little water was found between the rivers during times of drought. Thus, Wagon Bed Springs became a particularly important stopping point. Lack of water wasn't the only danger; pioneers also encountered Plains Indians, particularly Comanches and Kiowas. Mountain man Jedediah Smith was killed near Wagon Bed Springs in 1831 by Comanches after he had lost his way and wandered four days without water.

The spring got its name from a wagon box that was set in the water and used as a tank. The wagon box is no longer visible, and the spring has long since dried up, probably the result of a lower water table caused by pumping for irrigation. Grant County and much of southwestern Kansas is underlain by thick deposits of sand, silt, and gravel that contain ground water and are known collectively as the High Plains Aquifer. This aquifer is composed of the Ogallala Formation, an important water source in much of western Kansas, as well as younger Pleistocene deposits that cover the Ogallala in southwestern Kansas. These Pleistocene deposits are similar to those of the Ogallala, and the entire sequence, which is up to 700 feet thick, is often treated as a single aquifer and is tapped for irrigation in much of southwestern Kansas.

Irrigation has caused ground-water levels to decline, particularly in Grant County, where the water table dropped more than 60 feet from 1966 to 1980. Still, all the wells here are not on the verge of going dry. Measurements show that the High Plains Aquifer maintains an average saturated thickness of 220 feet in Grant County, which means that an average of 220 feet of sand and gravel in the aquifer are still saturated with water. But the declines have been enough to dry up springs, reduce the flow in many creeks and rivers, and make some of the shallower irrigation wells go dry. Even when wells don't go dry, lower water levels mean that the water must be pumped farther to reach the surface, which requires additional energy. As energy prices climbed and water levels dropped during the 1970s, irrigators became more conscious of their water consumption and began to take measures to conserve the remaining ground water.

41.6 *Lakin Draw* is the only named stream crossed by U.S. 160 between the Colorado border and Spring Creek, west of Meade, an indication of the scarcity of surface water in southwestern Kansas. Bear Creek, which begins 50 miles west of the Kansas/Colorado border, disappears on the plains of northern Grant County. During rare periods of flooding, Bear Creek sometimes overflows into Lakin Draw, sending water south toward Ulysses. Lakin Draw joins the North Fork of the Cimarron River a short distance south of the highway.

Low trees with feathery leaves grow along Lakin Draw on both sides of the highway. These are tamarisks, or salt cedars, which are native to the Mediterranean region and were introduced to the United States in Texas in the late 1800s. They quickly spread up river systems. Today they are found in many riverbeds of the American southwest, including western Kansas. Tamarisk trees thrive in arid and semiarid conditions; they also exhibit a high tolerance to saline or alkali conditions—hence the name salt cedar.

Tamarisks are in a group of plants called phreatophytes (pronounced free-at'-oh'-fights) that includes willows, cottonwoods, and alfalfa. The name comes from the Greek; it means "well plants," because phreatophytes obtain water by sinking deep roots and behaving like a water well, continually pumping ground water from the water table to the surface, where it is transpired into the atmosphere. Tamarisks can send their roots to considerable depths in search of water; for example, during the construction of the Suez Canal, excavations encountered tamarisk roots at depths of nearly 100 feet.

Tamarisks have few beneficial uses, and since they clog stream beds, they are considered a nuisance. However, in water-short areas such as southwestern Kansas, tamarisks can be more than a nuisance; they can also deplete the water supply.

47.2   The town of *Hickok*.

49.1   About 1.5 miles south are a railroad siding called *Columbian Track* and a nearby carbon-black plant. Carbon black is manufactured by the incomplete combustion of natural gas. It is used for manufacturing rubber (especially automobile tires), printer's ink, typewriter ribbons, and paint. Gas for the plant comes from the Hugoton gas field, which is the source of gas for most of the wells in this area. That gas is trapped in Permian rock formations called the Chase Group, the same limestones that are exposed at the surface in the Flint Hills of eastern Kansas. In southwestern Kansas they are between 2,200 and 2,600 feet underground. Geologists believe that much of the gas in the Hugoton was originally formed in the Anadarko Basin south of here, then migrated north until it was trapped in these limestones.

52.8   *Compressor station* south of the highway.

53.0   *Helium plant* south of the highway. Natural gas from the Hugoton field contains just under 1 percent helium, which makes it helium-rich. Helium, refined out of the gas at this plant, is used in the spacecraft industry for rocket fuel and for helping to achieve low temperatures in cryogenic research. Helium is produced in several Kansas plants. It was first discovered in natural gas in a well about 20 miles southeast of Winfield.

53.9   *K–190* runs five miles south to the town of Ryus and 12 miles southeast to the town of Satanta, which was named after a Kiowa chief. The name means "white bear." Satanta, an Indian spokesmen at the Medicine Lodge conference in 1867, was later captured by Gen. George Armstrong Custer and imprisoned in Texas, where he committed suicide.

Along a county road nine miles south of here is a deposit of volcanic ash. Ash is composed of tiny pieces of congealed lava, blown out of volcanoes during eruptions, such as the one at Mount St. Helens in Washington State in 1980. The deposits here are much older. They came from volcanoes west of Kansas, which erupted, throwing tiny ash particles into the air. That ash—most of which is from volcanoes in New Mexico, Wyoming, and California—was blown over Kansas by prevailing winds. In places it blanketed the ground and was then washed into streams and ponds, where it formed deposits up to 20 feet thick that occasionally contain fossils of snails and other aquatic life.

Beginning in the early 1900s, ash was mined and put to a variety of uses. The most common was as a cleanser; under a microscope, ash looks like tiny pieces of glass, and its abrasive powers were good for cleaning sinks, for example. Later the ash was used as a filter material, in ceramic glazes, and in making concrete. Perhaps its major use in southwestern Kansas was as a dressing material over the tops of asphalt roads.

54.9 *Grant/Haskell county line.* Haskell County is arguably the flattest county in Kansas. The elevation of U.S. 160 as it enters Haskell County is 3,013 feet; where it exits from southern Haskell County the elevation is 2,898, a difference of slightly more than 100 feet.

55.7 *High Plains depression* south of the highway. This depression doubles as a tailwater pit for irrigation. Some of the water applied to fields during irrigation runs off, rather than soaking in. In an attempt to conserve water, irrigators capture that excess water in features called tailwater pits. Usually those pits are dug exclusively for that purpose, but in this case the irrigator is using a natural depression in the ground to capture water, which is pumped out of the tailwater pit and used again in irrigation.

56.5–60.0 The wells in this area are part of the *Eubank oil and gas field*. Discovered in 1958, the Eubank has produced over 11 million barrels of oil and 38 billion cubic feet of natural gas from Pennsylvanian and Mississippian rocks that are between 4,100 and 5,400 feet deep. Like other southwestern Kansas counties, Haskell is a major producer of natural gas; it was responsible for 28 billion cubic feet of gas in 1984. It also produced 787,000 barrels of oil.

66.0 *Natural-gas compressor station.*

66.9 *Feed lot* north of the highway.

67.0 Junction with *U.S. 83* and *K–144.* U.S. 83 goes 29 miles north to Garden City and on to Oakley, Oberlin, and northern Kansas; the geology along its route in Kansas is described in chapter 4. At this point on eastbound U.S. 160, the numbering system changes to correspond with that of U.S. 83. Because U.S. 83's numbering system runs from south to north, the figures on the milepost markers grow smaller on eastbound U.S. 160 for the next 20 miles.

The town of Santa Fe, once located at this intersection near the center of Haskell County, had a population of 1,000 and was the county seat from 1887 to 1920. Even though the town bore the same name as the railroad, the Santa Fe by-passed this location and established a city at

Sublette in 1912. Eight years later, county residents voted to move the county seat to Sublette.

37.0–39.0 *Lemon Northeast oil and gas field.* Discovered in 1965, this field produces both oil and gas from depths of 4,600 to 5,200 feet.

38.3 Tailwater pit in a small *High Plains depression.*

36.5 To the west 0.5 miles is an *alfalfa dehydrating plant.* Alfalfa is a major feed crop for cattle in this area. In years past, alfalfa was usually baled into small square bales that were stacked and fed to livestock. Today, to make handling easier, farmers often bale alfalfa into large round bales; chop it and store it in silos; or bring it to plants like this, where it is dehydrated, stored, and fed to livestock. Depending on precipitation, alfalfa quickly grows back after it is cut; it can produce three or four crops of hay every year. At the same time it adds nitrogen to the soil, so farmers often plant fields to alfalfa, harvest hay for six or seven years, and then return the ground to wheat, corn, or some other nitrogen-consuming crop.

35.6 Junction with *U.S. 56,* which runs southwest to Satanta, Hugoton, Elkhart, and points west and is described in chapter 9. To the northeast it goes to Dodge City. One mile east is Sublette, the seat of Haskell County, which is named after frontiersman William Sublette.

29.5 *Haskell/Seward county line.* Haskell County is named for Congressman Dudley C. Haskell; Seward County is named after Senator William H. Seward, an ardent opponent of slavery from New York who was later secretary of state under Abraham Lincoln and Andrew Johnson. Seward is probably best remembered for his role in the 1867 purchase of Alaska from Russia for $7.2 million. Critics of the deal called Alaska Seward's Icebox.

28.4 *K–190* runs west to Satanta and Ryus.

24.8 West edge of a *High Plains depression.*

23.1 East of the road is a *natural-gas pipeline compressor station.*

87.0 At this point, eastbound U.S. 160 changes back to its original numbering system. Westbound U.S. 160 heads north, joining U.S. 83 and adopting its numbering system. Just a few miles south, U.S. 83 crosses over the Cimarron River, which cuts a path through the High Plains and provides some rare relief in the landscape. Liberal is 18 miles south on U.S. 83.

The town of Springfield once stood near this intersection, about three miles north of a rival town, Fargo Springs. The two fought bitterly to be the seat of Seward County, and the honor passed back and forth between them until the Rock Island Railroad was built through Seward County and by-passed them both. The towns soon disappeared, and with their decline, a town called Arkalon grew up where the Rock Island crossed the Cimarron River. To avoid floods, the Rock Island later built a new bridge over the Cimarron three miles downstream, leaving Arkalon to meet the same fate as Fargo Springs and Springfield. Many of the residents and businesses of these towns ended up in Liberal, which became the seat of Seward County in 1896.

90.1 and 91.3   *High Plains depression* north of the highway.

94.9   Six miles south on a county road is the town of *Kismet*. The name comes from a Turkish word meaning *divide*. The town reportedly was named by Rock Island Railroad workers who built a bridge over the nearby Cimarron River. The bridge, known as the Sampson of the Cimarron, is 1,200 feet long. This is also the approximate eastern edge of the *Hugoton gas field*. This huge gas field underlies almost 8,500 square miles, an area nearly five times as large as the state of Rhode Island.

99.3   *High Plains depression.*

99.9   *Seward/Meade county line.* During the 1930s, southwestern Kansas suffered severe damage from dust storms. Poor conservation practices and droughts took a toll in the early 1930s, and by the mid 1930s the area was experiencing the worst dust storms since it was settled in the late 1800s. Records from Garden City showed that for sixteen days during April 1935, dust restricted visibility to less than one-quarter of a mile. One of the worst storms came on Black Friday, April 10, when blowing dust cut visibility to less than 100 feet, drifted across highways and railroad tracks, and closed schools and businesses throughout western Kansas.

Increased precipitation and improved farming practices eventually helped bring an end to the dusty conditions of the 1930s, but not before many farmers in the area had been bankrupted. Those conservation practices, along with the water supplied by irrigation, have helped prevent a return of the large-scale dust storms in southwestern Kansas, although smaller storms may occur from time to time, particularly in winter and early spring.

101.9   *Wildhorse Lake* is six miles north on a county road. This is an ephemeral lake in a High Plains depression, holding water only during wet periods.

102.4   About 0.5 miles south is the town of *Plains*. The town was originally called West Plains, and the township retains that name.

A dust storm near Hugoton in 1935 (courtesy of the Kansas State Historical Society)

103.7 Intersection with *U.S. 54*. Plains is one mile to the southwest; Liberal is 28 miles southwest. To the east, U.S. 54 goes to Wichita, and eastbound U.S. 160 adopts the mileage numbering system of U.S. 54 for about the next 14 miles. Westbound U.S. 160 returns to its original numbering system here.

33.7 The town of *Hobart* is 0.3 miles north.

34.5 One mile north is the town of *Collano*.

37.0 This point marks the eastern edge of undissected High Plains along U.S. 160. To the west lies an unbroken landscape, marked only by shallow, saucerlike depressions and small tracts of subdued sandhills. To the east, U.S. 160 enters the first area of significant topographic relief in Kansas as it drops into the valley of Spring Creek. Spring Creek flows into Crooked Creek, which is one of numerous tributaries of the Cimarron River that have eroded headward into the High Plains, creating the Cimarron Breaks. The eroded valleys and canyons of the Cimarron Breaks contain exposures of Tertiary and Quaternary alluvial sediments that underlie the High Plains to the west and make up the High Plains aquifer.

37.4 *Pleistocene alluvial deposits*. These stream-laid sediments were deposited during the past two million years.

37.7 *Spring Creek*. This stream has cut through Pleistocene sediments into the underlying Ogallala Formation.

38.0 *Pleistocene deposits.* These recent alluvial deposits do not form distinct layers of rock, so these formations do not have specific names but are referred to by age.

38.7 The town of *Missler* is 2.2 miles north of the highway.

39.9 A tributary of Spring Creek has cut a path through the Pleistocene alluvium here. This valley, called *Hart Draw,* runs southeast two miles to Big Springs, a series of springs similar to the artesian valley north of Meade. Though most of the springs are dry now, they once flowed freely—the result of water that entered the ground to the west and moved through aquifers to the east. Where erosion exposed the aquifer, the water exited. Because the water moved in between impermeable layers of rock, it was under pressure and thus created flowing wells.

Near Big Springs was a quarry where paleontologists discovered vertebrate fossils from Ice Age animals, including the remains of ground sloths, camels, wolves, several varieties of horses, and large cats. Some of these cats were saber-toothed, some resembled modern-day mountain lions, and others were larger than today's African lions. The erosive work of Crooked Creek, the Cimarron River, and their numerous tributaries in Meade County has produced some of the best natural exposures of Tertiary and Quaternary sediments in the central Great Plains.

These exposures attracted vertebrate paleontologists, who came to scratch through the weakly cemented sands and gravels in search of animal bones. In each locality, paleontologists lumped the remains into groups of animals that lived at the same time, groups that they called faunal assemblages. The sequence of these animal groups was then determined by correlating and age-dating the rocks in which they were found. Changes in the fauna of this area reflect the changing climate before and during the Ice Age. For example, the appearance of large land tortoises and South American armadillos indicates that the climate during the periods between glaciers was much milder than it is today; at times it may have been subtropical.

41.8 *Pleistocene deposits* in the cliff south of the road.

43.2 West edge of *Meade,* the seat of Meade County. Both the town and the county are named after George G. Meade, a Civil War general and the commander of the Army of the Potomac at Gettysburg. Also located in the town is a museum dedicated to the Dalton Gang. Eva Dalton moved to Meade in 1887, where several of her brothers visited her. In 1892 four members of the gang were killed in a shootout in Coffeyville.

About six miles north of Meade there is an abandoned pit where

A map and cross sections of the Crooked Creek and Fowler faults

volcanic ash was mined. Volcanic ash is common throughout western Kansas, but some of the thickest deposits are here in Meade County. By the 1930s, Meade County had produced more volcanic ash than any other county in the state. Today, only two ash mines remain open in Kansas: one in Lincoln County and the other in Norton County. North of the ash pit is a site where alluvial deposits have yielded the remains of Ice Age horses, camels, and mammoths.

44.4   East edge of *Meade*. The elevation is slightly less than 2,500 feet.

44.7   *Crooked Creek* and the *Crooked Creek fault*. Crooked Creek is aptly named. From its source in northeastern Haskell County, it flows southeast to a point 10 miles north of Meade. It then veers to the northeast, into southwestern Ford County, where it makes a U-turn and heads southwest in linear segments, which appear to be controlled by faults. The Crooked Creek fault begins in this vicinity and controls Crooked Creek's course to the southwest. The Fowler fault, a few miles to the east, controls Crooked Creek's course to the north, near the town of Fowler. Geologists have found that faults often change the shape of the land and thus influence the course of rivers.

Faults have also shaped the geologic history of this area during the last few million years and even into the present. To the west, the faults have moved down, dropping the top of the Cretaceous and Permian bedrock 200 feet lower on the west sides of these faults. This down-dropped basin was then filled in by stream-borne sediment during the late Tertiary and Quaternary periods. The Crooked Creek and Fowler faults also cut downward through salt and gypsum in the Permian red beds in the subsurface. Ground water probably moved down along these faults, dissolving away the highly soluble evaporites, causing the collapse of overlying rocks, and lowering the surface near these faults even more.

The dissolution of salt and gypsum along the Crooked Creek fault probably continues today. About 1.5 miles south of Meade is the Meade salt sink, which opened up suddenly in March 1879. About 60 feet deep and 610 feet in diameter, it swallowed up part of the Jones and Plummer Trail, a wagon and cattle trail of the period. After the sink collapsed, it slowly filled with salty water to about 15 feet from the surface, and for a time, salt was produced by evaporating the water from the sinkhole. Today the sinkhole is generally dry and is gradually filling with sediment. In early 1973 a similar but smaller sinkhole developed 0.2 miles northeast of the Meade salt sink, in alignment with the Crooked Creek fault. Geologists expect sinkholes to continue to develop in this and other areas where underground soluble rocks are being actively dissolved.

The Crooked Creek valley northeast of Meade, by the way, is much different from the other stretches of Crooked Creek. It is called Artesian Valley because of a series of flowing wells. Measurements in 1939 showed more than 200 artesian wells in the area, with flows ranging from a dribble to 100 gallons per minute. The pressure in some of these wells was strong enough to raise the water 17 feet above the ground. Flowing wells and springs fed lakes and marshes in this area, including Sealock Lake, six miles to the northeast. However, with the development of irrigation, the water table has dropped, and many of the wells and springs have quit flowing. The lakes only occasionally hold water.

These lakes may be the vestiges of a much larger lake that occupied Artesian Valley until recent geologic time. Evidence for this lake is found in the topography of Artesian Valley, which is broad and flat compared to other stretches of the Crooked Creek valley. Along its borders, Artesian Valley has sandy ridges that resemble beaches. Along the western edge of this valley, the bluffs have a profile similar to the wave-cut cliffs that form in weakly cemented strata along the shores of oceans and large lakes. This lake basin was formed by the Crooked Creek and Fowler faults, as well as by subsidence from the dissolution of underground salt and gypsum. Crooked Creek and its tributaries carried in sediment that partially filled this basin until recent geologic time. The lake disappeared when the stretch of Crooked Creek below Meade eroded headward, creating an outlet about a mile north of U.S. 160 and draining the lake.

118.0 *Junction with U.S. 54.* Eastbound U.S. 160 and U.S. 54 divide, with U.S. 160 returning to its original numbering system. Westbound U.S. 160 joins U.S. 54 at this point, adopting U.S. 54's numbering system for the next 14 miles. Although the numbering system changes, the milepost numbers of U.S. 160 continue to grow smaller from east to west.

122.4 Here the highway passes along the trace of the *Fowler fault.* Subsurface formations on the west side of the fault have been dropped below those on the east. Although the Fowler fault is not exposed at the surface here, its trend to the north-northeast controls the course of Crooked Creek.

122.8 The elevation is 2,500 feet.

123.0–125.0 The highway passes along the north edge of a large *elliptical depression.* Such features are common in the flat uplands in this area; they probably reflect subsidence caused by the undergound dissolution of gypsum or rock salt. This depression is only about 30 feet deep, but it may be the beginning of a much larger structure, like Big Basin, 14 miles to the southeast on U.S. 160.

124.4 The town of *Fowler* is 7 miles north of here.

127.4 *Ogallala Formation* exposed in the creek to the south.

128.8 *Ogallala Formation,* overlain by loess. The Ogallala Formation is composed of silt, sand, gravel, clay, and other debris eroded off the face of the Rocky Mountains and deposited by streams on the plains of western Kansas during the past 5 million years. In much of the western third of the state, the Ogallala has since been covered by more recent material—in this case a dustlike silt called loess—and the formation serves as an aquifer where it is saturated with ground water. The Ogallala underlies all or parts of eight states. In Kansas it is generally thickest here in the southwest corner—as much as 600 feet in places—although not all of that thickness is saturated with water.

129.0 *Gyp Creek.* The Ogallala Formation is exposed in the valley walls north and south of the highway. Gyp is short for gypsum, or calcium sulfate, which occurs in the Permian red beds that this creek cuts through a short distance downstream.

129.1 *Ogallala Formation.*

129.3 *Indian Treaty Boundary.* This north-south line marks the western edge of the Osage Indian Reservation, as established in 1825, when the Osage signed away their lands in the rest of Kansas for a reservation that started 25 miles west of the Missouri border and extended to this line near the 100th meridian. The Osage generally lived in eastern Kansas, but the western part of this reservation allowed them an outlet to hunt on the plains of western Kansas during the summer.

This boundary also served as the western edge of Kansas for part of the state's history. Before 1804 all of Kansas belonged to France; it was sold to the United States in 1804 as part of the Louisiana Purchase. In 1819 the United States ceded the corner of Kansas west of this line and south of the Arkansas River—along with other lands in the West—to Spain in exchange for Florida. When Mexico gained its independence from Spain in the 1820s, this corner of Kansas became part of Mexico. In 1835, Texas claimed the area when it declared its independence from Mexico, making this area a part of the Republic of Texas.

Texas became a state in 1845, and this section of Kansas once again became part of the United States, although it was then part of the state of Texas, which sold southwestern Kansas, along with other lands outside its present border, to the United States in 1850 for $10 million. In short, southwestern Kansas changed hands five times betwen 1804 and 1850, during which period it was owned by France, the United States, Spain, Mexico, the Republic of Texas, and, finally, the United States again.

130.1 *Ogallala Formation* on the south side of the highway. Where it is exposed, the Ogallala sometimes forms hard, dense layers that geologists call mortarbeds. In these locations the particles that make up the Ogallala have been naturally cemented together by caliche, a type of calcium carbonate that was deposited as ground water moved through the rock.

130.3 *Meade/Clark county line.* Clark County is named after a Civil War captain, Charles F. Clark.

130.4 *Ogallala Formation* mortarbeds north of the highway. Some geologists believe that the Ogallala once formed a continuous sheet from the Rocky Mountains all the way across Kansas to the Flint Hills. Most of the formation has since been eroded away in central Kansas.

131.0 *Ogallala mortarbeds* exposed in road cut.

131.2 *Johns Creek.* Ogallala mortarbeds are exposed in a cliff 0.7 miles south of the highway.

133.3 *West Indian Creek.* In this creek, the Big Basin Formation is overlain by the Ogallala Formation. The location where these two layers come into contact is marked by the contrasting colors of the Big Basin Formation, which is red, below the Ogallala, which is white to buff. The Big Basin Formation is composed of red shale and siltstone that was deposited at the end of the Permian Period, about 240 million years ago. During the latter part of the Permian, an inland sea covered much of south-central Kansas. Occasionally an arm of that sea was cut off from the rest of the Permian ocean; when it evaporated, it left behind shales and siltstones of the Big Basin Formation, along with salt, gypsum, and other materials. Because of occasional uplifts and downdrops in the land, sea water would cover the area and then would be cut off again, leaving behind layer after layer of evaporites, rocks that remained as the sea water evaporated away. In some parts of central Kansas, salt deposits alone are up to 400 feet thick.

This is also the westernmost exposure of a Permian formation along U.S. 160, marking the west edge of the Red Hills physiographic region. The hills south of the highway are capped by bright white layers of the Ogallala Formation.

134.3 *East Indian Creek.* The Ogallala Formation is exposed in cliffs and valley walls south of the highway.

Big Basin, a sinkhole formed by the dissolution of underlying beds of salt and gypsum, at milepost 139 along U.S. 160 in Clark County

135.4 *Junction with U.S. 283,* which runs 16 miles north to Minneola and 32 miles north to Dodge City.

137.6 *Ogallala Formation* exposed in the road cut.

137.6–139.0 For 1.4 miles, the road runs along the floor of *Big Basin,* a large sinkhole that was probably caused by the dissolution of salt and gypsum beds several hundred feet below the surface. The walls of Big Basin are composed of the Permian-age Big Basin Formation, overlain by Ogallala Formation. From the age of the rocks, it is clear that Big Basin developed relatively recently in geologic history—probably within the past few thousand years, but before the days of frontier settlement. About 0.3 miles east of the eastern edge of Big Basin is a similar feature called Little Basin. About 0.5 miles in diameter, Little Basin apparently formed in much the same way as Big Basin. In the floor of Little Basin is a spring called St. Jacob's Well, one of many springs that are scattered in the hills east of the highway.

139.0 Road to *Little Basin* and *St. Jacob's Well.*

139.5 *Ogallala Formation* in road cut.

140.0 The view to the south is of the *Ashland-Englewood Basin,* a broad lowland area stretching from Englewood, more than 20 miles to the east, and underlying the town of Ashland. This lowland is occupied by the Cimarron River and its tributaries. The floor of this basin is 500 feet or more below the crest of the Red Hills, a few miles to the northeast. This basin is believed to have formed from a large number of sinkholes, such as Big Basin 1.5 miles north, that grew together by erosion and sedimentation. As in Big Basin, the dissolution of salt and gypsum in the subsurface caused subsidence in the Ashland–Englewood Basin. In fact, Big Basin and Little

St. Jacob's Well, a spring formed in the Little Basin, immediately east of Big Basin in Clark County

Basin may represent the latest episode in the northward growth of the Ashland–Englewood Basin. Erosion is attacking the walls of Big and Little basins. Soon, geologically speaking, these walls will be breached, and Big and Little basins will no longer be separate features but will be continuous with the Ashland–Englewood Basin to the south.

    140.5   *Ogallala Formation* south of the highway.

    141.0   *Antelope Creek.* The valley of this creek is more than a mile wide, but the creek is only 1.5 miles long. The valley becomes narrower both upstream and downstream, suggesting that this area may be an old subsidence feature similar to Big Basin, two miles to the north. But unlike Big Basin, stream erosion has breached the walls of this feature, creating a broad valley. One mile north, Antelope Creek valley widens again, indicating another old area of collapse caused by solution. The tributaries of Antelope Creek are now eroding away the south wall of Big Basin, and someday it will be captured and become part of Antelope Creek's drainage.

    141.9   Junction with *U.S. 283,* which runs 11 miles south to Englewood.

142.3   *Whitehorse Formation* overlain by *Ogallala Formation.* The Whitehorse is a Permian formation that includes layers of sandstone, siltstone, and shale. The Whitehorse was deposited late in the Permian, about 250 million years ago, while the Ogallala Formation was deposited within the last five million years. Thus, more than 200 million years elapsed between the deposition of these two formations.

143.6 and 143.8   *Ogallala Formation* in the road cut. Along westbound U.S. 160, this is the first exposure of the Ogallala Formation, a layer of sand, gravel, silt, and clay that eroded off the face of the Rocky Mountains and was deposited on the High Plains by rivers during the past five million years. While this is the easternmost exposure of the Ogallala along U.S. 160, the formation is found as far east as McPherson County in central Kansas. Along eastbound U.S. 160, this point provides the first clear view of the Red Hills to the east.

144.4   The rocks exposed here are part of a series of Permian deposits that geologists call *red beds.* These bright red rock layers form the Red Hills of Clark, Comanche, and Barber counties. Composed of shale, siltstone, and sandstone, along with interbedded layers of gypsum and dolomite, the red beds are colored by iron oxide, or rust, which turns bright red when exposed to oxygen.

145.4   *Keiger Creek.* The path of Keiger Creek has created some of the most spectacular views of the red beds along U.S. 160. Birds have built mud nests on the bridge's pillars and abutments; because the birds use local mud, even the nests are bright red.

145.4–145.6   *Whitehorse Formation* overlain by *Big Basin Formation.* Many of the rock layers in the red beds are part of the Whitehorse Formation. North of the highway the red beds have been extensively dissected by drainage into Keiger Creek, creating a series of bright red cliffs and canyons.

146.5   About 0.9 miles north of the highway is a hill called *Beckys Knob.* Its elevation is 2,180 feet, about 100 feet above the elevation at the highway. Near Beckys Knob are hills of 2,300 feet and higher.

147.0   *Little Sandy Creek.*

150.0   *Red beds* north and south of the highway. The Permian salt plains that covered this area were not particularly hospitable to most forms of life. Fossils are relatively rare, although paleontologists occasionally find remains of a fin-backed reptile or a lungfish that burrowed into the mud of pools of water. The scarcity of fossils made it difficult to determine

Bird nests cover the abutments of a bridge crossing Keiger Creek at milepost 145 along U.S. 160 in Clark County.

the age of these deposits, because geologists generally use fossils as a guide to the age of rocks. Geologists placed the rocks in the Permian, Triassic, Jurassic, and Cretaceous periods before finally classifying them as Permian in 1896.

150.8 *Redhole Creek* and *red beds* north and south of the highway.

151.3 *Whitehorse Formation* red beds north and south of the highway.

152.6 *Mount Casino,* elevation 2,130 feet, is 3 miles to the northeast.

153.1 West edge of *Ashland.* Ashland oil field is a mile north. Discovered in 1951, the field has produced less than half a million barrels of oil. Clark County is not a major oil-producing area; it pumped less than a million barrels in 1984.

Ashland, the seat of Clark County, was named by a group of Kentuckians from Winfield, Kansas, who organized a company to promote the town. The name came from Kentucky Congressman Henry Clay's home in Lexington, which was called Ashland. This town is near the east end of the Ashland–Englewood Basin, a broad lowland area that was created by the solution of subsurface evaporites and the collapse of overlying rocks. Permian deposits in the subsurface of western Kansas include evaporites, layers of salt and gypsum. The total thickness of these Permian evaporites is greatest here in Clark County, where it reaches approximately 1,400 feet.

154.1 East edge of *Ashland.* The county road at the east edge of Ashland runs north past Mount Casino, over Bluff Creek, and eventually to Clark County State Lake, about 15 miles to the north. The lake is set in the midst of a rugged canyon carved out of the Ogallala Formation and underlying Cretaceous shales and Permian red beds. The highway also crosses the *West Branch of Bear Creek* here.

154.2 *Roadside park.*

154.5 *East Branch Bear Creek.*

158.0 Five miles north, along a divide overlooking Dugout Creek, there is an isolated Ogallala-capped hill called *Mount Jesus.* With an elevation of 2,350 feet, it stands 350 feet above nearby Bluff Creek and commands a panoramic view of the surrounding countryside. The hill was named in 1868 by members of General Custer's command as they blazed a trail southward from Fort Dodge to Camp Supply in Oklahoma. During the Indian wars, the Plains Indians built signal fires on this hill and called it Fire Mountain.

159.0 *Day Creek,* the only creek with that name in Kansas. Day Creek Dolomite, named for exposures along this creek, is a few feet thick and is composed of dolomite, or magnesium carbonate, which looks like calcium carbonate or limestone, the most common carbonate rock exposed in Kansas.

160.1 Junction with *K–34* and *U.S. 183.* K–34 runs 26 miles north to Bucklin. To the south it goes one mile to Sitka and 26 miles to Buffalo, Oklahoma.

160.2 *Whitehorse Formation* red beds in the road cut.

162.2 *Harper Ranch oil and gas field* on both sides of the highway. Discovered in 1953, this field produces oil and gas from formations that are 4,900 to 5,400 feet deep.

162.7 Elevation is 2,000 feet. Clark County is second only to Wallace County in the amount of topographic relief within its borders. The elevation ranges from a low of 1,730 feet in the southeast corner along the Cimarron River to a high of 2,600 feet in the High Plains of the northwest corner, a difference of 870 feet, compared to 890 feet in Wallace County.

163.7   The hill one mile south of the road is called *Sugarloaf*. Its elevation is 2,004 feet, more than 100 feet above the elevation at the highway.

166.0   *Clark/Comanche county line.* Comanche County is named after the Plains Indian tribe that lived in the area until the mid 1800s. The Comanches were known for their horsemanship and their nomadic reliance on the buffalo. In the seventeenth century their population was estimated at 7,000, making them one of the largest tribes in western Kansas. They weren't the only Indians in this part of the state. The Kiowa, southern Cheyenne, and Arapaho also considered this part of Kansas in their range.

166.5   The *Overocker oil and gas field,* discovered in 1976, produces oil from wells about a mile deep.

168.3   *Bluff Creek,* one of six Bluff creeks in the state. This one probably gets its name from the high bluffs that it carves through the red beds, particularly to the north. It flows south into the Cimarron River, just above the Oklahoma border.

169.4   The town of *Protection* is immediately north of the highway.

170.2   *Kiowa Creek.*

170.7   *Cavalry Creek.* A small area of sand hills is visible on the east side of the creek.

172.7   *Bird South oil field.*

174.8   The soft, light-colored sandstone here is part of the *Cheyenne Sandstone,* which was deposited early in the Cretaceous Period, about 100 million years after the Permian formations that are more common in this area. Cretaceous formations underlie most of the western third of the state, but they are exposed mostly in west-central and north-central Kansas. However, the Cheyenne crops out in several counties in south-central Kansas, overlying the Permian red beds of the Red Hills. Outcrops are up to 94 feet thick, but the Cheyenne may be as thick as 300 feet where it is deeply buried in western Kansas.

177.0   At this point, eastbound U.S. 160 briefly leaves the Red Hills physiographic province as it climbs onto the undissected tableland of the *High Plains* to the east and north. For the next 20 miles the countryside is more level as the highway passes over a plateau capped by Cretaceous sandstone and shale that has been mantled with Pleistocene river-laid sands and gravels, along with more recent windblown sand and sediment that are typical of the High Plains.

178.0   Junction with *K–1,* which runs six miles to the town of Buttermilk.

180.9   *Roadside park* east of the highway.

182.5   This county road runs one mile west to *Lake Coldwater,* a small lake constructed on Cavalry Creek.

183.0–184.0   *Coldwater.* Named after a town in Michigan, Coldwater is the seat of Comanche County, one of the leading cave-exploration areas in Kansas. The Kansas Speleological Society has catalogued 528 caves in 37 counties. Comanche County leads the list with 128. Probably the most common source of those Red Hills caves is gypsum, which can be dissolved by water to form caves that extend several hundred feet. These caves often shelter large populations of bats, including some species that in Kansas are found only in the southwestern part of the state. One cave in southern Comanche County also harbors several Indian pictographs, figures that have been drawn on cave walls with charcoal or some other pigment.

184.0   Junction with *U.S. 183,* which diverges from eastbound U.S. 160 at this point and heads north to Greensburg. U.S. 183 joins westbound 160 here for the next 36 miles.

185.0–188.0   *Sand hills.* This area of windblown sand measures roughly 12 miles north-south and 3 or 4 miles east-west. Its stabilized dunes are typical of those found throughout the High Plains region of southwestern Kansas.

187.3   *West Branch Nescatunga Creek.*

188.9   *Nescatunga Creek.* This Indian name probably means Big Salt Fork or Rock Salt River. Either way, the Indians recognized the creek's salt content, caused by the dissolution of Permian beds in this area. The Nescatunga flows south to the Salt Fork of the Arkansas River, which in turn joins the Medicine Lodge River in Alfalfa County, Oklahoma. There the two rivers form the Great Salt Plains Reservoir, a sprawling salt flat that is a good source of brown crystals of selenite, a form of gypsum.

189.9   The town of *Wilmore* is four miles north, along a county road. The county road that runs east and north out of Wilmore is part of the Gyp Hills Trail, a road that encounters some of the scenic geology in northeastern Comanche and northwestern Barber counties. Immediately east of Wilmore, for example, the road passes an artesian well that flows into Mule Creek and then passes along the southern edge of Snooks Hollow.

190.5   *East Branch Nescatunga Creek.*

193.5   *Indian Creek.* The brownish soils developed on the alluvial deposits here in the High Plains contrast sharply with the reddish soils formed along the much older Permian red beds in the Red Hills to the east and west along U.S. 160.

197.0   The elevation is 2,000 feet. About a mile to the northwest, a branch of Mule Creek has carved a draw in the Red Hills, which is called Baker Canyon. This point marks the eastern edge of the High Plains as they dip into Comanche County. Westbound U.S. 160 leaves the Red Hills and passes over High Plains topography for the next 20 miles before reentering the Red Hills southwest of Coldwater.

About 13 miles to the north the small town of Belvidere is located in a scenic area of highly eroded remnants of Cretaceous and Permian formations. North of Belvidere, for example, are Osage Rocks and Cheyenne Rock, an outcrop of sandstone that supplied the name for the Cheyenne Sandstone. South of Belvidere is Champion Draw, famous for Cretaceous fossils that come from a formation called the Kiowa Shale, which is slightly younger than the Cheyenne Sandstone. The Kiowa contains a shell-filled layer of limestone that is sometimes called the Champion shell bed, named after this draw.

199.7   *Mule Creek.*

200.1   *Red beds* in the road cut.

200.3   *Walnut Creek.*

200.5–201.8   *Perry Ranch oil and gas field.* This field was discovered in 1958, but today it has only two producing wells.

202.0   *Comanche/Barber county line.* About seven miles north, in northeastern Comanche County, is a highly eroded area of Cheyenne Sandstone called Hell's Half Acre.

205.2   *Alluvial deposits* in road cut. These stream-laid sands and gravels were deposited in the Pleistocene, about a million years ago. They cap this ridge between the Medicine Lodge River drainage, to the northeast, and the Mule Creek Basin to the southwest. These deposits become much thicker to the west along U.S. 160. Where they are saturated by water, they are economically important aquifers.

Five miles north is a mine that produces gypsum from the Blaine Formation. Gypsum is found primarily in Permian rocks that crop out from northeast Kansas, southwestward in an ever-widening band. Three varieties of gypsum are found in Kansas. Crystalline forms called selenite and satin spar are found in these hills; some specimens are prized by rock collectors. The mine produces massive, or rock, gypsum that is layered throughout these hills. Gypsum is used in cement, sheet rock, and plaster of Paris.

Butte and mesa topography is characteristic of the Red Hills of south-central Kansas (Barber County).

205.9   *Sun City* is eight miles north along this county road, which is part of the Gyp Hills Trail. In a pasture south of Sun City are the remnants of a natural bridge that was formed by the erosion of a gypsum layer in the Blaine Formation. Before its collapse in 1964, the bridge was 35 feet wide and 55 feet long and stood 12 feet above the stream that cut through it. Because gypsum is susceptible to erosion, many smaller natural bridges have formed in the Red Hills.

A natural bridge often begins as a cave that has an underground stream issuing from its mouth. The dissolution of the rock containing the cave—in this case gypsum—continues and can cause part of the cave to fall in but leave part of the cave roof standing as a natural bridge. Close to the old Natural Bridge site is Natural Bridge Cave, one of 117 caves in Barber County.

207.0   The town of *Deerhead* is three miles south. This county road is also part of the Gyp Hills Trail, which cuts through the hills south of the highway before rejoining U.S. 160 west of Medicine Lodge.

207.5   The elevation is 2,000 feet.

This sandstone, at milepost 210 along U.S. 160 in Barber County, slumped after the natural dissolution of an underlying layer of gypsum.

208.0 Wells from the *Stumph oil and gas field,* discovered in 1952, are south of the highway.

209.0 *Dog Creek.* The Dog Creek Formation, named for exposures along this stream, is composed of maroon shale, siltstone, and sandstone. It overlies the Blaine Formation in this area.

209.2 Gypsum layers in the *Blaine Formation.* The Blaine is the source of the massive gypsum that is mined near Sun City, and its outcrops are the source of the name Gyp Hills, which is often applied to this part of the Red Hills.

209.6 and 209.9 *Blaine Formation,* which is up to 50 feet thick and includes several layers of gypsum, dolomite, and red shale. The prominent gypsum bed along U.S. 160 is the Medicine Lodge Gypsum Member of the Blaine Formation.

210.5 *Dog Creek Formation.* The sandstone on the north side of the highway forms a long downward arch that is evidence of solution and collapse that took place after the formation was deposited. Ground water probably dissolved a cavity in the Blaine Formation gypsum, which is about 30 feet below the low point of the arch. This cavity allowed the rocks above

Satin spar from the Red Hills

it to slump down and thus formed this long solution feature. Caves and solution channels are common in the gypsum of the Red Hills. Some layers are so honeycombed with cavities that by stamping your foot on top of an outcrop, you will produce a hollow sound.

210.8 *Flower-pot Shale.* Named for the hill 2.5 miles east of here, the Flower-pot is composed of reddish-brown shale. Up to 180 feet thick, outcrops of this formation are often strewn with clear crystals of selenite and with red, pink, and white crystals of satin spar.

211.2 *Oak Creek.* At 6 a.m. on 6 January 1956, an earthquake struck this part of south-central Kansas and northwestern Oklahoma. The quake was felt as far away as Great Bend and Dodge City, although the center seems to have been here in Barber County. The strongest effect was felt in the small town of Coats, about 15 miles north of here, and in Pratt, where a guest in the Hotel Roberts reported that his bed bounced across the floor. Since 1977 the Kansas Geological Survey has recorded six microearthquakes, too small to be felt, in southern Barber County. Geologists speculate that the earthquakes may be caused by movement along the Pratt Anticline, a subsurface feature that runs along the western edge of Barber County.

Flowerpot Mound, at milepost 213 along U.S. 160 in Barber County, is capped by Permian-age gypsum.

212.0–212.6 *Flower-pot Shale.* The sediments that make up these bright red formations were deposited on vast alluvial plains and coastal lagoons that were the remnants of a saline Permian sea.

213.2 *Little Bear Creek.* Immediately northeast of the highway is *Flowerpot Mound,* elevation 1,800 feet. Like many hills in Barber County, this mound is capped by layers of Blaine Formation gypsum that are somewhat more resistant to erosion than shales and siltstones are. Layers of gypsum are responsible for much of the butte-and-mesa topography that is common around Medicine Lodge.

213.4 This thick road cut goes through the *Flower-pot Shale.* Flowerpot Mound is immediately north of the highway.

214.2 Massive gypsum beds in the *Blaine Formation.* The draws and valleys of these hills are covered with cedar trees, and the area is sometimes called the Cedar Hills.

214.6 *Flower-pot Shale,* including numerous interlacing layers of gypsum.

215.5 *Bitter Creek,* which probably got its name from the taste of its water. Sulfate, derived from the abundant gypsum (or calcium sulfate) in this stream's watershed, may occur in the water of Bitter Creek in the form of magnesium sulfate (Epsom salts) or sodium sulfate (Glauber's salt). In sufficient quantities, these salts give water a bitter taste.

216.3 *Flower-pot Shale,* overlain by gypsum from the *Blaine Formation.*

218.4 The two sharp hills south of the highway are called *Twin Peaks.* The peak closer to the road is 1,785 feet high; the south peak is slightly lower. The elevation along the road is 1,631 feet.

219.4 *Flower-pot Shale.*

220.0 Layers of sandstone and siltstone here are part of the *Cedar Hills Sandstone.*

220.4 *Cedar Creek.*

221.0–221.2 These layers of red sandstone are part of the *Cedar Hills Sandstone.*

221.5 *Cedar Hills Sandstone.*

222.4 *Walnut Creek,* elevation 1,500 feet.

222.5 Cross-bedded sandstone in the *Cedar Hills Sandstone.*

224.0 The southbound county road is part of the *Gyp Hills Trail,* which winds through the rugged scenic hills south of the highway before rejoining U.S. 160 about 17 miles to the west.

224.3 *Junction with U.S. 281.* The town of Hardtner is 23 miles to the south. The oil wells to the south are part of the Bloom North oil field. Discovered in 1960, this field has produced about 500,000 barrels of oil from Mississippian formations.

224.5 *Roadside park.* To the north 0.5 miles is the plant where the gypsum mined near Sun City is processed. The gypsum industry in Kansas got its start in Medicine Lodge in 1888, when Prof. Robert Hay, an early-day Kansas geologist, wrote a magazine article describing the gypsum-capped buttes and mesas of Barber County. Among the article's readers were a group of Englishmen who manufactured Keene's cement, a gypsum-derived cement used to make columns, figurines, statuettes, and other ornamental features on expensive buildings. The group sent a man from England to investigate Barber County's gypsum deposits, and his glowing report resulted in a branch factory of the Keene Cement Company. Gypsum has been produced here ever since. The original mines were in the gypsum-capped hills a few miles southwest of Medicine Lodge. Quarrying operations later sprang up in northwestern Barber County, near Sun City, where mining began in 1930 and continues today.

224.8 *Medicine Lodge River.* For years the Plains Indians believed that this river had healing properties, and they came here to bathe in and to drink the water. In a sense, they were probably right, because these streams and springs are high in calcium and magnesium sulfates and in other natural salts dissolved from layers of gypsum and dolomite. Magnesium sulfate, or Epsom salts, has long been used as a solution for soaking. Present in sufficient quantities in drinking water, Epsom salts may also have a laxative effect.

225.0   Junction with *U.S. 281*. To the north, this highway runs along the west edge of Medicine Lodge. Pratt is 27 miles north.

225.0–225.9   *Medicine Lodge,* the seat of Barber County. Before the town existed, this was the site of the 1867 Medicine Lodge Peace Treaty, negotiated between the United States Government and five tribes of Plains Indians. The area was settled by whites during the 1870s, and its economy was based primarily on the cattle business. It was the home of "Sockless" Jerry Simpson, a Populist leader who won a seat in Congress in 1890, and of Carry Nation, who was famous for her opposition to alcohol. Nation began her saloon-smashing career in nearby Kiowa and Wichita, but she soon moved on to Kansas City and Topeka. Her fight against alcohol brought her national recognition and may have been partially responsible for restrictive liquor laws in several southwestern Kansas counties. Nation's home is preserved as a museum, just off U.S. 160, west of Main Street in Medicine Lodge.

Kansas prohibited the sale of hard liquor even before Nation's activities. In 1880 the state voted a constitutional prohibition on hard liquor—the first state in the Union to do so—although the sale of beer was still allowed and the law was difficult to enforce. Even after national prohibition was repealed in 1936, Kansas maintained its prohibition until 1948, when, in the wake of World War II, Kansans voted to allow liquor stores. Even Barber County voted wet in that election.

225.9   *Elm Creek.*

227.1   The road to the south runs to a *natural amphitheater,* where, every three years, the Medicine Lodge Peace Treaty pageant is held. The pageant reenacts the 1867 signing of a treaty between the United States and the Plains Indians who roamed the American Southwest—the Kiowa, Comanche, Arapaho, Apache, and Cheyenne. Among the negotiators were Kit Carson, William Mathewson, Jesse Chisholm, Little Raven (of the Arapaho), and Satanta (of the Kiowas). After two weeks of negotiating, the tribes ceded away their claims to Colorado, North Dakota,

South Dakota, New Mexico, and Arizona (Kansas was already a state at the time) and agreed to move south of the southern border of Kansas, an area that was called Indian Territory. The treaty, in essence, helped to clear the way for white settlement in much of the West.

228.6   *Antelope Creek.*

229.6   The former town of *Pixley* was 0.2 miles south of the road. All that remains of it is a grain elevator.

230.0   Wells from the *Highway oil field,* discovered in 1956, produce oil from rocks in the Douglas Group of formations, which are about 3,500 feet deep here. The Douglas Group was named for Douglas County, where some of these rocks are exposed.

232.0–234.0   *The McGuire–Goemann oil field.* These wells produce oil from formations that were deposited in the Pennsylvanian Period. They are 4,100 to 4,300 feet underground, significantly deeper than the wells in the nearby Highway field.

231.6   *Sand hills.*

231.9   *Little Sandy Creek.*

234.9   The town of *Sharon.*

235.2   *East Branch Little Sandy Creek.*

236.5   Red beds in the *Salt Plain Formation* are exposed in the road cut. This formation is up to 265 feet thick and includes the Cedar Hills Sandstone and the Flower-pot Shale formations.

237.3   *Sharon oil field* south of the highway. These wells produce oil from Mississipian rocks that are more than 4,300 feet deep.

237.5   *Spring Creek.*

238.2   *Sharon oil field* north of the highway. Most of the oil wells drilled in Kansas are either development wells or exploration wells. Development wells, drilled to produce more oil from a known deposit, are the most common oil wells in Kansas; oil or gas is discovered in about 70 percent of them. Exploration wells, commonly called wildcats, are drilled to find oil in areas that are not known to have production. To qualify as an exploration well, the well must be at least a half-mile from an existing field or be completed in a stratigraphic layer below or above a rock formation that is currently producing. About 20 percent of the wells drilled in Kansas each year are exploration tests, and commercially recoverable amounts of oil are discovered in about 22 percent of them.

238.7   *Barber/Harper county line.* Thomas W. Barber was a native of Ohio who was killed in 1855 while defending Lawrence during a raid by proslavery forces. The Kansas Legislature honored Barber by naming this county after him. For several years it was spelled Barbour, before it was formally corrected in 1883. Harper County is named for Sgt. Marion Harper, who was killed in 1863 during a Civil War battle near Waldron, Arkansas.

238.8 *Gene Creek.*

239.7 *West Sandy Creek.*

240.5 *Bachelor Creek.*

241.3 *Cottonwood Creek.* This creek and the next three streams to the west drain the Red Hills Upland, north of the highway. The hilltops are 200 to 300 feet higher than the highway at this point. All four of these creeks converge to form Sandy Creek a few miles to the south.

243.6 Red beds in the *Salt Plain Formation* in the road cut on both sides of the highway. The elevation is 1,500 feet.

244.5–245.0 *Sullivan oil field* on both sides of the highway. This field was discovered in 1961; the wells are about 4,600 feet deep.

245.0 *Camp Creek.*

245.3–246.3 *Attica.* Natural gas was discovered in wells drilled on city property in this small town during the 1970s. The wells, several of which are visible along U.S. 160, have provided income that has helped provide a variety of city services.

247.0 The wells along the highway are part of the *Sullivan East oil field,* discovered in 1977. This field produces oil from a sandstone formation that is about the same age as rocks that crop out along U.S. 160 west of Independence, Kansas.

248.5 *West Branch Bluff Creek.*

249.3 Red beds in the lower part of the Permian *Nippewalla Group* of rocks are exposed in the road cut north of the highway.

249.5 *East Branch Bluff Creek.*

249.5–250.0 Immediately north of the highway is a classic example of a Kansas *shelter belt.* The strip of trees includes cedars on one side and taller deciduous trees on the other. After the dust storms of the 1930s, which were due in part to poor conservation practices, Kansas farmers began to plant shelter belts along the edges of their fields in order to slow down the wind and help keep the topsoil in place. Many of these shelter belts are mature today, serving their original purpose and supplying much-needed cover and food for area wildlife.

250.0 Here the highway passes through *Crystal Springs oil field,* which produces oil from Mississippian rocks at depths of 4,400 feet. Oil-producing zones are deeper here than in much of Kansas, in part because Paleozoic sediments are thicker here. Harper County and the surrounding counties in south-central Kansas overlie the Sedgwick Basin, a broad bowl-shaped depression that has been filled with thick accumulations of marine sediments. Like the Hugoton Embayment of southwestern Kansas, the Sedgwick Basin is a northward extension of the Anadarko Basin, a much deeper basin to the south in Oklahoma. In the Sedgwick Basin, the depth to crystalline Precambrian rocks reaches a maximum of 5,500 feet along the southern Harper County border. The depth to these crystalline Precambrian rocks in the Hugoton Embayment is 9,500 feet.

The Sedgwick Basin is bounded on the east by the Nemaha Uplift and on the west by the Pratt Anticline, an arch of rocks that separates the Sedgwick Basin from the Hugoton Embayment. The Pratt Anticline dies out in northern Oklahoma, where the Sedgwick Basin and the Hugoton Embayment merge to form the northern flank of the Anadarko Basin. These basins began to drop down after the Mississippian Period, and that down warping continued through the Pennsylvanian and Permian periods, during which the thickest layers of sediments were deposited. Pennsylvanian and Permian rocks are up to 20,000 feet thick in the deepest part of the Anadarko in western Oklahoma, but in southern Kansas these same rocks are relatively thin, and the area is sometimes called the northern shelf of the Anadarko Basin.

251.3 The town of *Crystal Springs* is two miles to the north.

253.3 *Rush Creek.*

255.4 *Spring Creek.*

256.3 Junction with *K-2 and K-14,* which go six miles south to the town of Anthony. Horse and dog races are held in July at Anthony Downs, a race track north of the town.

258.8 *Roadside park.*

259.2 Junction with *K-14,* which runs 26 miles north to the town of Kingman.

259.2–260.8 *Harper,* the seat of Harper County, is the county's largest town.

260.4 *Junction with K-2,* which runs northeast 47 miles to Wichita. Along the way it passes near the former site of Runnymede, a town established by Englishmen in 1889 and named after Runnymede, England, where King John was forced to sign Magna Carta in 1215. Young Englishmen came to Runnymede, Kansas, to learn farming, but by 1892 most of them had returned to England, leaving behind a ghost town.

262.3 *Sand Creek,* which joins the Chikaskia River southwest of Argonia.

265.2 The present-day town of *Runnymede* is five miles north of here. After the English left old Runnymede, the town's buildings were moved several miles to this new site.

267.4 The town of *Danville* is north of the highway. In this vicinity, the Red Hills physiographic region to the west merges with the Wellington Lowlands to the east, although the boundary between the two regions is indistinct here. Topography characteristic of the Red Hills is found north and west along U.S. 160, while the Wellington Lowlands extend eastward to the Flint Hills in Cowley County. This area is covered by sand, silt, and gravel of Pleistocene age, carried here by streams that have eroded slightly older deposits in the High Plains to the north.

269.2 The town of *Freeport,* six miles south, has only twelve residents. It is the smallest incorporated city in Kansas.

271.4 The wells to the south are part of the *Rex oil field,* discovered in 1961. This field produces from rocks in the Simpson Group, which were deposited during the Ordovician Period and are 4,600 feet deep. One important bed in the Simpson Group is the St. Peter Sandstone, which is found throughout the Mississippi Valley. The St. Peter is a very clean sandstone, composed almost entirely of quartz grains, or silica, and thus is valuable for making glass. The nearest outcrops of St. Peter Sandstone are in central Missouri and northwest Arkansas.

272.0 *Chikaskia River,* which flows southeast, joining the Salt Fork of the Arkansas west of Ponca City, Oklahoma.

272.4 *Harper/Sumner county line.*

272.0–272.8 Area of *sandhills* east of the Chikaskia River.

274.5 *Argonia Creek.*

274.5–275.0 *Argonia,* which claims to have elected the first woman mayor in the United States. Mrs. Susana Salter was elected in 1887, even before women had the right to vote in general elections in Kansas.

276.0 *Spring Creek.*

277.3 *Sand Creek.* The highway crosses four creeks in the next four miles, all of which flow south into the Chikaskia River.

278.0 *Silver Creek.* This creek's name must have come from the water's reflection, because silver has never been commercially mined within today's Kansas borders. The only silver that might be found in Kansas would come from rivers that drain the Rocky Mountains, such as the Arkansas, and that is a long shot.

278.3   The wells here are part of the *Love-Three oil and gas field.* Discovered in 1969, this field produces oil from wells that range from 1,400 to 4,300 feet deep.

279.3   *Shore Creek.*

279.7   Junction with *K-205.* The town of *Milan* is one mile south of here.

281.7   Junction with *K-49,* which runs eight miles north to Conway Springs, another Kansas town that produced spring water for use as a medicine and for drinking.

282.7   *Beaver Creek.*

284.5   This is the approximate point of contact between the *Ninnescah Shale,* to the west, and the older *Wellington Shale* to the east. These are both soft shales, so their contact is not expressed topographically; rather, it is characterized by a change in soil color. To the west the soil is a subtle red, caused by the presence of iron oxide in the reddish-colored Ninnescah Shale. To the east the road encounters brownish soils that are part of the Wellington Shale. Both shales were deposited in the Permian. Drained by the Arkansas River, most of the land here is flat, and outcrops of rock are rare because the common formations, such as shale, are easily eroded.

284.6   Junction with *K-49.* This road runs 18 miles south to Caldwell, one of the last of the Kansas cow towns.

286.6   Junction with *K-271,* which goes south to the town of Mayfield.

287.6   The park a mile south of the road contains a marker commemorating the path of the *Chisholm Trail,* which came north out of Oklahoma and passed through this area on its way to Wichita and, eventually, Abilene.

288.9   *East Prairie Creek.*

291.0   *O'Hara oil field* on both sides of the highway. This field, discovered in 1955, produces from rocks in the Kansas City Group, about 3,200 feet deep. Rocks in this group crop out about 100 miles to the east, forming the hills in the Cherryvale area.

291.8   A railroad siding is all that remains of the town of *Roland,* immediately south of the highway.

292.7   Shale in the *Wellington Formation* on the north side of the highway and in the hillsides north of the highway. The shale, salt, and limestone that make up the Wellington were deposited early in the Permian Period, before the Permian red beds further west. Though the Wellington is composed of shale here, the formation also includes the Hutchinson Salt Member, the thick layer of salt that underlies central Kansas. Rock salt was discovered in Kansas in 1887 in nearby Wellington, when a hole drilled in search of coal struck the Hutchinson salt at 240 feet.

293.1   Shale in the *Wellington Formation,* on the north side of the highway.

293.7   *Slate Creek.* Slate is a metamorphic rock that is formed when a shale is subjected to intense heat and pressure. It maintains the slabby nature of shale but is much harder. Slate does not occur in this area, but shale is common, and hard shales are sometimes mistaken for slate.

293.7–296.5   *Wellington.* Wellington is the seat of Sumner County, which regularly leads the state in wheat production and is among the leaders in oil production. In 1984, Sumner County wells produced 1.4 million barrels of oil. The Wellington oil field begins two miles northwest of town.

295.0   Junction with *U.S. 81* which heads south. South Haven is 15 miles south; Caldwell is 27 miles southwest; and the Oklahoma border is 19 miles south.

295.7   Junction with northbound *U.S. 81.* Wichita is 33 miles north. At Wichita, U.S. 81 joins I–135 until they reach Salina, where I–135 terminates and U.S. 81 continues north to the Nebraska border and beyond. Before the Interstate system, U.S. 81 was one of the major north-south highways in the state.

296.0   *Hargis Creek.*

297.4   *Beaver Creek.*

298.0   Wells to the south are in the *Interchange oil field,* discovered in 1959. Oil is produced from the Simpson Group of formations, about 4,000 feet deep.

298.6   *Deer Creek.* Shale from the Wellington Formation is exposed along the creek on the south side of the highway.

298.8   *Kansas Turnpike interchange.* At this point, U.S. 160 passes over the Kansas Turnpike and I–35. The South Haven exit from the turnpike is 16 miles south of here, and the south Wichita interchange is 24 miles north. The route along the turnpike is described in chapter 8 of this book.

300.3   *Antelope Creek.* Kansas has eleven Antelope creeks, none of which is east of the Flint Hills. This probably reflects the range of pronghorn antelope before Kansas was settled.

302.5   The small town of *Dalton* is 0.5 miles to the south; *Belle Plain* is 8.5 miles north.

302.9   *Winser Creek.*

304.5   The *Oxford West oil field,* south of the highway, was discovered in 1926 and has since produced more than a million barrels of oil. The field is beginning to play out, however; 1984 production was only 12,000 barrels.

307.2   Nine miles south of here and four miles northwest of Geuda Springs there is an outcrop of gypsum in the Wellington Formation. This massive, or rock, gypsum is composed of a compact intergrowth of small crystals. Because of its hard, crystalline nature, this gypsum was called marble by local residents. It was quarried in the late 1800s and was used to construct a block of business buildings in Wellington that was known as the ''marble block.''

307.5   *Lost Creek* is parallel to and south of the highway.

307.5–308.5   The town of *Oxford,* named after the university in England. English names are common along this stretch of U.S. 160. Thirty miles east the highway passes through Cambridge. Oxford is located on the banks of the Arkansas River. About a mile north of town, a mill race diverted water out of the river and into a channel where it ran an electrical powerhouse, which has since been abandoned. About three miles north of Oxford the Arkansas and Ninnescah rivers join together.

South of Oxford is the Oxford oil field. The Churchill oil field is 3 miles north. These fields are located over the Nemaha Anticline, which extends from southeast Nebraska into central Oklahoma. These two fields have produced 38 million barrels of oil since they were discovered in the late 1920s.

Twelve miles south of Oxford is the small town of Geuda Springs, one of the mineral-springs resorts that was established in Kansas in the late 1800s. These springs produced saltwater, and around them were built a bathhouse, a hotel, and a bottling plant. Later a dam was built, creating a 30-acre pond of slightly saline water called Salt Lake.

308.5   *Arkansas River.* The Arkansas River is 1,459 miles long, the third-longest river in the continental United States. The river officially gets its start in Lake County, Colorado, though today it is dry through much of eastern Colorado and western Kansas. It joins the Mississippi in the state of Arkansas; the river drains more than 160,000 square miles.

In Kansas, tributaries of the Arkansas River are home to the

U.S. 160 crosses the Arkansas River west of Winfield at milepost 308.

alligator snapping turtle, the largest species of turtle in the state. Alligator snappers look like ordinary snapping turtles, although they grow much larger, weighing up to 200 pounds. Despite their size, alligator snappers are seldom seen because they are rare, secretive creatures that spend much of their time on the bottoms of rivers or pond. These turtles eat fish, frogs, snails, and other aquatic life; they attract small fish into their mouth by using a small wormlike appendage on their tongue as a lure.

    308.8   Small area of *sand hills*.

    308.9   *Sand pit* south of the highway.

    309.2   *Sumner/Cowley county line.* Sumner County is named after Senator Charles Sumner, a slavery opponent from Massachusetts. During debate over the Kansas–Nebraska Act of 1854—the debate to determine if slavery should be allowed in the new territories—Sumner was attacked by a South Carolina congressman who beat Sumner with a cane until he was bloody and unconscious. Nearly two years passed before Sumner recovered sufficiently to return to the Senate, where he remained a staunch supporter of Kansas and an opponent of slavery. Cowley County was named after Lt. Matthew Cowley, who was killed in a Civil War battle at Little Rock in 1864.

309.5 Small *oxbow lake* on both sides of the highway. This lake formed along an old channel in the Arkansas River. To the east of the lake, the highway climbs onto a low terrace that was formed during the Wisconsinan Ice Age.

310.5 *Slick–Carson oil field* south of the highway. This field, which extends southwest into Sumner County, has produced more than 8 million barrels of oil.

311.4 *Spring Creek.* Spring Creek is the most common name for streams in Kansas, but rarely are three Spring creeks as close together. This Spring Creek drains south into the Arkansas River. Another Spring Creek begins five miles southeast of here, flowing south into the Arkansas River east of Geuda Springs. Four miles east, still another Spring Creek drains east into the Walnut River.

312.0 The town of *Kellogg* is a mile north of here.

314.0 *Henderson oil field* on both sides of the highway. Oil in this field comes from Pennsylvanian limestones in the Kansas City Group, which are about 2,700 feet deep, and from Arbuckle Group rocks deposited in the Cambrian and Ordovician periods, which are about 3,400 feet deep.

315.5 This point marks the boundary between the *Arkansas River Lowlands*, to the west, and the *Flint Hills Upland* to the east. To the west, the landscape is a level or gently rolling plain that is underlain by the soft shales of the Wellington Formation, along with much younger alluvial deposits laid down by the Arkansas River. To the east is a hilly upland formed on layers of limestone and shale that dip to the west. Some of these limestones contain chert or "flint," which gave these hills their name.

315.8 *Strother Airport* is about five miles south of here.

316.7–320.0 *Winfield.* Located at the junction of the Walnut River and Timber Creek, Winfield is the seat of Cowley County and home to Southwestern College (a private Methodist school) and a state hospital, which is located on a hill north of town. Many of the town's buildings are constructed out of native limestone.

317.0 A short distance south of the road there is a small cave, located in a thick layer of limestone, probably the Cresswell Member of the Winfield Limestone. The cave appears to be mostly natural, although some man-made supports have been added to it. Because limestone layers here are particularly thick, caves are not unusual in this part of Kansas. In addition, an area near the town of Rock, north of Winfield, is known as a

A geode from the Cresswell Limestone Member of the Winfield Limestone

collecting spot for geodes, which sometimes form in the Cresswell. Geodes are crystal-lined cavities in rocks. Ground water has deposited minerals on the cavity walls, and the resulting crystals point toward the center of the cavity. Geodes, which are found occasionally in Kansas, particularly in the limestones of the Flint Hills, are prized by rock and mineral collectors.

317.4 *Walnut River.* It rises in northern Butler County and drains the western edge of the Flint Hills before joining the Arkansas River near Arkansas City.

317.7 Every autumn a festival of bluegrass music is held in the *fairgrounds* south of the road. Near the fairgrounds, in a bend of the Walnut River, is an area of river bottom called the Kickapoo Corral, where Indians reportedly kept horses.

318.5 Junction with *U.S. 77* and *K–15.* U.S. 77 forms Main Street in Winfield. To the north it runs to a city park on an island in Timber Creek. The town of Augusta is 32 miles to the north, and Arkansas City is 14 miles south.

318.7 *Cowley County Courthouse,* south of the highway, is built of native Fort Riley limestone, locally called Silverdale limestone, after the town where it is quarried. The courthouse's most distinctive feature is a map of Cowley County, carved into stone on the north side of the building.

320.0   The area 1.5 miles to the south is called the *Cup and Saucer Hills.* Two prominent limestones—the Winfield and the Fort Riley—produce the unusual topography in this area. The Winfield Limestone, which caps many of these hills, often erodes into isolated rounded buttes that have the appearance of overturned cups. The layer below the Winfield is the Fort Riley, which is riddled with caves and solution cavities, some of which have collapsed to form shallow saucer-shaped sinkholes.

320.6   *Black Crook Creek.*

320.7   *Fort Riley Member* of the Barneston Formation, on both sides of the highway. This is one of the limestones that is commonly used for building stone.

321.9   *Gage Shale Member* of the Doyle Shale, overlain by the *Winfield Limestone.* North and south of the highway are oil wells in the Winfield oil field, which is located along the Winfield Anticline, a linear, upward fold of rocks that is parallel to but smaller than the Nemaha Anticline, about 15 miles to the west. The Winfield Anticline affects the position of the Winfield Limestone and underlying rocks. For example, this outcrop is 1,300 feet in elevation, yet at the west end of the Walnut River bridge in Winfield, the formation is exposed just above bank level, at an elevation of 1,120 feet. Natural gas was discovered in this field in 1902, the first such discovery in Kansas west of the Flint Hills. For a time, gasoline was produced from this natural gas; 1,000 cubic feet of natural gas yielded one-third to one-half gallon of gasoline.

322.5   *Winfield Limestone.* About 25 feet thick, the Winfield is one of the prominent limestones in the western flank of the Flint Hills, appearing at the surface as far north as Washington County.

322.9   *West Badger Creek.* Badgers are found throughout Kansas, although they may be more common in the central part of the state. During the summer, badgers are primarily nocturnal and are predators of gophers, prairie dogs, ground squirrels, mice, and rabbits.

323.5 and 323.7   *Winfield Limestone.*

325.1   *East Badger Creek.* Four miles to the southwest, East and West Badger creeks converge to form Big Badger Creek.

326.4   The town of *Tisdale.*

327.0   *Fort Riley limestone,* another of the rock layers common in the Flint Hills. The formation ranges from 30 to 45 feet in thickness and often forms the rim rock in Flint Hills pastures.

327.3 *Snake Creek.*

328.2 *Fort Riley limestone.*

328.6 *Silver Creek.* About 15 miles downstream is the town of Silverdale, where Fort Riley limestone is extensively quarried for use in construction. At Silverdale the Fort Riley is nearly 60 feet thick, and in places it is necessary to drill and blast blocks of rock away from the ledge, rather than simply cutting them away from the exposure. Once the blocks of stone are removed, they are taken to a plant and cut into slabs on a gang saw—a series of long, parallel saw blades that cut through the rock, which is then cut into desired shapes and dressed.

328.6–328.9 *Fort Riley limestone.*

330.9 Junction with *K–15,* headed east, and a *roadside park.* The town of Eaton is one mile south; Dexter is six miles east and south, on K–15. In 1903 a well near Dexter produced large volumes of natural gas. Anticipating the economic boom the gas would bring to the area, Dexter's town fathers decided to hold a public lighting ceremony, flaring off the gas and making it visible for miles around. When a crowd assembled, officials tried three times to light the well, and each time it blew out their flame. H. P. Cady, a professor at the University of Kansas, analyzed the gas and found that it contained slightly less than 2 percent helium, the first time that helium had been discovered in natural gas. A plant was constructed to extract the helium, but it was later closed, although helium is still extracted from natural gas at several other locations in Kansas. Helium is used in balloons and missiles.

333.8 *Fort Riley limestone.*

333.9 *Plum Creek.*

336.4–336.9 The town of *Burden.*

338.2 The *Florence Limestone Member* of the Barneston Limestone. The Florence is named after a small town in Marion County; it is one of the chert-bearing limestones common in the Flint Hills.

338.4–338.6 *Kinney Limestone Member* of the Matfield Shale.

338.8 *Threemile Limestone Member* of the Wreford Limestone.

338.9 *Blue Branch of Grouse Creek.*

339.1 and 340.2 *Threemile limestone.*

340.5 *Grouse Creek.* This creek was probably named for the wild grouse that inhabit the native grasslands of the Flint Hills. These grouse are more commonly called prairie chickens. Kansas boasts the largest population of these birds in North America. Male prairie chickens put on extravagant displays during middle and late March. Spreading his tail feathers, the male fills himself with air, puffs up to twice his normal size, then expels the air, creating a booming sound. This ritual takes place on so-called booming grounds, which are usually located on prominent wind-swept hilltops, where the grass is shorter and the male is more visible to interested females.

A smaller relative of the prairie chicken is the ruffed grouse, which was found in the woodlands of eastern Kansas until it disappeared around 1900 because of the destruction of its habitat. Today, the ruffed grouse is being reintroduced to the woodlands of Jefferson and Douglas counties in northeastern Kansas.

340.8   *Funston Limestone.*

341.7   The town of *Cambridge.*

343.0   The hills south of the road are capped by the cherty *Wreford Limestone.*

345.0   *Crouse Limestone.*

345.5–348.5   For these three miles the road follows *Cedar Creek*, which flows to the southwest and is crossed by the highway at milepost 348.5.

349.0   The former town of *Grand Summit* was one mile to the north. The town site is now only a railroad siding, where a spur of the railroad was built alongside the main track. At this site, a branch of the Santa Fe Railroad crests the Flint Hills.

350.0   *Crouse Limestone.* This formation ranges from 6 to 18 feet in thickness. The elevation here is 1,500 feet, the highest point along U.S. 160 in the Flint Hills. Twelve miles north of here the elevation is 1,680 feet, the highest point in the Kansas Flint Hills.

350.3   *Spring Creek.*

350.4   *Cowley/Elk county line.* Elk County is named after the Elk River, which flows diagonally to the southeast across the county. Before the arrival of whites in the 1800s, elk were common in Kansas, particularly in the prairies and along the edges of woodlands in eastern Kansas. Settlers generally eliminated elk from the state, but a small herd has been reintroduced in the Cimarron National Grasslands of southwestern Kansas.

350.6   The elevation is 1,400 feet.

352.2   *Foraker Limestone.* The Foraker Limestone forms a prominent east-facing slope, or escarpment, in this part of the Flint Hills. The elevation is 1,300 feet.

352.5 This is the approximate boundary between Permian rocks to the west and Pennsylvanian rocks to the east. The shales, limestones, and cherts to the west were deposited early in the Permian, about 290 million years ago; they make up the Flint Hills. The shales and limestones to the east were deposited slightly earlier, during the Pennsylvanian Period.

353.6 *Channel sandstone*, in the French Creek Shale Member of the Root Shale. Sandstones were deposited at the floor of streams that drained this area during the Pennsylvanian Period.

353.7 *Caney River.* The Caney flows south, along the western edge of Elk County, into Chautauqua County, Kansas, where it creates a waterfall at Osro Falls. After crossing into Oklahoma, it is dammed up to form Hulah Reservoir; it then flows south, joining the Verdigris River near Tulsa. The elevation here is 1,100 feet, 400 feet lower than the crest of the Flint Hills, a few miles to the west. The Caney River marks the eastern edge of the Flint Hills along U.S. 160. To the east are the Osage Cuestas.

354.4 *Stotler Limestone.*

354.7 The town of *Grenola* is immediately south of the highway.

355.8 *Corum Creek.*

357.0 The *Flint Hills escarpment* dominates the view to the west of the highway.

358.6 The wells north of the road are part of the *Starr oil field*, discovered in 1937.

360.2 *Walker oil field*, which produces oil from rocks in the Lansing and Kansas City groups, which were deposited during the Pennsylvanian and are about 1,550 feet deep. These same rocks crop out in Montgomery County, the next county to the east on U.S. 160. This field also produces oil from Mississippian rocks at a depth of 2,200 feet.

362.0 *Wildcat Creek*, on the north edge of Moline, is crossed by what is reportedly the oldest suspension bridge in Kansas.

362.6–363.5 The town of *Moline*, named after Moline, Illinois, which is famous for the manufacture of farm machinery. About 4.5 miles southwest of Moline was the town of Boston, established by a group of Irish settlers in the 1870s. Boston competed with Elk Falls to be the county seat, and at one point, Boston residents stormed Elk Falls, captured the county records, and declared Boston the county seat. After losing a legal battle over the question, Boston also lost most of its population and had become a ghost town by 1885.

Erosion along natural joints in limestone creates Osro Falls along the Caney River south of U.S. 160 in Chautauqua County.

363.9   *Topeka Limestone.*

364.1   Junction with *K-99,* running to the south. This highway heads 19 miles south to Sedan, the seat of Chautauqua County and the birthplace of circus clown Emmett Kelly. South of Sedan, near the Oklahoma border, is the small town of Elgin. The railroad reached Elgin in the late 1800s, and it became a major point for shipping cattle out of Indian Territory. By the turn of the century, Elgin had turned into a rough, lawless boom town, with a population of 1,100, which shipped out more than 6,000 carloads of cattle per year. When the railroad pushed further south, the boom came to an end. The depression provided the *coup de grâce,* and today the town has only 139 residents. Elgin continues to hold on, however; on Main Street a sign proclaims that Elgin is "A Town Too Tough to Die." North of Elgin, on a hillside overlooking the town, in the Elgin cemetery, is the grave of Romulus Hanks, a first cousin of Abraham Lincoln.

364.6   *Wildcat Creek.*

364.7   Junction with *K-99,* headed north. Howard, the seat of Elk County, is 7 miles north. The county courthouse, completed in 1908, is east of the town's business district.

364.8   *Coal Creek Limestone Member* of the Topeka Limestone.

365.0–366.5   *Moline Quarries.* Taking limestone out of the Ervine Creek Member of the Deer Creek Limestone, this is one of the largest quarries in Kansas. Most of the limestone from this quarry is ground into gravel for use in road and highway construction. The Ervine Creek limestone is from 5 to 32 feet thick in other places, although it averages 20 feet in these quarries and is overlain by 20 feet of Topeka Limestone, which is treated as overburden and is removed to reach the Ervine Creek. The Ervine Creek is light gray to nearly white and is very pure, its composition averaging 95 percent or more calcium carbonate. It also contains a variety of fossils, including fusulinids, corals, crinoids, bryozoans, brachiopods, and mollusks.

369.5–370.4   *Elk Falls.* This town, named for a waterfall in the Elk River, is located at the junction of Wildcat Creek and the Elk River. When Elk Falls was settled in 1870, Elk County and Chautauqua County to the south were part of one large county called Howard County. Elk Falls competed with several other towns to be county seat, and in the furor, the county was split. Elk Falls was no longer near the center of the new county, so Howard became the county seat.

370.4   *Oread Limestone.*

370.5   *Elk River.*

371.6 *Oread Limestone*, overlain by the *Kanwaka Shale*.

373.0 *Oread Limestone*. This formation crops out in Douglas County. It is named for Mount Oread, site of the University of Kansas.

374.6 *Snyderville Shale Member* of the Oread Limestone.

376.1 Oil wells in this vicinity are part of the *Longton oil field*, which extends to the north and east. This field was discovered in 1902; it produces oil from the top of Mississippian limestones that have been draped over a 25-mile-long anticline called the Longton Ridge. This ridge is not apparent at the surface, but the Pennsylvanian rocks on top of this ridge are 250 feet higher than they are a short distance to the east.

376.7 The town of *Longton*.

377.3 *Hitchen Creek*.

378.7 *Painterhood Creek*.

379.5 Wells here are part of the *Longton oil field*. Geologically recent stream-deposited gravel blankets some of the older formations in this location. This point also marks the boundary between the Osage Cuestas, to the west, and the Chautauqua Hills, to the east. The Chautauqua Hills constitute a wooded area, formed by thick limestone and sandstone outcrops of the Douglas Group, deposited during the Pennsylvanian Period.

379.7 The hills here are capped by the *Ireland sandstone*, a thick Pennsylvanian sandstone that was deposited in an ancient river valley that extended as far north as Douglas County. The Ireland is an important aquifer in Elk County, where many rural areas are dependent on ground water. Six miles north, near the small town of Busby, water in the Ireland is confined between impervious shales above and below, forming a sort of sandstone sandwich. In places, the pressure on the water forces it to the surface in artesian wells.

380.7 The *Elk River* is on the west side of the highway as the road follows the river's flood plain.

381.3 *Hickory Creek*. Because of rock outcrops in this area, the hills are generally not cultivated but are left for pasture. Many are heavily wooded by stands of blackjack oak, post oak, and other hardwoods. This mix of medium-tall grasslands and scattered stands of deciduous trees is called the Cross Timbers by scientists who map vegetation. In Kansas, the Cross Timbers closely corresponds to the Chautauqua Hills physiographic region. This marks the northern extension of the Cross Timbers, which extend south into central Texas.

382.3 The town of *Oak Valley*.

383.5 On the hillside to the north, the *Tonganoxie sandstone* is overlain by the *Robbins shale*. The Tonganoxie is similar to Ireland sandstone, which is exposed a few miles to the west. Both are sandy deposits that filled a large, ancient valley during the Pennsylvanian Period. These two sandstones cap the Chautauqua Hills in this area.

384.9 *Bachelor Creek.*

385.1 *Elk/Montgomery county line.* In the Elk River, 0.7 miles south of the road, the elevation is 820 feet, the lowest point in Elk County. The highest point in Elk County is 1,645 feet, in the Flint Hills in the northwest corner of the county, which is also where the Elk River gets its start. Despite its small size, Elk County has 825 feet of topographic relief between its high and low points, one of the greatest amounts of relief of any Kansas county. Because of this change in altitude, the Elk River has completed half its vertical drop to sea level even before it leaves Elk County.

386.0 *Elk City oil and gas field.* The porous rocks in this field are used for the storage of natural gas.

389.0 The hill to the west is capped by the *Tonganoxie Sandstone Member* of the Stranger Formation. The elevation at the top of the hill is 1,070 feet, compared to 850 feet here along the road. The Tonganoxie is named for a small town in Leavenworth County.

389.7 Duck Creek flows west into the Elk River, slightly to the west of Elk City.

390.3 *K-39.* To the west the highway goes into Elk City.

392.2 *Elk River.* East of here the Elk River flows into the Elk City Reservoir, before joining the Verdigris River.

392.3 *South Bend Member* of the Stanton Limestone.

393.0 Wells in the *Sorghum Hollow oil and gas field,* which extends west of the highway, produce oil from a sandstone in the Bandera Shale. The Bandera is 800 feet deep here but crops out to the east in Labette County.

393.2 *Weston Shale Member* of the Stranger Formation, on the east side of the road. Named for the Missouri River town in northwest Missouri, this fossiliferous shale is as much as 120 feet thick.

393.8 The oil wells to the west are part of the *Sorghum Hollow oil and gas field.* Several of the wells produce oil from the Arbuckle Group of formations, which were deposited in the Cambrian and Ordovician periods of geologic history.

394.0 *Card Creek.*

395.5 The wells to the south are part of the *Coleman oil field.*

Like other fields in this area, the Coleman produces oil from the Arbuckle Group of formations from a depth of about 1,700 feet. The Arbuckle is found only in the subsurface in Kansas, although it does crop out in Missouri. Composed mostly of dolomite, the Arbuckle is up to 1,200 feet thick and was deposited about 500 million years ago. In addition to producing oil, the Arbuckle is used for the disposal of hazardous waste and of the brine that is produced along with petroleum. In southeastern Kansas, where it is closer to the surface, the Arbuckle is also an important fresh-water aquifer because of the relative scarcity of ground water. Because the Arbuckle dips to the west, brine can be returned to the formation in central and western Kansas without threatening the fresh water to the east.

395.7   *South Bend Limestone Member* of the Stanton Limestone Formation.

395.8   *Coon Creek.*

395.9   *South Bend limestone.*

396.5   To the north is the hill *Bald Mound,* elevation 975 feet.

397.5   *Timber Hill* is north of the highway. Its elevation is 1,050 feet, compared to 880 at the road. The hill is composed of shale and sandstone from the Stranger Formation.

398.0   Shale in the *Stanton Formation.* This point marks the approximate boundary between the Chautauqua Hills physiographic region, to the west, and the Osage Cuestas to the east.

398.8   *Chetopa Creek.* This creek is named after a chief of the Osage Indians, who occupied a reservation in this area.

399.0   *Lane Shale* overlain by the *Bonner Springs Shale.* The Wyandotte Limestone, a prominent formation in the Kansas City area, separates these two shales in east-central Kansas. However, it pinches out or disappears to the south in Anderson County. South of there, these two shales are difficult to separate and therefore are usually mapped as one formation.

400.0   South four miles is the small town of *Bolton.* The first oil produced in Montgomery County came from wells drilled near Bolton in 1903. These wells were the first in the state to produce more than 1,000 barrels per day.

400.5   The dam for *Elk City Lake* is four miles to the north.

400.7   *Squaw Creek.* Walker Mound, elevation 950 feet, is 0.8 miles south. The elevation at the road is 800 feet.

401.1   Here the road crosses a *levee* that is used as flood storage for Elk City Lake. Table Mound, elevation 1,000 feet, is 2.5 miles to the north. This hill is capped by Stanton Limestone.

402.0 Junction with *U.S. 75* running south. Four miles south of here, along U.S. 75, is the Independence Municipal Airport, formerly a base for the Army Air Force during World War II. The town of Caney is 19 miles to the south.

403.8 West edge of *Independence,* the seat of Montgomery County. Independence was the home of Alfred Landon before he was elected governor of Kansas in 1933. In 1936 the former oil man was the Republican nominee for president. Independence was also the home of Martin Johnson, who, along with his wife, Osa, produced motion pictures about Africa, the South Seas, and Borneo. The couple made photographic expeditions from 1917 to 1935, providing glimpses of unfamiliar wildlife for movie audiences in the United States. Today the work of Osa and Martin Johnson is commemorated in a museum at Chanute. Independence is also the birthplace of the Pulitzer Prize–winning dramatist William Inge, author of *Picnic, Bus Stop, Come Back Little Sheba, Dark at the Top of the Stairs,* and *Splendor in the Grass.* Many of his stories were set in Kansas.

405.4 Junction with northbound *U.S. 75* and *K-96.* Along U.S. 75, 13 miles north, is the town of Neodesha. Near this town was the site of Norman No. 1, a well that was drilled in 1892 and struck oil at 832 feet. It opened up oil exploration throughout the mid continent, including this part of Kansas, Oklahoma, and Texas. A replica of the drilling apparatus is on display at a Neodesha park.

Drilling was common around here in the early 1900s, and the area produced more than its share of "gushers" that blew drilling tools completely out of the hole. Today, most oil wells are drilled by rotary rigs, which are equipped with bits that gnaw their way through rock. In the early days of Kansas oil exploration—and even today in some of the shallower fields—drillers used cable rigs that lifted drilling tools up and then dropped them down, pounding holes into the rock.

406.2 East edge of *Independence.* This is one of the southeastern Kansas towns that benefited from the discovery of natural gas and the subsequent economic boom in the late 1800s and early 1900s. With the lure of plentiful gas supplies, industries rushed to relocate in this area and in the counties to the north and east. Cement plants and glass plants, both of which depended on steady energy supplies, were common; zinc smelters were established in this area to refine ore from lead and zinc fields in the tri-state district. As the gas played out, many of those industries closed up or moved on. Three glass factories and two cement plants were located in Independence between 1890 and 1930. Today, only one cement plant remains.

406.9 *Verdigris River.* The elevation at the riverbed is 720 feet, the lowest point along U.S. 160 in Kansas. From here the river cuts south, and just before entering Oklahoma, it reaches the lowest point in Kansas at 680 feet. A mile south of here are a quarry and cement plant that takes rock out of the Corbin City Member of the Drum Limestone.

407.0 *Corbin City Member* of the Drum Limestone. The Corbin City is a limestone layer that, at 50 feet, is thickest around Independence. The same formation is only about a foot thick near Kansas City.

407.8 *Mouse Creek.* This area was part of the Osage Indian Reservation, which was established in 1825 and extended west to near Ashland. In 1863, a group of twenty Confederate soldiers, on their way to recruit troops in the West, were intercepted here by Osage Indians, who chased them onto a gravel bar in the Verdigris River three miles to the north. All but two of the Confederates were killed, in what has since been called the Battle of Rebel Creek.

407.9 *Corbin City Member* of the Drum Limestone to the south of the highway. A creek called Bloody Run, which drains into Mouse Creek, is north of the road.

409.6 *Noxie Sandstone Member* of the Chanute Shale. This is another sandstone that was deposited in the channel of a Pennsylvanian river. It is up to 100 feet thick.

409.7 *Noxie sandstone.*

410.2 *Chanute Shale.* This shale ranges from about 12 feet thick near Kansas City to 200 feet thick in southeastern Kansas.

410.8 *Drum Creek,* elevation 750 feet, which drains to the southwest, emptying into the Verdigris River near Independence. On a site along Drum Creek in 1870, Osage Indians agreed to sell their lands in southeastern Kansas and to move to Indian Territory, south of the Kansas border. Ironically, the Osages moved onto oil-rich land in eastern Oklahoma, which made them one of the wealthiest Indian tribes in the West. It also made them targets for con artists and swindlers, and for years, Indian oil scandals in eastern Oklahoma were legendary.

412.0 Junction with *U.S. 169* and *K-96,* to the south. Coffeyville is 15 miles south on U.S. 169. In 1892, the Dalton Gang attempted to rob two Coffeyville banks at the same time, and four gang members were killed in a subsequent gun battle that is commemorated at a Coffeyville museum. The only surviving member of the Dalton Gang, Emmett, served a prison stretch and then moved to Hollywood, where he helped make westerns. Coffeyville was also the long-time winter home of major-league pitching star Walter ("Big Train") Johnson, who was born in nearby Humboldt.

The hills to the south and east, with an elevation of about 900 feet, are capped by Drum Limestone. About 14 miles south of here, in southeastern Montgomery County, is the Treaty Rocks site, where Indian petroglyphs were carved onto native rock. These drawings, probably done by the Osage Indians whose reservation covered this area, include pictures of elk, horses, and buffalo.

414.5 To the southeast 0.5 miles is a *shale pit* in the Cherryvale Shale Formation, which produced the raw material for bricks. Here U.S. 160 is about 15 miles from the Oklahoma border. This far south, the highway encounters the habitat of animals that are not normally found in other parts of Kansas. For example, armadillos are often found in southern Kansas. Mainly nocturnal, armadillos thrive on insects, worms, fruits, and berries. According to some sources, armadillos appear to be extending their range northward into Kansas—one specimen was reported in Ellis County— although it is unclear if the animals are naturally moving north or if specimens are being transported and released by people.

415.5–416.7 *Cherryvale.* This town was the home of Vivian Vance, who is famous for her portrayal of Ethel Mertz on the "I Love Lucy" television show.

417.8 *Cherry Creek.* The wells west of the highway are part of the Coffeyville–Cherryvale oil and gas field. This field, discovered in 1892, was part of the gas boom that enveloped southeastern Kansas in the late 1800s and provided the gas for the zinc, brick, and cement industries that flocked to the state.

419.2 *K–37* runs 16 miles north to Neodesha.

419.5 The hills to the west are topped by the *Drum Limestone;* their sloping sides are formed on the Cherryvale Shale. The Drum Limestone, named after the creek two miles to the west, runs south across the Oklahoma border. In Oklahoma, the Drum is called the Dewey Limestone, and the Cherryvale Shale is called the Nelie Bly Shale. Generally, geologists in different states try to give formations the same name, no matter where the rocks are found. However, sometimes they are unable to decide which name should be given priority, so formations may have different names in different states.

420.1   Wells to the west are part of the *Coffeyville–Cherryvale oil and gas field*. The Dennis Limestone is on the west side of the road.

421.0   Junction with *U.S. 169*, headed north. Chanute is 25 miles north of here. The seat of Neosho County, Chanute is named for Octave Chanute, an engineer who helped to construct a bridge over the Missouri River in Kansas City in 1869. Twenty miles northwest of Chanute is the small town of Piqua, the birthplace of silent-film star Buster Keaton.

421.6   *Montgomery/Labette county line*. Montgomery County is named for Gen. Richard Montgomery, a hero of the Revolutionary War who died at the Battle of Quebec. Montgomery County was originally part of Wilson County, to the north, but it became a separate entity in 1867. While the name for Labette County is clearly French, the exact source is not certain, although it may be named after French explorer Pierre Le Bete.

424.0   These hills, capped by Drum Limestone, were originally called the Labette Mounds, but the name was changed to *Bender Mounds*, after a homicidal family that lived a mile north of here. The Bender family— probably the most famous member was daughter Kate, who was known locally as a spiritualist—built a house in 1871 along a trail that ran from Independence to the Osage Indian mission northeast of here, near the town of St. Paul. The Benders served meals to travelers and operated a small trading post. After several missing people were traced to this area, local residents decided to check on the Benders. The family had gone, leaving behind eleven bodies in their garden. Apparently the Benders seated guests in their house with their backs against a curtain that served as a partition. The Benders then smashed their victim's head with a hammer. The family was never found, although their crimes are remembered in a Cherryvale museum.

424.2   Wells in this area are part of the *Dartnell oil and gas field;* they produce oil from a sandstone formation in the Pennsylvanian-age Cherokee Group, which crops out to the east, in Cherokee County.

424.5   *Roadside park.*

425.0   The hills to the north 1.2 miles are called *Twin Mounds.*

425.4   *Big Hill Creek*. Five miles south, this creek is dammed to form Big Hill Lake. The lake is formed in a scenic valley that is carved by Big Hill Creek and is called the Labette County Ozarks. These forested hills are capped by the Swope Limestone.

427.5   *K-133* leads 0.5 miles north to the town of Dennis. The

Dennis Limestone Formation was named after this town. The type section for the Dennis Limestone is found nearby. A type section is a location where the layers that make up a formation are found in their typical sequence. A type section is important to geologists in the same way that a type specimen, showing a typical plant or animal, is important to biologists.

429.5   The town of *Fern* is 0.5 miles north.

432.4   *Little Labette Creek.* The shale exposed on the south side of the highway is part of the Tacket Formation, which includes layers of sandstone, shale, and bluish-gray limestone. It was named for exposures on the slopes of Tacket Mound, six miles to the southwest.

434.0   West edge of *Parsons,* the seat of Labette County, which was named after railroad official Levi Parsons.

435.0   *Parsons State Hospital* is 0.5 miles north. Originally constructed for the treatment of epileptics, it is now a state hospital for the mentally retarded.

436.0   Junction with *U.S. 59.* To the west, U.S. 160 splits around a mall-like downtown area. The town of Erie is 16 miles north, and Oswego is 20 miles southeast of here.

436.6   *Labette Creek.*

437.5   *East edge of Parsons,* which is the birthplace of movie star Zasu Pitts.

438.5   South of the road is the *Kansas Army Ammunition Plant.*

443.2   The county road here leads 0.5 miles south to *Laneville* and 4.5 miles south to Montana.

444.3   *The Neosho River,* which produced the state record flathead catfish. Caught by Ray Weichert of Brazilton in August 1966, the fish weighed 86 lbs. 4 oz. and was 55 inches long. The Neosho is also home to the paddlefish, or spoonbill, which in Kansas is found only in the northeastern and southeastern parts of the state. Paddlefish have a long spoon-shaped snout, which gives them their name; they look like no other fish found in North America. The largest paddlefish ever taken in Kansas was caught in the Neosho near the town of Chetopa, south of here. Taken in 1973, the fish weighed 73 pounds and was 69 inches long. Paddlefish feed entirely on plankton, microscopic organisms that they strain out of the water. However, paddlefish can be caught by snagging—dragging a heavily weighted hook along the bottom of the river until it snags the fish— particulary during their spring spawn. Because they have no bones or scales, paddlefish are easy to clean and are thus a popular game fish in southeastern Kansas.

444.5   *Hickory Creek,* which drains into the Neosho immediately south of the highway. The Neosho is one of the major rivers in southeastern Kansas, carrying more water out of the state than any other stream except the Kansas River. Because the surrounding country is flat, the river has a large flood plain. During spring rains, the Neosho often widens and covers several miles. During droughts it can dry up completely.

445.5   The town of *Strauss.*

446.2   *Litup Creek,* which also marks the point at which Crawford, Cherokee, and Labette counties come together. Labette County is west of this point, while eastbound U.S. 160 forms the boundary between Cherokee County, to the south, and Crawford County to the north. This point also marks the former border between the Osage Indian Reservation, to the west, and the Cherokee Neutral area to the east. The Cherokees were given this ground by treaty, though they never lived here. Today they live primarily in northeastern Oklahoma, where they are one of the largest Indian tribes in the country.

449.0   This small break in the landscape is formed by the outcrop of the Fort Scott Limestone and marks the approximate boundary between the Osage Cuestas, to the west, and the Cherokee Lowlands to the east. The Osage Cuestas are a series of east-facing escarpments formed by outcropping limestones and shales of Pennsylvanian age. These erosion-resistant limestones form the rims of these escarpments, or cuestas. The rocks in the Cherokee Lowlands are also Pennsylvanian in age, but they are soft shales, siltstones, and coals, which are easily eroded and have been worn down to a gently undulating plain that stretches eastward to the Ozark Plateau.

449.5   Junction with *K–126,* running north. McCune is one mile north of the road.

451.1   *Mulberry Creek.*

452.2   *Lightning Creek.*

453.0   The hill to the south of the highway is capped by the *Fort Scott Limestone.*

453.5   The town of *Monmouth* is 1.5 miles north.

454.2   *Strip pits* are apparent north and south of the highway here, as well as scattered throughout the countryside on both sides of the road. The Bevier (pronounced buh-veer') coal bed was mined in this area. The Bevier is one of the uppermost coals in the Cherokee Group of formations; it averages 14 to 18 inches in thickness.

454.5   The town of *West Mineral* is four miles south of here on a

Big Brutus, a coal-shovel-turned-museum south of the town of West Mineral, four miles south of U.S. 160 in Cherokee County

county road. A mile southwest of the town is Big Brutus, an abandoned coal shovel that was among the largest in the world. Today it is a museum, with exhibits related to coal mining and southeastern Kansas.

455.5 The *quarry* south of the highway is in the Fort Scott Limestone.

455.8 *Little Osage shale* overlain by the *Higginsville limestone,* both members of the Fort Scott Limestone. At this location, Little Osage shale is a dark-gray platy shale, which contains marble-size nodules that are rich in phosphate. In places along the outcrop, the nodules have weathered out of the host rock and are scattered on top of the ground. Studies have shown that these nodules contain as much as 30 percent phosphate and up to 0.03 percent uranium. Little Osage shale has also produced up to 23.3 gallons of oil per ton of shale. Thus, it may someday be possible to mine such shales for their oil, uranium, and phosphate, which could be used as

Marble-size phosphate nodules in the Little Osage shale in Cherokee County

fertilizer. By recovering all three minerals, mining might be profitable, although it is not now.

456.1   *Strip pits* north and south of the highway are in the Bevier coal bed.

456.6   *Wolf Creek.*

457.0   To the north 0.3 miles is a *clay pit* in the Cabaniss Formation. Four miles south is an old mining town called Carona. During the days of prohibition, southeastern Kansas was famous for producing moonshine, which became known as deep shaft whiskey because it was allegedly manufactured in old mine shafts in order to avoid the authorities. Liquor produced from the Carona area was particularly well known; it was even given the brand name of Carona.

458.5   *Wolf Creek.*

458.8   To the south is *strip-mined land* that has been reclaimed.

459.5   Junction with *K-7*. To the north, this highway runs to Girard, 13 miles away; to the south it goes four miles to Scammon and 12 miles to Columbus.

459.8   South of the highway is a *gob pile,* and beyond that are abandoned strip mines. These are the visible signs of the coal-mining industry that flourished in this area from the 1870s into the 1970s. Gob piles are composed of waste material that was separated from the coal brought up via underground mining. Gob piles mark the site of nearby mine shafts. The abandoned strip mines are made up of linear piles of soil, shale, and siltstone, which was removed by large mechanical shovels and draglines to reach the underlying coal. In this area, the Mineral coal bed was strip mined at the surface, above deep mines in the Weir–Pittsburg coal bed. Although strip mines are the most visible reminder of coal mining in Cherokee and Crawford counties, most of the coal was removed from underground, generally in the Weir–Pittsburg coal seam.

460.0   The town of *Cherokee* is north of the highway. A pond created by strip mining, about a mile south of the highway, is called Cherokee Tank.

461.3   *Strip mines,* to the north and south, take coal from the Mineral coal bed. These mines mark the outcrop of the Mineral coal along this portion of U.S. 160.

462.0   A *gob pile* is visible south of the highway.

462.5   The county road here goes two miles south to the town of *Weir.* Daisy Hill is 0.5 miles south of the road. Weir is located along the outcrop of the Weir–Pittsburg coal bed and provides half of its name. In addition to coal mines, a brick plant was once located at the south end of Weir. In the 1870s the first zinc smelter west of St. Louis was erected here. Similar industries sprang up in Pittsburg a little later, taking advantage of cheap energy provided by surrounding coal fields.

463.0   *Strip mines* in the Weir–Pittsburg coal bed to the north of the highway. From here the Weir–Pittsburg extends northeast to Pittsburg and southwest towards Columbus.

Strip pits, once the source of coal, now provide fishing in Cherokee County.

463.4 *Brush Creek.*

466.4   Junction with *U.S. 69*. At this point, eastbound U.S. 160 joins U.S. 69 and adopts the numbering system for U.S. 69 for the next nine miles. Westbound U.S. 160 departs from U.S. 69 here and returns to its original numbering system. Along U.S. 69, Crestline is 11.5 miles south, and Miami, Oklahoma, is 41 miles south.

475.2   Junction with *U.S. 69*. Eastbound U.S. 160 departs from U.S. 69 here and heads east. Westbound U.S. 160 joins U.S. 69 here and adopts the latter's numbering system for the next nine miles. The geology along that part of U.S. 160 is described in chapter 6. This is also the west edge of the town of Frontenac, named after a governor of New France, which included the Mississippi River valley and Canada.

476.5–476.8 *Strip mines* in the Mineral coal bed east of the highway. Coal is composed of the organic debris that compacted in a dense swamp during the Pennsylvanian Period of geologic history. It took about 10 feet of leaves, tree trunks, and other organic matter to produce a one-foot layer of coal.

477.4   North edge of the town of *Frontenac*.

477.5–477.8   The land south of the highway has been reclaimed after strip mining in the Mineral coal bed. These mines mark the outcrop of the Mineral coal bed along U.S. 160. A slight dip makes the Mineral coal bed deeper to the west.

479.0   *East Cow Creek.*

479.4   The old mining town of *Yale* is one mile north of the highway. Large gob piles dot the landscape here, indicating the presence of shafts leading to underground mines in the Weir–Pittsburg coal bed.

480.2   The old mining camp of *Cornell* is 0.5 miles south of here. In eastern Kansas, U.S. 160 passes near four towns with names celebrated in the annals of higher education. Cornell is south of here; Yale is northwest; Cambridge is further west, in Cowley County; and Oxford is west of Winfield in Sumner County.

480.8   A *gob pile* is located 0.2 miles north of the highway.

480.9   *Kansas/Missouri state line.* The town of Mindenmines, located in the midst of strip mines in the Weir–Pittsburg coal bed, is 1.5 miles east. Lamar, Missouri, the birthplace of Harry Truman, is 20 miles to the east. Westbound U.S. 160 passes through the coal fields of southeast Kansas. These coals, which were deposited in the Pennsylvanian Period, are in the Cherokee Group of rocks, which is composed of shale, siltstone, and coal beds, several of which have economic value. The rocks in the Cherokee Group are soft and easily eroded, so that the area where they crop out has been leveled by erosion to a gently undulating plain known as the Cherokee Lowlands.

70·INTERSTATE·70

From Denver to Kansas City

0.0 *Kansas/Colorado state line.* Kansas measures 411 miles from east to west along its southern border. The trip across Kansas on I–70 is 424 miles long, because the interstate doesn't travel directly across the state. Today's state line has not always marked the western edge of Kansas, however. In the days of the Kansas Territory, the border extended to the Continental Divide in Colorado, including Denver and much of the Rocky Mountains. When Kansas became a state in 1861, its organizers lopped off the western portion of the state.

Although the Rocky Mountains begin 150 miles west of the state line, their presence affects the geology and climate of western Kansas. Moisture-laden air, as it travels across the Rockies, is lifted to high elevations and cooled, causing the moisture to condense and fall as rain and snow over the mountains. This air then enters the plains of eastern Colorado and western Kansas much drier than before, creating a "rain shadow" of low precipitation over the High Plains. Rainfall in Kanorado averages less than 18 inches per year. Precipitation gradually increases across the state to more than 35 inches per year in Kansas City. The windiness, low humidity, and abundant sunshine of western Kansas further reduce the effectiveness of the small amount of precipitation that does fall, creating a semiarid climate. Crops can be grown only with irrigation or dry-land farming techniques, which require leaving fields unplanted for a year to build up moisture in the soil. The natural vegetation of western Kansas reflects the arid climate. Short grasses (which are drought tolerant) are dominant, trees are scarce, and desert-type plants, such as cactus and yucca, are common.

Surface water is scarce in western Kansas. Porous soils allow rapid infiltration of the scant precipitation, and the streams are intermittent, carrying water only briefly after periods of rain. However, much of western Kansas is covered with a thick deposit of sand and gravel, called the Ogallala Formation, which was carried out of the Rockies by streams over geologic time. The aridity of the plains caused these streams to dry up and deposit their loads over a wide area from South Dakota to Texas. The Ogallala has acted like a sponge, soaking up rainfall for millions of years and storing it below ground, providing abundant ground water in much of the High Plains. Just the opposite is the case in eastern Kansas. Plentiful precipitation has eroded much of the landscape, exposing rocks with poor water-storage properties. Perennial streams and reservoirs are common; however, plentiful and good-quality ground water is found only in the stream deposits of major river valleys.

0.5 *Kanorado.* A number of Kansas place-names come from combining parts of two names into one. The name Kanorado comes from the combination of Kansas and Colorado.

1.4 *K-267 interchange.*

1.7 *Middle Fork of Beaver Creek.*

2.4 *Ogallala Formation* on the south side of the highway.

2.5 North of the highway, in the bed of Middle Beaver Creek, is a *sand and gravel pit.*

2.8 *Ogallala Formation* (poorly exposed). The Ogallala Formation is composed mostly of sand, gravel, and silt, although in some locations it is naturally cemented together. The Ogallala underlies most of the western third of Kansas and is up to 600 feet thick. The Ogallala is important because in much of western Kansas it is a water-bearing rock formation—an aquifer. This water-saturated rock underlies not only Kansas, but parts of seven other states as well; its average thickness is about 100 feet. Water in the Ogallala has accumulated over millions of years as rainwater seeped into the ground, but today the aquifer is being rapidly depleted, primarily because of irrigation. Many of the irrigation wells that are visible along I-70, particularly in Sherman County, tap the Ogallala aquifer.

3.4 At this point, *I-70 reaches its highest elevation in Kansas,* 3,910 feet, or 3,150 feet above the eastern border of Kansas. The highest point in Kansas is Mount Sunflower—4,039 feet above sea level—located about 22 miles south of I-70, near the Colorado border in Wallace County.

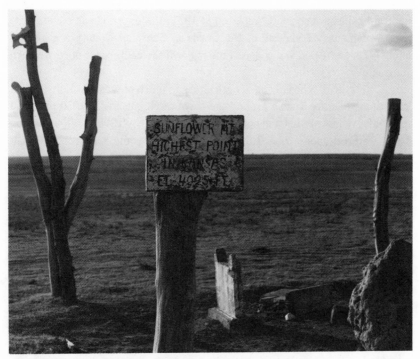

Mount Sunflower. At 4039 feet in elevation, this is the highest point in Kansas, south of I-70 in Wallace County.

3.7 The elevation here is 3,900 feet.

7.0–8.0 *Rest areas.* The rest area on eastbound I-70 contains a tourist information center.

8.6 The town of *Ruleton* is 0.7 miles to the north.

11.6 *South Fork of Beaver Creek.*

12.2 The *Ogallala Formation,* overlain by windblown loess deposits, is exposed south of the highway.

12.3 *Caruso exit.* The town of Caruso, 0.7 miles to the north, is named after the famed operatic tenor from Italy.

12.5 The *sand and gravel pit* north of the highway is in the bed of South Beaver Creek.

14.5 *Middle Fork of Sappa Creek.* This stream, though often dry, is a tributary of the Republican River. The name comes from the Sioux word meaning "black" or "dark."

17.3 *K–27 interchange.* Just to the north is *Goodland,* the seat of Sherman County. The Sherman County Courthouse is located east of the central business district. Completed in 1931, the exterior has several art-deco ornaments. Today all of the towns in Sherman County are located along I–70, and Goodland is one of the major trade centers for northwestern Kansas.

Irrigation has changed agriculture in much of western Kansas, and irrigation wells are visible around the town of Goodland. In 1950 only 100 irrigation wells were used in northwestern Kansas; by 1978, 3,400 irrigation wells watered more than 3.5 million acres. In some parts of Sherman County, the water table dropped an average of three to four feet per year during the 1970s.

Throughout western Kansas, huge center-pivot irrigation systems have done much to change agriculture. Center-pivot systems consist of large sprinklers mounted on wheels; they irrigate circular patches of ground. Because the wheels can roll over uneven terrain, center-pivot systems eliminate the need for extensive leveling of ground, which is necessary in other types of irrigation. Thus, center pivots opened up large new areas for irrigation. Although most of the irrigated crops in western Kansas are used to feed cattle, some crops are grown for human consumption, including sugar beets, pinto beans, and sunflowers.

20.3 The area around Goodland is underlain by the *Goodland–Niobrara Gas Area,* one of several natural-gas fields in northwestern Kansas and adjacent parts of Colorado and Nebraska where natural gas is found at shallow depths in the Niobrara Chalk Formation. Outcrops of the Niobrara form the chalk badlands of western Kansas, where numerous vertebrate fossils from the Cretaceous seas have been found. Around here the Niobrara has been folded and fractured, allowing natural gas to accumulate beneath the overlying Pierre Shale, which acts as an impermeable cap on the gas field.

27.3 *K–253 interchange.* The town of Edson is 0.5 miles north.

35.3 *Sherman/Thomas county line.* Along I–70, this marks the dividing line between Central and Mountain time zones. Much of northwestern and west-central Kansas is in the Mountain time zone. Both Sherman and Thomas counties, by the way, are named for Civil War generals who fought in the Atlanta campaign: William Sherman and George Thomas.

36.4 *K–184 interchange.* The town of Brewster is 1.5 miles north. South of the highway is the South Fork of Sappa Creek, which runs parallel to I–70 for the next 4 miles to the east.

40.4 *South Fork of Sappa Creek.*

42.6 *Loess* exposure in the quarry south of I-70. Loess (pronounced luss) is a finely ground dust that covers much of Kansas. This windblown silt is responsible for much of the rich soil in the state; in some places it is several hundred feet thick. Thick loess exposures can be seen in the Kansas City area (at milepost 417). Geologists theorize that loess was originally carried by streams when glaciers melted at the end of the Ice Age. As these streams dried up, they dropped their sediment, and strong winds whipped up the finer particles into clouds of dust and dispersed them over large areas of the central United States.

46.0 Road to *Levant.* U.S. 24 exits to the east.

48.4–49.0 *Rest area.* West of the rest area on the north side of I-70 is a depression in the ground that occasionally holds water. These features, common in western Kansas, show areas where the ground has subsided slightly. Scientists are not sure of the exact cause of these features, although they suspect solution of soluble rocks, wind erosion, and settling of near-surface silt deposits. Even buffalos have been suspected of causing these depressions because of their habit of wallowing in mud. That is why local residents often call these depressions buffalo wallows.

49.7 *Prairie Dog Creek,* named by John Charles Frémont when he explored Kansas in 1843. It is impounded near Norton to form Keith Sebelius Lake.

54.0 *K-25 interchange.* Just north is the town of *Colby,* the seat of Thomas County. The county courthouse, located east of the Colby business district, was built in 1907 of red brick with limestone trim.

57.4 *North Fork of the Solomon River.* Within 20 miles, I-70 crosses the beginnings of both the Solomon and Saline rivers, two major tributaries of the Kansas River. The highway crosses the same rivers in central Kansas, after they flow across much of north-central Kansas and grow much larger.

62.9 The town of *Mingo* is one mile west of I-70.

63.9 *South Fork of Solomon River.*

70.2 *U.S. 83 interchange.* This highway runs north and south across western Kansas; its geology is described in chapter 4. Four miles south of here is *Oakley,* named after Eliza Oakley Gardner Hoag, the mother of the town's founder. Oakley is the home of the Fick Fossil and History Museum, which contains displays of fossils taken from the Pleistocene sediments and Cretaceous chalk formations of western Kansas. Oakley, by the way, is the county seat of Logan County. Most county seats are situated in the central part of their counties, but Oakley is in the far northeast corner of Logan County. After an especially bitter fight during the early 1960s, the county seat was moved here from Russell Springs.

70.6 *North Fork of Saline River.* The Indians called the Saline the "salt river," and French explorers later gave it the name of Saline. The name comes from the salt content that the river gains as it drains north-central Kansas. This river also gave its name to Saline County and the city of Salina.

72.8–73.2 *Rest area.*

74.0 *South Fork of Saline River.* The Ogallala Formation is exposed east of the highway.

74.4 Wells in the *Campus North oil field* are situated on either side of the highway.

74.9 *Logan/Thomas county line.* Logan County was once named St. John County, after one of the state's early governors. That was before the Kansas Legislature decided to change the name to honor the Civil War general John A. Logan.

75.6 *Gove/Logan county line.*

76.0 *U.S. 40 interchange.* Oakley is two miles to the west. Westbound U.S. 40 exits from I–70, passing through western Kansas and eastern Colorado before rejoining I–70 near Limon, Colorado. To the east, U.S. 40 and I–70 are joined as far as Topeka.

78.0 The elevation here is 3,000 feet above sea level. Between Hays and this point on I–70, the highway gains about 1,000 feet in altitude in only about 85 miles.

80.0 One mile to the north is the town of *Campus.* This name came, not from a college, but probably from camps that formed in the area when the railroads were being built. Monument Rocks, a highly eroded outcrop of Niobrara Chalk, is about 21 miles south of here.

81.6 Here I–70 crosses the upper reaches of Big Creek, which parallels I–70 a short distance to the south for nearly 100 miles before joining the Smoky Hill River southwest of Russell.

86.0 K–216 leads 0.5 miles north to the town of *Grinnell.*

89.6 *North Fork of Big Creek.*

94.0 *K–23 interchange.* Just to the north is the town of *Grain-field.* Here it is easy to see why this area is called the High Plains. The ground is flat and seemingly featureless, although the Saline River cuts through the earth just a few miles to the north.

96.6–97.0 *Rest area.*

99.8 *K–211* leads north one mile to *Park.* This town was originally named Buffalo, then Buffalo Park, and finally simply Park. During the late 1800s, fossil collectors often used this town as a supply base as they moved into the Cretaceous chalk to look for fossils. At the time, western Kansas was a paleontologist's paradise, offering numerous samples of previously unknown species. Swimming reptiles called mosasaurs; flying reptiles called pterosaurs; shark's teeth; toothed birds; and other fantastic fossils that came out of the Cretaceous—all lured paleontologists to western Kansas.

The scientists weren't always at home on the western Kansas range. Prof. O. C. Marsh of Yale University came to Kansas in 1870 and began to dig for fossils in Wallace County the day before Thanksgiving. That night the coyotes scared away the party's mules, which returned to nearby Fort Wallace. When the mounts straggled into the fort, the commander surmised that "the Indians have jumped the professor. We must save the party if possible." He sent soldiers to Marsh's dig; they found the scientists safe, although "somewhat short of mules."

With the episode settled, Marsh invited the soldiers to supper and provided the following description of that Thanksgiving night on the plains. "Our bill of fare for the game courses included buffalo tongue, steak, and roast rib; antelope meat in various forms, and stewed jack rabbit; with pork and beans, canned fruit, and vegetables as side dishes. Our beverages were limited but good army coffee and the wine of Kentucky were in abundance. As the feast progressed, the fun became fast and furious. I was requested to make a speech, thanking the lieutenant and his comrades for our rescue, and he replied in feeling terms. Yale songs were sung by our party and western stories were told by the army officers. The November wind howled through our camp and the coyotes, sniffing the feast, serenaded us from the orchestra bluff above, but we heeded them not for we were safe and happy, and that Thanksgiving dinner on the plains of Kansas will, I am sure, not be forgotten by any one present." Marsh's visit may mark the first time that the strains of a nineteenth-century version of "Boola Boola" were heard in western Kansas.

The competition among paleontologists to find and record new fossils became so intense that it was later known as the Fossil War. Collectors resorted to bribery and to covering up new finds to keep their rivals from discovering them. And they lived in fear of attack by Indians. Geologist F. V. Hayden, who explored Kansas during the 1850s and 1860s, was given the Indian name of "The-man-who-picks-up-rocks-running." According to one report, Hayden was once stopped by the Indians, but "finding him armed only with a hammer and carrying a bag of rocks and fossils, they concluded he was insane and let him alone."

Yale's Marsh wrote that he was worried about Indians during his trip to Kansas in 1870. "We kept a sentinel posted on the high part of the bluffs, so that our enemies, the Cheyennes, could not steal upon us unawares." But other geologists took issue with Marsh's concern about Indians. Samuel Wendell Williston, a noted paleontologist and early member of the Kansas Geological Survey, later wrote that "his (Marsh's) reference to the personal dangers encountered from hostile Indians is amusing in the extreme to all those who know the facts. I think that I can say without fear of dispute . . . Prof. Marsh never ran any greater danger from Indians than when he entertained Red Cloud at his home in New Haven."

107.5 *K-212* leads north to *Quinter.* This is also the exit for Castle Rock Road, a county road that runs about 14 miles south to the chalk formation called Castle Rock. Castle Rock overlooks the valley of Hackberry Creek, while nearby is a heavily eroded outcrop of the Niobrara Formation, the same formation that makes up both Castle Rock and Monument Rocks in western Gove County. The old Butterfield stage line ran past Castle Rock. The ruts are still visible in the grass just north of the formation.

113.1 *Gove/Trego county line.* These counties are named for two captains in the Civil War, Grenville Gove and Edgar P. Trego.

115.1 *K-198* leads a short distance north to *Collyer.* The county road that runs south from here winds near Castle Rock. Visitors to Castle Rock should keep in mind that they are on private property.

About a mile north of Collyer is Coyote Creek. Coyotes, the largest member of the dog family still found in Kansas, live throughout the state and are often seen in fields and along creeks. Coyotes are generally solitary, nocturnal animals whose howls and yips can be heard at night. Resembling underfed German shepherds, coyotes live on a variety of plants and animals, eating everything from mice to birds and berries. In his book *Roughing It,* Mark Twain described a prairie encounter with a coyote: "The coyote is a long, slim, sick and sorry-looking skeleton, with a gray wolf-skin stretched over it, a tolerably bushy tail that forever sags down with a despairing expression of forsakenness and misery, a furtive and evil eye, and a long, sharp face, with slightly lifted lip and exposed teeth. He has a general slinking expression all over. The coyote is a living, breathing allegory of Want. He is *always* hungry. . . . When he sees you he lifts his lip and lets a flash of his teeth out, and then turns a little out of the course he

Castle Rock, a chalk outlier south of Quinter in eastern Gove County

was pursuing, depresses his head a bit, and strikes a long, soft-footed trot through the sage-brush, glancing over his shoulder at you, from time to time, till he is about out of easy pistol range, and then he stops and takes a deliberate survey of you; he will trot fifty yards and stop again—another fifty and stop again; and finally the gray of his gliding body blends with the gray of the sage-brush, and he disappears."

118.5   To the north are wells in the *Hladek oil field*. These wells produce oil from the Lansing and Kansas City groups of formations at a depth of about 4,000 feet. These rocks, deposited in the Pennsylvanian Period when a shallow sea left limestones and shales across the state, crop out at the eastern end of I–70 in Kansas.

120.1   *Voda Road* exit. The town of Voda is 2 miles to the north of the highway.

121.7   The elevation here is 2,500 feet.

127.2   *U.S. 283 interchange*. North of the highway is the town of WaKeeney, a place-name that was created from combining two names: Albert Warren and James Keeney, owners of a Chicago business firm that bought land here in the 1870s. According to the United States Postal Service, WaKeeney is still the only town in the country by that name. This is also about the halfway point between Kansas City and Denver.

131.8–133.0   *Rest area.*

132.5   Oil well in the *FCS field*.

134.6   The wells here are part of the *Ogallah Northwest oil field*, discovered in 1957. In 1984, wells in Trego County produced about 1.8 million barrels of oil.

135.4   *K–147 exit*. To the north this highway goes to Ogallah. To the south, Highway 147 goes about 13 miles to Cedar Bluff Reservoir, a lake built in 1951 on the Smoky Hill River by the Bureau of Reclamation. Chalk bluffs and outcrops are visible along many of the banks of Cedar Bluff.

136.0   Just south of Ogallah is the *Ogallah oil field*. Since oil was discovered here in 1951, this field has produced more than 14 million barrels. In 1984 the field included 40 wells that were producing from the same Pennsylvanian rocks that are visible on the surface in eastern Kansas. From here east to the area around Russell, I–70 passes through numerous oil fields that are located on the Central Kansas Uplift, a broad, northwest-southeast trending upwarp of the earth's crust. This uplift occurred in the early Pennsylvanian Period and has several smaller folds and faults superimposed upon it, which have localized many of the oil accumulations. Oil was first discovered in the Central Kansas Uplift in 1923, northwest of Russell.

Since then, the area has been the site of continued oil exploration. Today it is the most densly drilled geologic feature in the United States.

138.0 This is the approximate boundary between Tertiary rocks, to the west and older Cretaceous rocks to the east. Chalks and chalky shales are common to the east, while the Ogallala Formation is more common to the west. The Cretaceous Period was named from the Latin word *creta,* which means chalk. Limestones and chalks were common deposits in the seas of this age; they are found over large parts of the world, including the island of Crete in the Mediterranean. The famous white cliffs of Dover, England, are thick Cretaceous chalk deposits, similar to the Niobrara Chalk that is widespread in western Kansas. The Niobrara Chalk contains famous fossil beds that have yielded skeletons of fish, marine reptiles, and even flying reptiles from the age of dinosaurs.

139.5 The well south of the highway is in the *Spring Creek oil field.* Wells in this field are up to 3,800 feet deep.

139.8 *Spring Creek.* According to the U.S. Geological Survey, Spring Creek is the most popular name for rivers and creeks in Kansas.

140.5 *Riga Road.* This town was settled by Volga-Germans who moved westward from Ellis County to Trego County. Now just a railroad siding, Riga is a mile south of I-70.

141.0 About 3.5 miles south of I-70 is a hill called *Round Mound* (elevation 2,323 feet). Oil wells in the Ridgeway South oil field are also visible from here. These wells produce from formations deposited in the Cambrian and Ordovician periods of geologic history.

143.0-145.0 Numerous wells in the *Ellis oil field* are visible along both sides of the highway in this area. These wells produce oil from four different horizons, or layers, of Paleozoic rocks.

143.7 *Ellis/Trego county line.* Ellis County was named for George Ellis, a Civil War lieutenant who was killed in the Battle of Jenkins' Ferry in Arkansas. Lieutenant Ellis happened to be in the Eldridge House in Lawrence when it was burned by Quantrill's raiders in 1863. To the north of Ellis County is Rooks County, named for a Civil War private, John C. Rooks. Forty-seven Kansas counties are named for soldiers, but only Rooks County is named for a private.

145.6 *K-247* exits south to *Ellis,* the boyhood home of Walter Chrysler, of automobile fame. Although Chrysler grew up in Ellis, he was born in Wamego, in northeast Kansas.

147.7 Wells north of the highway are in the *Werth field.* Because of its location astride the Central Kansas Uplift, Ellis County has long been the top oil-producing county in Kansas. In 1984, Ellis County wells produced 5.7 million barrels of oil, down slightly from the previous year's production.

148.4 Oil wells to the south of the highway are in the *Werth Southeast oil field*.

149.2 *Fort Hays limestone.* This massive limestone bed often contains fossilized clamshells. It marks the base of the Niobrara Chalk.

149.5 *Fort Hays limestone.*

150.0 South of the highway are wells in the *Pfeifer North oil field*, discovered in 1982. Nearby is the Pfeifer Northwest oil field, discovered in 1983. During the late 1970s and the early 1980s, rising oil prices led to a drilling boom in Kansas, which resulted in new oil fields such as these.

A mile south of the highway is a railroad siding named Hogback, which probably got its name from a sharp bluff formed by an outcrop of the Fort Hays limestone along the Big Creek valley. In geological terms, a hogback is a sharp ridge formed by a resistant rock unit that dips steeply. A less sharp ridge formed by rocks that dip more gently is called a cuesta. Although the Fort Hays limestone dips very gently here, the ridge was still called a hogback. Mount Oread, the hill that is the site of the University of Kansas in Lawrence, was originally called Hogback Ridge.

153.0 *Yocemento Road.* The U.S. Portland Cement Company, owned by I. M. Yost, once operated here, and the town's name may have come from the combination of Yost and cement. The geological expertise for the plant was provided by Erasmus Haworth, a director of the Kansas Geological Survey around the turn of the century.

154.0 The wells here are in the *I–70 oil field*, producing from the Topeka Limestone and rocks in the Lansing and Kansas City groups.

155.3 *Gatschet oil field.* The Toronto Limestone Member of the Oread Limestone is one of the producing horizons for this field. The Toronto crops out along I–70 around Lawrence.

156.0 Wells here are in the *Gatschet Southeast field*.

159.3 *Hays,* named for the fort that was established here, is today a regional trade center for much of western Kansas. It is the home of Fort Hays State University; many of the school's buildings are constructed out of native limestones, particularly the Fort Hays limestone seen at milepost 149.5. On the campus of Fort Hays State is the Sternberg Memorial Museum, which includes a fine collection of fossils from the Cretaceous chalk. Perhaps the museum's best-known specimen is a large

Cretaceous fish that died with the well-preserved remains of another, smaller fish in its stomach.

Fort Hays was established here in 1865, one of a string of forts designed to protect the trail to Denver across western Kansas. The fort was abandoned in 1889. Gen. George Armstrong Custer was once stationed at Fort Hays, as well as at Fort Riley and Fort Leavenworth, which are also in Kansas.

160.1 *Chetolah Creek.* The name Chetolah was once applied by Indians to the Smoky Hill River.

162.0 *Rest area.*

163.0 *The Fort Hays escarpment,* a long, continuous ridge, is visible on the western horizon; it is formed where the Fort Hays limestone comes to the surface.

164.0–165.6 Wells here are in the *Younger oil field.* Discovered in 1944, it has produced over 2.5 million barrels of oil.

165.0 The elevation is 2,000 feet.

166.0 *North Fork of Big Creek.* To the north is the town of Catharine, named for Catherine the Great, an empress of Russia. Catherine, who was of German ancestry, offered free land in southern Russia to German immigrants. These German–Russians, or Volga-Germans, later immigrated to the United States, many of them settling in Ellis, Russell, Trego, and Rush counties.

166.1–166.6 The wells here are in the *Sugar Loaf Southeast oil field,* named after a rounded hill capped by the Fort Hays limestone, southwest of the town of Catharine.

167.2 Oil wells in the *Toulon field* are south of the highway. This field contains 66 wells producing from depths of 3,300 to 3,500 feet.

168.6 *K–255* exits south to the town of *Victoria.* Like Catharine, Victoria was named for royalty. Victoria was settled by German–Russians and by Englishmen, who named the town after their queen. The English were not particularly successful farmers or ranchers; they were more interested in chasing rabbits and coyotes across the prairie. There are even reports that they dammed up a portion of Big Creek and imported a steamboat to float up and down the river. Like the English at a similar community called Runnymede in south-central Kansas, most of the English in Ellis County went back to England.

The twin towers of the church of St. Fidelis in Victoria are visible to the south. This church, also known as the Cathedral of the Plains, was built out of 17 million pounds of native Fencepost limestone between 1908 and 1911.

168.8   The wells here are in the *Herzog oil field,* which, since it was discovered in 1940, has produced over 3.8 million barrels of oil. It would be the largest producer in most Kansas counties, but here in Ellis County, where huge fields are common, the Herzog is just one of many major producing fields. At least twelve other fields in Ellis County have out-produced the Herzog field during their history.

170.5   *Mud Creek,* one of nineteen Mud creeks in Kansas.

172.5   North of I-70 are the *Blue Hills,* which include the Fort Hays escarpment. The Blue Hills are formed by hard Cretaceous limestones such as the Fort Hays. Major east-flowing streams, such as the Saline and the Solomon, have cut deep valleys through these rocks, forming many picturesque side canyons. Exposures of bluish-gray shales, interbedded with the limestone, probably gave the region its name.

173.0   *Walker.* This town is the site of another attractive prairie church, St. Ann's, which is also built of native limestone. Walker was the site of a major airfield in World War II. The runways and hangars are still present about two miles north of I-70.

174.6   *Walker Creek.*

175.0   *Ellis/Russell county line.* Of the 105 counties in the state, Ellis and Russell are regularly among the leaders in oil production. These two counties are located atop the Central Kansas Uplift, a dome of rock that runs from Barber County, in south-central Kansas, to Norton County in the northwestern part of the state. About one-third of the wildcat wells in Kansas are drilled in the Central Kansas Uplift area. In 1983, 45 percent of the state's oil production came from this geologic feature.

176.0   *Gorham* is one mile north, on K-257. Near the town of Fairport, about 13 miles north of here along this road, are several locations that have produced fossilized shark's teeth from the Fairport Chalk Member of the Carlile Shale Formation. Sharks were common in the Cretaceous seas, and they have left behind fossil vertebrae as well as teeth. The teeth are especially common because they were continually being replaced by row after row of new teeth. When a shark lost a tooth, it was quickly replaced by a new one. In addition to shark's teeth, the Fairport chalk has produced fossils of clams, ammonites (a relative of today's *Nautilus,* a chambered, shelled animal), fish scales, bryozoans, barnacles, and worm tubes.

176.0–182.0   Extending several miles to the north and south is

The Crawford sinkhole, formed by the dissolution of underground salt beds, at milepost 179 along I-70 in Russell County

the *Gorham oil field,* discovered in 1926. This is one of the older oil fields in central Kansas and one of the largest in the state. By 1984 it included 514 wells and had produced 90 million barrels of oil from six different geologic horizons.

178.5 *Witt sinkhole.* At this point, I-70 passes through the Witt sinkhole, which is forming much like the Crawford sinkhole, described below. Here the highway once had a sharp dip in it, which was considered a hazard, because a car parked in the bottom of the dip was invisible to oncoming traffic. In 1984 the Kansas Department of Transportation raised the highway to its original grade, thus removing both the dip and the hazard.

179.0 *Crawford sinkhole,* which creates a small pond and a dip in the highway just east of the underpass. Even in dry weather, the pond contains water. Twenty years ago, this stretch of the highway was flat, but the sinking roadbed has created a steep dip in the road. This subsidence is caused when ground water dissolves away underground salt beds that are 1,300 to 1,600 feet beneath the surface. In some places, the ground water washes out cavities that suddenly collapse. In other cases, such as the Witt sinkhole, the earth subsides more slowly. Seismic studies indicate that there appears to be little chance of a catastrophic collapse, though the roadbed is dropping about seven inches per year and may be putting stress on the bridge that crosses I-70.

A catastrophic collapse did occur in 1978 at a sinkhole in a pasture a few miles northwest of here. That sink, about 100 feet in diameter, opened up overnight and continues to grow. Geologists believe the collapse was caused by water leaking out of a saltwater-disposal well, which dissolved away space in the salt. When the overlying rocks could no longer support the weight above, they collapsed, and the sinkhole opened up.

179.4 A slight dip in the highway here marks the *Rouback sinkhole,* which, like the Crawford and Witt sinkholes to the west, is related to oil production and was caused by the solution of salt around abandoned oil wells. The sinkhole was discovered in 1979 while the Crawford and Witt sinkholes were being studied. With slow subsidence, the sinkhole became more noticeable over the years.

182.8 *Fossil Creek,* which parallels the south side of I-70 for three miles, was named for the fossil beds in the area. Russell was originally called Fossil Creek Station, which was shortened to Fossil, then finally changed to Russell.

183.8 Wells here are in the *Russell oil field,* discovered in 1934, which has produced more than 17 million barrels of oil. During the 1970s, Russell was regularly the second-ranking oil-producing county in Kansas, trailing only neighboring Ellis County. In 1984, Russell County's production was 4.4 million barrels, and it fell to fourth place. Second place was taken by Barton County, and third by Rooks County, both of which produced 4.7 million barrels of oil in 1984.

184.8 The active wells here are in the *Russell oil field;* north of the highway there is a display of old oil-field equipment, including a wooden derrick. Derricks were once erected to support the equipment used in drilling oil wells. Most of these old wells were drilled with cable-tool rigs, which over and over again raised and dropped a bit so as to punch a hole in the ground.

185.0 *U.S. 281 exit.* Directly south of I-70 is Fossil Lake. To the north is Russell, the seat of Russell County and a center for much of the area's oil and gas business. It is also the hometown of Kansas' Senator Robert Dole.

187.2 *Rest area.*

189.0 South of I-70 is the *Hall-Gurney oil field,* a large field that extends south into Barton County and includes more than 1,000 wells.

Stone fence posts from the Fencepost limestone bed guard a Smoky Hills pasture near Wilson Lake in Russell County.

Discovered in 1931, it has produced more than 137 million barrels of oil, which is nearly twice the average annual production of the entire state of Kansas today. In 1984, Kansas produced 75.7 million barrels of oil, which made it the eighth-largest oil-producing state in the nation, following Texas, Alaska, Louisiana, California, Oklahoma, Wyoming, and New Mexico. The value of Kansas oil production amounts to several billion dollars annually.

    193.0   *Bunker Hill Road.* The town of Bunker Hill, 1 mile to the north, is situated on a divide between the Smoky Hill River, 5 miles south, and the Saline River, 6 miles to the north. The Smoky Hill River valley is visible to the south.

    198.4   *Greenhorn Limestone* is exposed on the north side of the highway.

    199.6   *K–231* leads 0.5 miles south to the town of *Dorrance*.

    200.3   The wells here, in the *Dorrance oil and gas field,* are about 2,300 feet deep.

    200.9 and 202.9   *Greenhorn Limestone Formation* (poorly exposed). Fenceposts and buildings made of native limestone line the roads in this area. Most of that limestone comes from a thin but widespread and

An abandoned bridge constructed out of native Fencepost limestone south of Wilson Lake, north of I-70 in Lincoln County

uniform limestone bed at the top of the Greenhorn Limestone called, for obvious reasons, the Fencepost limestone.

203.0   These wells are in the *Plymouth Northwest oil and gas field*. These wells produce from rocks in the Wabaunsee Group of formations, which crop out just east of the Flint Hills in eastern Kansas.

203.9   About 4.5 miles south of here, along the county line, *Coal Creek* empties into the Smoky Hill River. This creek gets its name from lignite, which was mined in the walls of the creek valley. Lignite, the lowest rank of coal, differs from the higher ranks, bituminous and anthracite, by having a brown color, a woody appearance, and a high moisture content. It occurs in the upper part of the Dakota Formation, and it crops out from Hodgeman County, north of Dodge City, to Washington County, at the Nebraska border. Along its outcrop belt are subtle scars left by small mines that operated during the late 1800s and early 1900s. The Coal Creek Mining district, south of here, was one of the largest; it included fourteen mines. Some of these mines completely undermined a large hill along the Russell County/Ellsworth County line, so that it was possible to go underground in one county and come out in the other.

205.0   *Russell/Ellsworth county line.*

The route of the Wilson Channel, an ancestral course of the Saline River

205.4 *Wilson Creek.*

206.0 *K–232 exit,* to Wilson and Lucas. A mile south of I–70 is Wilson, a town that was settled primarily by Czechoslovakians and calls itself the Czech capital of Kansas. Every July the town holds an After Harvest Czech Festival. North of I–70 about 7 miles is Wilson Lake, a man-made lake on the Saline River. The Wilson Lake area offers deep canyons and steep scenic hills, which belie the picture of western Kansas as being flat and featureless. Farther north, in Lucas, is the Garden of Eden, a series of unique cement statues built by a disabled Civil War veteran, S. P. Dinsmoor. The statues depict Biblical characters, a flag, and other items and events. Dinsmoor is buried in a coffin with a glass top; for a price, visitors can see his remains.

207.8 *Dakota Formation sandstone.* Just north of I–70 there are several unusual erosional features composed of Dakota Formation sandstone. These features are not easily visible from the eastbound lane. The Dakota was deposited along the edge of a Cretaceous sea. Cretaceous seas were among the most widespread in all of geologic history, covering much of North America. The Dakota Formation occurs throughout the north and central Great Plains. From here on west, it is in the subsurface of western Kansas and eastern Colorado, coming to the surface as a sharp sandstone ridge just east of the Colorado Front Range.

209.0–212.0 *Wilson channel.* At this point, I–70 passes between the Smoky Hill and Saline rivers. However, geologists have discovered that at about this point the Saline River once meandered south and joined the Smoky Hill. This flat, valleylike area represents that old stream channel. Volcanic ash deposits in this channel were quarried a few miles north of the highway.

211.2 *Cow Creek.*

214.6 *Buffalo Creek.*

219.0 *K-14 exit.* This road runs 7 miles south to Ellsworth, the seat of Ellsworth County and a former cow town. James Butler ("Wild Bill") Hickok, one of the town's early residents, later moved to Hays and became town marshal. Wyatt Earp reportedly got his start as a law officer when he made two arrests in Ellsworth.

220.6 *Fencepost limestone,* on frontage road north of I–70, is not easily visible from the eastbound lane.

222.0 *K–14 exit* and exposures of the *Greenhorn Limestone.* This highway goes about 13 miles north to Lincoln, the seat of Lincoln County. Many of the town's buildings are made out of native Fencepost limestone, and the road north of Lincoln is lined with stone fence posts. In addition, calcite-cemented sandstone from the Dakota Formation is quarried in the vicinity of Lincoln; it is known locally as Lincoln Quartzite. The cemetery on the east edge of the town contains a small tombstone in the shape of a suitcase, which marks the grave of a salesman who made his last stop in Lincoln.

South of here is the small town of Kanopolis, which was once promoted, because of its central location, as the capital of Kansas. Several city lots were reserved for that purpose, and the Populist party promoted the move in 1893; nevertheless, the capital remained in Topeka. Kanopolis is also the site of the oldest continuously operated salt mine in Kansas. Work began at the mine in 1914. Today a 10-foot section of the Hutchinson Salt Member of the Wellington Formaton is being mined at a depth of 850 feet. This salt bed underlies most of central and western Kansas.

223.2 *Graneros Shale.*

223.7 *Dakota Formation sandstone.* In addition to sandstone, the Dakota Formation contains large amounts of clay, which is mined and used in the manufacture of brick in a plant at nearby Kanopolis.

224.0–225.0 The eastbound lane has a *rest area,* which contains several mushroom-shaped sandstone concretions that are similar to those found at Mushroom Rocks State Park, north of Kanopolis Reservoir, and at Rock City, southwest of Minneapolis.

225.5 *Dakota Formation sandstone* outcrops on hill south of I–70.

Toadstool-shaped concretions weather out of the Dakota Formation at Mushroom Rocks State Park in Ellsworth County

225.6   *K-156* goes southwest to Ellsworth and Great Bend.

225.7   *East Elkhorn Creek.*

225.8   *Dakota Formation sandstone* southeast of the highway. The Dakota has produced many fossilized plants. For example, in this area, it was the source of a fossilized pine-cone-like structure that may have grown on a Cretaceous relative of today's redwood tree.

226.5   *Dakota Formation sandstone* south of the highway.

228.2   *Ellsworth/Lincoln county line.*

228.5   *Graneros Shale,* overlain by the *Greenhorn Limestone.* Because shale is easily eroded, shale outcrops often seem only to color the road cut, rather than to stand out as prominently as does a bed of limestone or sandstone.

229.0   *The Smoky Hills* dominate the view to the east, as I–70 crosses the eastern edge of the uplands, which are capped by the Greenhorn Limestone.

232.5   The elevation is 1,500 feet.

Palmer's Cave, an erosional feature in Dakota Formation sandstone, north of Lake Kanopolis in Ellsworth County.

233.2   *Dakota Formation* on the north side of I–70.

233.5   *Carneiro exit.* The small town of Carneiro is about 10 miles to the southwest, and the town of Juniata is 2.5 miles to the north. In a pasture a few miles south of here, in an outcrop of Dakota sandstone, there is an opening in the rock, which is called Palmer's Cave. The cave consists of a passage about 15 feet long, eroded into the sandstone by wind and water. Carved inside and outside of the cave are Indian petroglyphs, including an enigmatic reclining figure, tally marks, and geometric designs. Because it is relatively soft and easily carved, Dakota sandstone was a favorite medium for Indian artists in central Kansas. Petroglyph sites include pictures of horses, buffalo, people, birds, and other figures. Archaeologists don't know which Indian tribes carved the figures, although they speculate that it could have been the Wichitas, the Pawnees, or any of eight other tribes that inhabited Kansas during the 1700s and the 1800s. The purpose of the carvings is less clear, although in some cases they provide directions to the nearest water source, a sort of permanent paleolithic road map. Because buffalo bones have been found near Palmer's

Cave, scientists speculate that the drawings at the cave may be related to hunting.

233.8 and 234.3   *Dakota Formation.*

235.5   *Lincoln/Saline county line.*

236.5   *Dakota Formation.* This exposure shows the angled patterns that geologists call cross-bedding, which is caused by the deposition of sand by moving water.

236.7   *Dakota Formation sandstone* on south side of I–70. The hills south of the road are capped by sandstone in the Dakota, making them more resistant to erosion.

237.0   *Mulberry Creek.*

237.7   *Dakota Formation.*

238.0   *Dakota Formation* on south side of I–70.

238.5   *Brookville exit.* Many of the buildings in Brookville, 7 miles to the south, are built out of Dakota sandstone. This town is also the site of the Brookville Hotel, a well-known Kansas eatery, which opened in the 1870s.

242.9   *Mulberry Creek.*

244.2   *Dakota sandstone.*

246.0   *Soldier Cap.* About 11 miles south of I–70 is a flat-topped hill called Soldier Cap (elevation 1,578 feet). This hill is located in the Smoky Hill Bombing Range, a site where planes from McConnell Air Force Base in Wichita make practice bombing runs.

247.0   *Iron Mound,* another Dakota-capped hill, is visible to the southeast.

248.0   The *Smoky Hill Buttes* are on the horizon to the south. These Dakota-capped hills, which include Coronado Heights, rise above the Smoky Hill River valley north of Lindsborg.

250.8   *U.S. 81/I–135 interchange.* The geology along these north-south highways is described in chapter 5. Salina, at the intersection of I–70 and I–135, is a major trading center for north-central Kansas. In 1980, its population was more than 41,000, and it was the sixth-largest city in the state. Salina, the seat of Saline County, is located west of the confluence of the Smoky Hill and Saline rivers. The Smoky Hill once cut through the heart of Salina, but it was the source of several severe floods during the early 1900s and has since been diverted, so that it skirts the east edge of the city. Salina was founded in 1858 by William A. Phillips, a newspaperman from New York. Phillipsburg, in northwestern Kansas, is named after Salina's founder.

251.5 *North Pole Mound* (elevation 1,466 feet) is visible 5.5 miles north of I-70.

253.0 *Mulberry Creek.*

254.1 *The Saline River,* elevation 1,200 feet, which, shortly after it crosses under I-70, joins the Smoky Hill.

256.1, 257.4, and 258.4 Shale in the *Dakota Formation.*

258.5 *Iron Mound* (elevation 1,497 feet) is visible about 7 miles south of I-70. This hill is capped by Dakota sandstone, which allows it to resist erosion. Iron Mound marks the eastern extent of this Cretaceous sandstone.

259.0 *Kiowa Formation.* Like the Dakota Formation, the Kiowa is a Cretaceous deposit, although it is slightly older than the Dakota. The Kiowa includes layers of shale, sandstone, and limestone. In places the shale contains a scaley concretion known as cone-in-cone; sandstone in the Kiowa contains selenite crystals in some locations.

259.5 This is the approximate boundary between the Permian and Cretaceous formations in Kansas. To the east, the rocks are older, often containing limestone and flint. To the west, the most common rocks are the clays, shales, and sandstones of the Dakota Formation. The Permian Period was named after Perm, an area in Russia where rocks of this age were first described. The Permian rocks in Kansas were deposited in a vast shallow sea, where the environment was continually changing; thus, diverse sediments were deposited. The rocks found along I-70 generally are alternating beds of limestone and shale. Some of these limestones contain chert or "flint" and are resistant to erosion, forming the Flint Hills to the east. South-central Kansas contains brightly colored Permian red beds composed of shale, silt, sand, and occasional beds of gypsum.

264.3 *Solomon River.* Around this point, salt springs near the river mark an area where Permian salt beds lie near the surface. Much of central Kansas is underlain by Permian salt deposits in the Hutchinson Salt Member of the Wellington Formation. Mined in Reno, Ellsworth, and Rice counties, this salt bed is several hundred feet thick in places. However, where it approaches the surface, as it does in this area, the salt dissolves away, often contaminating local ground water and surface water. Also at this point, Iron Mound is visible to the southwest.

265.5 *Rest area.*

266.0 *Saline/Dickinson county line.* This boundary is also the location of the Sixth Principal Meridian. This longitudinal line runs through

Wichita, providing the name of Meridian Street in Wichita. The Sixth Principal Meridian is a survey line that was used to lay out the public lands of the early Kansas and Nebraska territories. It marks the division between eastern-range townships and western-range townships and is used in determining legal descriptions of land in Kansas, Nebraska, Wyoming, Colorado, and part of South Dakota. Its location does not correspond to any important longitudinal meridian but marks the point where the original survey of the Kansas–Nebraska line was halted due to fear of hostile Indians.

266.5 *K–221* leads south to *Solomon,* about a mile south of which the confluence of the Solomon and Smoky Hill rivers is visible from a bridge on a county road.

269.0–272.0 *Sand-dune topography.* In recent geologic times, this area was covered by sand dunes. While grass has since covered the hills, the rolling land surface still displays that sand-dune topography. The sand in this area was blown out of the valley of the Smoky Hill River, and the resulting sand dunes resemble those common today around the Arkansas River in central and western Kansas.

271.0 About 1.5 miles south, just south of this area of sand hills and north of the Smoky Hill River, is a place called *Sand Spring.* The porous and permeable dune sand in these hills allows precipitation to percolate downward into equally porous alluvial terrace deposits. Eventually the ground water enters solution channels in the Herington limestone. These channels act like water mains; where the Herington is exposed, as it is just above the bed of the Smoky Hill River, these solution channels bring ground water to the surface in the form of springs. The quality of this water is much better than that derived from the alluvium of the nearby Smoky Hill River, and in 1881 Abilene developed Sand Spring as a source of water. The loose, sandy soil of these hills is also conducive to growing melons, and this area is sometimes locally referred to as the Sand Spring melon district.

274.3 *Mud Creek.*

275.2 *K–15 interchange* and *Abilene.* Another former cow town, Abilene is today the site of the Eisenhower Museum and Library, one of the leading tourist attractions in Kansas. Abilene was first settled in 1858 but got its start as a cow town in 1867, when the Kansas Pacific Railroad reached the area. The Illinois cattleman Joseph McCoy promoted Abilene as the northern terminus of the Chisholm Trail, and in 1871 nearly 700,000 head of cattle reached the town. Violence from the cowboys grew to such proportions that Tom ("Bear River") Smith and "Wild Bill" Hickok were brought in to clean up the town. By 1872 the railroad had reached Ellsworth, and Abilene was finished as a cow town.

Abilene is also the home of the Greyhound Hall of Fame, which is built out of native Kansas limestone. Abilene is a noted center for breeding and raising greyhounds, which race throughout the midwest.

278.5 *Wellington Formation shale* (poorly exposed). This formation is named for the town in south-central Kansas. In places it is up to 700 feet thick and contains the Hutchinson Salt Member.

280.0 *Sand-dune topography.* See mileposts 269–272.

281.0 The towns of Detroit and Enterprise lie south of here on K–43. In 1901, Enterprise was the site of one of Carry Nation's saloon smashings. Her antialcohol antics weren't always popular, however. In Enterprise, she was arrested and chased out of town by women who pelted her with rotten eggs.

281.8 *Nolans Limestone* on the north side of I–70.

282.1 *Lone Tree Creek.* A cut-off meander loop in the Smoky Hill River south of here has created an oxbow called Terrapin Lake.

285.8 *Winfield Limestone.*

286.2 *K–206* exits to the south and leads to the town of Chapman.

286.4 *Chapman Creek.* To the north are the remains of a small town called Moonlight, which was named after a Civil War veteran, not the lunar reflection. Thomas Moonlight was later the governor of territorial Wyoming. Moonlight was settled by members of the Church of the Brethren, who moved here after finding Abilene too wild for their tastes.

289.5 *Geary/Dickinson county line.* John Geary was a governor of territorial Kansas. He was also, at various times, the governor of Pennsylvania, a Civil War officer, and mayor of San Francisco, where Geary Street is named after him. He is the only governor of territorial Kansas to have a county named after him.

290.6 *Fort Riley Member* of the Barneston Limestone.

292.5 Less than a mile south of the highway is an area called *Seven Springs.* These springs issue from outcrops of the Barneston Limestone, a formation that caps the Flint Hills in this area. Where it is fractured, or includes solution channels, it collects ground water. Springs that come out of this limestone occur in many of the valleys of this area and feed numerous small streams. Farther to the south is a stretch of the Smoky Hill River that was once called Kansas Falls, after a series of water falls in the river. A nearby railroad siding still bears the name.

294.0   *Rest area* and *Goose Creek*.

295.0   *Exit of U.S. 77,* which skirts the west edge of Junction City, the seat of Geary County. Northwest of Junction City is Milford Reservoir, which has over 16,000 acres of water and is Kansas' largest lake. The lake is named after a town on its eastern shore that became famous in the early 1900s as the home of Dr. John R. Brinkley, who offered to restore male virility with a goat gland transplant. Those operations made him famous. Brinkley opened a hospital, a drugstore, and his own bank; in 1922 he started the first radio station in Kansas. After his medical license was revoked, Brinkley became a politician, using his radio station to campaign for governor as an independent write-in candidate in 1930. Brinkley finished a close third in a disputed election. He ran again in 1932, when his name was on the ballot; but this time he lost to the Independence oilman Alf Landon.

298.7   *Smoky Hill River,* which, 2 miles north of here, joins the Republican River to create the Kansas River. Grand View Hill is south of the highway.

299.3   *Matfield Shale,* overlain by the *Florence Limestone Member* of the Barneston Limestone Formation, south of the highway. This massive limestone is common along roadcuts in I–70. It ranges in thickness from 12 to 45 feet and often contains a variety of fossils. Florence limestone is one of the chert- or "flint"-bearing limestones that gave the Flint Hills their name. The "flint" appears as bluish-gray beds in the otherwise buff-colored limestone.

299.8   *Franks Creek.* Franks Hill is east of the highway.

301.0–302.0   *Grant Ridge,* a series of hills on the southeast side of I–70, is capped by the Fort Riley Limestone Member. Limestone boulders, weathered from the Fort Riley, litter the hillside. Also called Military Ridge, these hills overlook the Smoky Hill River valley, which today is the Fort Riley Military Reservation. Immediately north and west of I–70 is Marshall Airfield.

Fort Riley was established here in 1853 as an army outpost to protect the Santa Fe and Oregon trails. The Seventh Cavalry, with Gen. George Custer as second in command, was organized here in 1866. Since then, Fort Riley has become an important training location for cavalrymen; 150,000 soldiers took their basic training here during World War II.

At Fort Riley is a building that housed the first territorial legislature of Kansas in July 1855. That legislature was unanimously in favor of slavery; it was elected mostly by Missourians who wanted to make

Florence limestone overlain by the Fort Riley limestone at milepost 305 along I–70

Kansas a slave state. The legislature met here, in what was then the town of Pawnee, within the Fort Riley Military Reservation, for four days before adjourning to Shawnee Mission, where it passed a number of proslavery acts and was labeled, by antislavery forces, the "Bogus Legislature."

    302.5   A short distance north of the highway is a cut-off meander loop of the Kansas River. This oxbow is known as *Whiskey Lake.*

    302.7   *Matfield Shale.*

    303.6   *Clarks Creek.*

    304.2   *Crouse Limestone* overlain by *Blue Rapids Shale, Funston Limestone, Speiser Shale,* and the base of the *Wreford Limestone.*

    305.0–305.3   *Florence limestone,* overlain by *Oketo shale,* overlain by *Fort Riley limestone,* all members of the Barneston Limestone Formation.

    305.6 and 306.1   *Fort Riley limestone,* which is named for the nearby fort.

    306.4–306.8   *Florence limestone.*

    306.8   *Matfield Shale,* overlain by *Florence limestone.* Matfield Shale comes in a variety of colors, from bright red to dull green; it is common along I–70 road cuts.

307.3 *Speiser Shale,* overlain by the *Wreford Limestone.*

308.2 *McDowell Creek.* About 20 miles to the north are a dozen volcanolike formations that geologists call kimberlites. Kimberlite is an igneous rock that has welled up from deep underground. Although the rock does not flow out onto the surface like lava, kimberlites do form craterlike structures. Kimberlite is of interest for several reasons. As the kimberlite comes to the surface it carries along pieces of other deeply buried rock. By drilling into kimberlites and examining the core, geologists can glimpse rocks that come from deep underground, perhaps as deep as 150 miles. These formations exploded to the surface about 100 million years ago, and they represent a few of the rare outcrops of igneous rock in Kansas.

Kimberlites are also interesting because they are the only known source of diamonds, although no diamonds have been found in the Riley County kimberlites. During one research project, the Kansas Geological Survey drilled into a kimberlite; when the drill bit was returned to the surface, it had been deeply grooved by something. Was it a diamond? Geologists don't know for sure, but the possibility remains, and at least one company thought the odds were good enough to dig deep trenches through several kimberlites and examine the rock for diamonds.

308.6 A thick road cut south of the highway exposes rock from the *Easly Creek Shale* up through the *Crouse Limestone, Blue Rapids Shale, Funston Limestone,* and the *Speiser Shale.*

309.3–310.0 *Rest area.* The rest area on the westbound lane is decorated by several red quartzite boulders brought south by glaciers. A stream called Pressee Branch runs nearby.

309.9 *Speiser Shale* north of the highway.

310.3 *Wreford Limestone,* a formation characterized by an abundance of chert.

311.0 *Konza Prairie.* Immediately north of I-70 is an 8,616-acre area of tall-grass prairie that was purchased by the Nature Conservancy and is used for research by Kansas State University. Much of the Prairie was originally part of a ranch owned by the Chicago industrialist C. P. Dewey; the ground is covered by bluestem grass that has never been plowed.

The Konza Prairie is named after the Kansa Indian tribe, which supplied the name for the state of Kansas. Konza is one of more than 100 variations in the spelling of Kansa. The word *Kansa* probably first appeared on a map in 1763, when Father Marquette, a French missionary, sketched the Missouri River, using information supplied by the Indians.

311.2–311.8 *Florence limestone* lines this stretch of the highway. The underlying Matfield Shale is exposed at both ends of the road cut.

312.2 and 312.5 *Wreford Limestone.*

312.7 *Swede Creek,* named for Peter Carlson, who settled in Riley County in 1857.

313.1–313.8 Several road cuts exposing the *Matfield Shale,* overlain by *Florence limestone.*

314.0 *K–177 exit* and exposures of the *Florence limestone.* At this point, the elevation is 1,450 feet above sea level. This is about the highest point along I–70 in eastern Kansas. From here west the highway begins to drop, following the Smoky Hill River valley. It does not regain that lost altitude until it reaches Lincoln County, 15 miles west of Salina. North on K–177 is Manhattan, the home of Kansas State University. Farther north is Tuttle Creek Reservoir, where a fault is visible in the shales in the spillway at the east end of the dam. South on K–177 is Council Grove, a historic spot on the Santa Fe Trail.

314.2 *Florence limestone.*

314.6 *Matfield Shale,* overlain by the *Florence limestone.*

315.1–315.7 Several exposures of the *Wreford Limestone.*

315.9 *Crouse Limestone,* overlain by *Blue Rapids Shale, Funston Limestone,* and *Speiser Shale.*

316.0 *Riley/Geary county line.* The chert in the limestone formations here in the Flint Hills garnered much attention when geologists first explored this part of Kansas. In 1819, a scientist named Augustus Edward Jessup accompanied Maj. Stephen Long in a reconnaissance of eastern Kansas. A portion of Long's party had reached Riley County before an encounter with Pawnee Indians sent them packing back to Council Bluffs. Before they left, however, Jessup noted outcrops of limestone, some of which "embrace numerous masses of chert. . . . This occurs of various colours, and these are arranged in spots and or strips. Some specimens have several distinct colours arranged in zigzag lines."

316.2 *South Branch of Deep Creek.* Along Deep Creek, south of the town of Zeandale, is Pillsbury Crossing, where Deep Creek flows over a ledge of Elmont limestone, forming a waterfall about 40 feet across, with a drop of about 5 feet.

316.5 *Beattie Limestone,* overlain by *Stearns Shale,* overlain by *Bader Limestone,* overlain by *Easly Creek Shale.*

316.8 *Eskridge Shale,* overlain by *Beattie Limestone.*

318.1 *Neva Limestone* on the south. Also scattered along the south side of I–70 are wells in the Yaege oil field. Oil production is relatively

Pillsbury Crossing, a waterfall along Deep Creek north of I–70 near Manhattan

uncommon in northeastern Kansas, although production and exploration have increased in recent years. This oil field was discovered in 1959.

318.2   *East Branch Deep Creek.*

318.8   *Bader Limestone.*

319.1   *Crouse Limestone,* overlain by *Blue Rapids Shale,* overlain by *Funston Limestone.*

319.7   *Threemile limestone* overlain by *Havensville shale,* both members of the *Wreford Limestone Formation.* Before it is exposed to erosion, Threemile limestone can be light gray to nearly white. In most places it is a massive thick layer of rock that contains some chert.

320.1 and 320.8   *Schroyer Limestone Member* of the Wreford Limestone.

321.0   About 2 miles north of I–70, obscured by hills, are *Tabor Valley* and *Tabor Hill,* named after A. W. Tabor, who later moved to Colorado and discovered gold at Leadville.

321.2   *Schroyer Limestone Member* of the Wreford Limestone.

321.6   *Threemile Limestone Member* of the Wreford Limestone.

322.0   *Riley/Wabaunsee county line.*

A geologic cross section along I-70 in eastern Kansas showing the Nemaha Uplift, the source of small earthquakes

323.7   Hendricks Creek.

324.6   Most rock layers in eastern Kansas dip to the west. That is, they are angled so that they grow increasingly deeper to the west. At this point, however, the beds of the Bader Limestone are slightly upturned, dipping to the east. This change is probably the result of the Nemaha Uplift, a zone of faulting associated with the *Humboldt fault zone*, a series of faults that run from Nemaha County in northern Kansas to Sumner County in south-central Kansas. The fault zone is associated with the Nemaha Ridge, a buried mountain range composed of igneous rock. The fault zone was probably responsible for producing one of the largest earthquakes recorded in Kansas. The quake, which occurred in April 1867 near Wamego, was felt as far away as Dubuque, Iowa. In nearby Manhattan, people fled to the streets, and several foundations were cracked. Farther west, near Solomon, a train shook so much that its engineers abandoned it. Less powerful earthquakes struck the area in 1906 and twice in 1929. While there have been no strong earthquakes in the vicinity since that time, the Kansas Geological Survey has recorded a number of smaller "microearthquakes" in the Wamego area, which are too small to be felt, measuring less than 2.5 on the Richter Scale.

324.8   Crouse Limestone.

325.3   Crouse Limestone, overlain by *Blue Rapids Shale,* overlain by *Funston Limestone,* overlain by *Speiser Shale.*

325.7 and 326.1   *Wreford Limestone.* This exposure has several thin layers of chert running through the limestone.

A field of glacial boulders atop a hill near Wamego, north of I-70 along Kansas Highway 99

327.5   *K-99 exit* and exposures of the upper part of the *Beattie Limestone.* Along this highway, about five miles north of I-70, is a large field of glacial boulders on the side of a hill. This probably represents an area where a retreating glacier began to melt and dropped much of its load of rock debris. That debris includes a number of red quartzite boulders—many today are stained green by accumulations of lichen—that were transported south from South Dakota or Minnesota, a distance of several hundred miles. To the east, I-70 is usually within a few miles of the farthest advance of the Ice Age glaciers in Kansas, which occurred during the Kansan Glaciation, about a million years ago. Glacial deposits are common north of this line.

Early explorers often commented on these glacial erratics. In fact, one of the first recorded references to Kansas geology came from the Frenchman Etienne Veniard de Bourgmont, who explored the Mississippi River valley in 1724. Somewhere between present-day Topeka and Manhattan, Bourgmont wrote: "Along rivers there is found a slate, and in the meadows, a reddish marble, standing out of the earth one, two, and three feet. Some pieces of it upwards of six feet in diameter." In his description of the reddish marble, the explorer was almost certainly referring to the quartzite glacial erratics.

327.9   *Eskridge Shale* overlain by the *Cottonwood Limestone Member* of the Beattie Limestone. Cottonwood limestone is massive, maintaining a nearly constant thickness of six feet wherever it is found. The rock's weathered surface is nearly white. These qualities make it valuable as a bulding stone; many buildings and stone bridges in the state are made of Cottonwood limestone. Just south of I-70 are abandoned quarries in the Cottonwood.

328.5   *Pretty Creek.*

328.7   Northwest of I-70 is a creek valley that may mark the ancestral course of the Kansas River before it took a more northerly route. Today, a railroad runs through this creek valley. Roads and railroads often follow creek and river valleys because they represent level, easy surfaces for travel. I-80 in Nebraska, for example, follows the Platte River much of its way.

330.0   *K-185 interchange.* The town of McFarland is 0.5 miles to the south.

330.2   *Paw Paw Creek.* Pawpaw trees are found throughout the eastern United States; eastern Kansas is near the western limit of their range. These trees often grow in small colonies. They are notable for their sweet, fleshy yellow-skinned fruit.

332.4   *Spring Creek.*

332.9   *K-138 interchange.* Paxico is 1 mile to the east.

333.3   *Mill Creek,* which drains parts of northern Wabaunsee County before joining the Kansas River. It is one of at least ten Kansas streams named Mill Creek.

333.6   *Hamlin Shale Member* of the Janesville Shale.

335.0   Exit to *Skyline Scenic Drive,* which runs south of I-70 and rejoins the Interstate at the Wamego–Alma exchange, just east of Manhattan. The drive traverses parts of the Flint Hills.

335.1   *Snokomo Creek.*

336.0–337.0   *Rest area.*

337.5   *Dog Creek.*

338.1   *West Branch shale,* overlain by the *Five Point limestone,* both members of the Janesville Shale. An unnamed coal bed in the West Branch shale crops out along the north-south county road just west of this outcrop. Thin layers of coal are unusual in these Permian deposits.

339.0   At this point the road passes an exposure of *Janesville*

Shale and crosses *Buffalo Mound.* At 1,273 feet elevation, this hill is more than 300 feet above Mill Creek valley, just to the north. It was reportedly named because its shape resembles a buffalo's back. Geologists consider Buffalo Mound a landmark on the eastern edge of the Flint Hills. It is capped by Grenola Limestone, and its outcrops are good locations for collecting fusulinid fossils.

339.8   *Brownville Limestone Member* of the Wood Siding Formation, overlain by the *Towle Shale Member* of the Onaga Shale. In this outcrop the Brownville limestone, which is actually bluish-gray, is stained red by the overlying red-colored Towle shale. The contact between these two units is defined as the boundary between rocks of Pennsylvanian age, below, and younger strata of Permian age, above. The contact between these two units appears conformable. That is, there is no drastic change from the rocks of Pennsylvanian age to those of Permian age, probably because there was no interruption of depositon. In other locations, Indian Cave sandstone may occur at the base of Towle shale and may cut down into the underlying Pennsylvanian rocks. Indian Cave sandstone, which marks the beginning of Permian depositon, is exposed about 8 miles southeast of here in the banks of Mission Creek at a location called Echo Cliffs.

340.4   *Brownville limestone,* overlain by *Towle shale.* This marks the easternmost exposure of Permian rocks along I–70. To the east, I–70 passes over Pennsylvanian rocks, which are generally alternating limestones and shales, with occasional sandstones and thin seams of coal. Thicker, economically important coal beds of Pennsylvanian age are found in southeastern Kansas.

341.8   *Dry Creek.*

342.0   *Keene Road exit.* Taken south, this road intersects with K–4, which meanders through the Flint Hills. K–4 passes through Eskridge—a town that bills itself as the Gateway to the Flint Hills—and past a pretty man-made reservoir called Lake Wabaunsee. Buffalo Mound is clearly visible on the western horizon.

342.2   *Pillsbury Shale,* overlain by the *Stotler Limestone.* The Pillsbury Shale often contains a thin bed of coal.

344.4   *Dover Limestone Member* of the Stotler Limestone.

345.4   *Post Creek.*

346.0   *Wabaunsee/Shawnee county line.* Wabaunsee County was named for a Potawatomi chief whose name meant "Dawn of Day." This area was once part of a 900-square-mile Potawatomi reservation that stretched 30 miles north and west from Topeka. A 120-square-mile

Potawatomi reservation still exists about 15 miles to the north, in Jackson County. The Potawatomi language has provided what is probably the best-known Indian phrase in America—"kemo sabé"—which means faithful friend and was used by television's Tonto in reference to the Lone Ranger.

347.3 *Willard Shale* overlain by the *Tarkio Limestone Member* of the Zeandale Limestone. This Pennsylvanian limestone is gray, but it weathers to a deep yellow-brown. It is characterized by an abundance of large fusulinids.

348.0 *Vassar Creek.*

348.4 *Willard Shale,* overlain by *Zeandale Limestone.*

348.8 and 349.1 *Zeandale Limestone,* which was named for a small town east of Manhattan. *Zea* is the Greek word for grain, and that part of the Kansas River valley was known as Zeandale Bottoms.

349.5 *Willard Shale.*

350.8 *Mission Creek.*

351.3 Here the road passes through a hill called *Hickory Knob.* Along the road, the Scranton Shale is overlain by the Bern Limestone.

351.4 *Blacksmith Creek,* a small creek that joins Mission Creek and drains into the Kansas River.

352.4 *Emporia Limestone.*

352.8 *Auburn Shale,* overlain by the *Emporia Limestone.*

353.1 *K-4 interchange.*

354.4 *Burlingame Limestone Member* of the Bern Limestone. Fusulinid fossils and algal remains are common in this limestone layer, which ranges in thickness from 1 to 25 feet.

355.2 *I-470 interchange.* This road loops around the western and southern edges of Topeka before intersecting with the Kansas Turnpike and I-70.

355.4 *The Kansas Museum of History,* operated by the State Historical Society, opened in 1984 with exhibits related to the history of Kansas.

356.1 Exit for *Wanamaker Road* and *U.S. 75 By-pass South.* The Kansas Museum of History and the Menninger Foundation are accessible from this exit.

356.6 North of I-70 is the Menninger Foundation, a nationally known center for the treatment of and research into psychiatric disorders. The main building, which is modeled after Independence Hall in Philadelphia, contains a collection of Sigmund Freud's papers.

357.1 On the hill to the north is Cedar Crest, the official residence of the governor of Kansas.

358.0 *Interchange for U.S. 75,* which exits to the north and joins I-70 for a short distance to the west. The eastbound exit for Gage Boulevard and Gage Park is at this interchange.

358.3 Just north of I-70 is a view of the *Kansas River.* This is also the westbound exit for Gage Boulevard. One mile south is Gage Park and the Topeka Zoo, which is noted for its tropical rain forest and its gorilla exhibit, which allows zoo goers to walk through a glass-enclosed tunnel in the gorillas' cage.

359.6 Here I-70 crosses *Ward Creek,* which flows a short distance north and empties into the Kansas River. The Kansas, or Kaw, drains almost all of northern Kansas, eastern Colorado, and most of Nebraska south of the Platte River. The Kansas begins at the confluence of the Republican and Smoky Hill rivers near Junction City and empties into the Missouri River at Kansas City. Other rivers that drain into the Kaw include the Solomon, Saline, Delaware, Blue, and Wakarusa. In all, the Kaw drains a watershed of more than 60,000 square miles, carrying an average of 4.75 million acre-feet of water per year. In northeastern Kansas, where groundwater supplies can be rare, the Kaw is an important water source, and it is no accident that many of the state's major cities—Topeka, Lawrence, Kansas City—are perched on its banks.

361.4 The twin spires of *St. Joseph's Catholic Church* are visible south of the highway.

362.5 Tenth Avenue exit. The *Kansas Statehouse* and other state office buildings are accessible from this exit. Construction on the Kansas Statehouse began in 1867, using Fort Riley limestone, quarried near Junction City (outcrops of the Fort Riley can be seen at milepost 306). Construction on the west wing began in 1869, using Cottonwood limestone, dug near Cottonwood Falls in Chase County. Located on a 20-acre square in the center of Topeka, the Statehouse features murals by John Steuart Curry on its second floor.

Since Kansas was organized as a territory in 1854, the legislature has met at four locations in addition to Topeka. The 1855 legislature, called the Bogus Legislature because many of its members were elected by proslavery Missourians, met at Pawnee in Fort Riley and at Shawnee Mission. Later legislatures met at Lecompton and Lawrence, before Topeka was voted the permanent state capital.

Construction on the east and west wings of the Kansas Capitol (courtesy of the Kansas State Historical Society)

362.9 *Shunganunga Creek.*

363.2 *Topeka Limestone.*

364.8 *Deer Creek,* which is dammed 1.5 miles to the south to form Lake Shawnee. The Deer Creek Limestone is named after exposures along this stream.

365.0 *Topeka Limestone.* This is the last milepost along non-turnpike I-70. Just east of here, I-70 joins the Kansas Turnpike and takes on a new numbering system, beginning with marker 182.

181.6 *Kansas Turnpike tollbooth.*

181.8 *East Topeka interchange.* To the east, I-70 and the Kansas Turnpike continue to Lawrence and Kansas City. To the southwest, the turnpike leads to Emporia, Wichita, and Oklahoma.

182.8 *Topeka Limestone* is exposed on the north side of west-bound I-70.

183.0 *Topeka service area.*

183.3 *Ervine Creek limestone,* which is the uppermost member of the Deer Creek Limestone, is exposed on the south side of the eastbound lane of I-70.

183.4 *Stinson Creek* is named after the founder of the town of Tecumseh, which is situated near the stream's mouth.

183.7–184.0 *Calhoun Shale* overlain by *Topeka Limestone.*

184.6 *Deer Creek Limestone.* The Ervine Creek Member of the

Deer Creek Limestone is 14 to 18 feet thick in this area. It has been extensively quarried south of the highway along Tecumseh Creek.

184.7 *Tecumseh Creek* flows into the Kansas River 1.5 miles north of here, near the town of Tecumseh, which was named after the most famous of the Shawnee Indian chiefs.

185.6 *Calhoun Shale,* overlain by *Topeka Limestone.*

185.8 *Deer Creek Limestone.*

186.0 *Whetstone Creek.*

186.3 *Deer Creek Limestone.*

187.0–187.3 and 187.9 *Topeka Limestone,* which is named for the Kansas capital, is 33 to 55 feet thick.

188.5 *Shawnee/Douglas county line* and the *Oregon Trail crossing.* Here I-70 passes over U.S. 40, which was one of the major east-west routes through Kansas before the days of the Interstate highway system. Between Topeka and Lawrence, U.S. 40 closely follows the path of the old Oregon Trail. When gold was discovered in California in 1848, traffic along the Oregon Trail increased. One branch of the trail ran from Independence, Missouri, to Topeka, then northwest to Marysville, and into Nebraska; its course would have crossed I-70 at about this point. Although wagon ruts are not visible at this location, they can be seen along a number of wagon trails and stage trails that cross the state.

188.6 *Topeka Limestone.*

188.8 *Spring Creek.*

189.4 *Calhoun Shale,* overlain by *Topeka Limestone.*

189.5 The town of Big Springs, south of the highway, is named after a series of springs that flowed near here. Also at about this point the road crosses a feature that geologists call the Big Springs Anomaly. According to precise measurements by geophysicists, magnetic levels in this area are significantly higher than in the surrounding countryside. These readings are intriguing because they give clues about the subsurface. For example, the Big Springs Anomaly indicates that the underground igneous rock contains more magnetite than in other areas, producing the higher magnetic readings. Geologists have mapped magnetic levels across Kansas; they use the information to explore for oil and gas and to understand the geologic history of Kansas.

190.2 *Topeka Limestone.*

193.0 *Lecompton* is visible to the northeast. This small town was once the capital of territorial Kansas and was known as a proslavery stronghold. Today, bald eagles often roost in the winter in the trees along the Kansas River near Lecompton—in fact, the town was once called Bald Eagle.

193.3 *Deer Creek Limestone.*

193.8 *Tecumseh Shale,* overlain by *Deer Creek Limestone.* This shale layer is up to 65 feet thick near the Kansas River.

195.0 *Tecumseh Shale,* overlain by *Deer Creek Limestone.*

195.6–196.8 Numerous outcrops of *Lecompton Limestone.*

196.3 *Oakley Creek.*

198.1 *Plattsmouth limestone.* The Oread Formation is made up of four smaller limestone layers. The Plattsmouth limestone is the thickest of those limestones.

198.6–199.0 *Oread Limestone,* one of the most prominent formations in eastern Kansas, averages about 52 feet in thickness in the northern part of the state. The Oread was named for Mount Oread, the hill that overlooks downtown Lawrence. Mount Oread, also the home of the University of Kansas, is capped by Oread Limestone. The hill's name came from the home of Eli Thayer, a Massachusetts resident who was one of the foremost promoters of the New England Emigrant Aid Society, which helped to settle Kansas in the 1850s. Mount Oread was the name of Thayer's home; he also operated the Oread Female Seminary near Worcester, Massachusetts.

199.7 *Baldwin Creek.*

201.9 *West Lawrence interchange.* Lawrence is the home of the Kansas Geological Survey, a division of the University of Kansas. Also, the Museum of Natural History on the KU campus displays excellent specimens of Kansas fossils, including a number taken from the Cretaceous chalks of western Kansas.

203.0 *The Kansas River* looks wider and deeper here than at other places along its course, primarily because it is impounded a little more than a mile downstream by Bowersock Dam. Bowersock is the only dam on the Kansas River and the only hydroelectric dam in the state. It has the capacity to produce just under two megawatts of electricity.

At Lawrence, the first bridge across the Kansas River was completed in December 1863. Construction was held up when eight members of the building crew were killed in Quantrill's raid. When the wooden bridge was finally finished, the construction company charged a toll of 25 cents per trip.

Because of its sandy bottom and its connection to the Mississippi

Former channels of the Kansas River near Lawrence

River via the Missouri River, the Kansas River is home to several species of fish that are rarely found elsewhere in Kansas. Sturgeon, lamprey, and eels have all been taken from the Kansas River, although such fish are rarely caught today.

203.5 *East Lawrence interchange* and junction with U.S. 24, U.S. 40, and U.S. 59. The Lawrence business district is 1.5 miles to the south.

204.3 The Kansas River has wandered across this area for centuries, regularly changing course. To the north is an *old river channel,* which the Kaw has abandoned, although it still often holds water.

204.6 Sandpits appear on the south side of the road. Mount Oread and the University of Kansas are visible to the southwest.

205.4 A *river channel* abandoned by the Kansas River. The slightly higher ground east of this channel is underlain by deposits known locally as the Newman Terrace. These deposits were laid down following the last Ice Age, when the Kansas River flowed at a higher level and carried more sediment than it does today.

205.8 *Douglas/Leavenworth county line.*

206.0 *Mud Creek.* Blue Mound (elevation 1,052 feet) is visible 6.5 miles to the south. This hill has resisted erosion because it is capped by Oread Limestone. Standing more than 250 feet above the nearby Wakarusa River, the hill was once the site of a ski slope. Although Blue Mound is shaped much like Iron Mound (the hill east of Salina, seen at milepost 258.5), Blue Mound appears to be more rounded and less steep because it is covered by deciduous trees rather than by short prairie grasses.

207.0 *Stranger Formation,* partially obscured by vegetation and best seen on the north side of the highway, is named for the creek that runs nearby.

208.2 *Kent Creek.*

209.0 *Lawrence service area.*

209.5 *Haskell Limestone Member* of the Lawrence Formation. This rock layer is best seen on the south side of the eastbound lane. It marks the base of the Lawrence Formation, separating it from the Stranger Formation (visible at milepost 212) below. Both of these formations are mostly sandstone and shale.

210.7 *Nine-mile Creek.*

211.6 Sandstone in the *Stranger Formation.*

212.0 *Stranger Formation,* overlain by *reddish Pleistocene deposits.* The Pleistocene Period is the most recent in geologic history, and these deposits are the rocks, gravels, and silts that were dropped by glaciers, streams, or wind.

212.2 *Cow Creek.*

213.4 On the south side of the highway there is an exposure of one of the thin *coal beds* in the Stranger Formation. These coals were once mined in this vicinity and in other parts of northeastern Kansas by means of small underground mines. Today, all Kansas coal is mined in surface pits in southeastern Kansas.

214.9 *Stoner Limestone Member* of the Stanton Formation. This limestone is generally 10 to 20 feet thick in this area but thickens to 50 feet in Montgomery County in southeastern Kansas.

215.9 *Stranger Creek.* Stranger is a translation of the Indian word *okeetsha,* which means "wandering aimlessly about," an apt description of the meandering habit of this stream.

218.2 To the west on the skyline is the *Oread escarpment,* which is capped by the resistant limestones of the Oread Formation. This prominent break in the landscape extends from Doniphan County, in the far northeast corner of Kansas, to Chautauqua County, on the Oklahoma border.

220.5–220.7 *Tonganoxie Sandstone Member* of the Stranger Formation on both sides of the highway. Several exposures of this sandstone are visible east of this point. Note the sandstone's cross-bedded appearance—thin, angling lines in the rock indicate that it was deposited by flowing water. The Tonganoxie occurs in a relatively narrow band, never more than 20 miles across, that extends to the southwest. This sandstone reaches a thickness of 160 feet and appears to be the valley deposit of a large southward-flowing river that cut downward into older rocks during the Pennsylvanian Period. This formation is named after a small town 8 miles to the west, in southern Leavenworth County. Tonganoxie sandstone is an important aquifer, supplying ground water to farms and small towns.

221.0   *Stanton Limestone.*

221.4   *Bonner Springs Shale,* overlain by *Plattsburg Limestone.* The Plattsburg averages about 25 feet in thickness in Kansas.

221.8   *Wolf Creek.* Gray wolves were once common throughout most of Kansas. Last reported in Kansas in 1905, wolves preyed almost entirely on buffalo.

222.4   *Leavenworth/Wyandotte county line.*

222.4–222.9   *Bonner Springs Shale,* upward through the *Plattsburg Limestone, Vilas Shale,* and the top of the *Stanton Limestone.*

223.2   *Tonganoxie Sandstone Member* of the Stranger Formation. Tonganoxie was a Delaware Indian chief.

224.2   *K-7 exit.* To the north are the Agricultural Hall of Fame and the city of Leavenworth; to the south is Bonner Springs. Leavenworth is the site of historic Fort Leavenworth. Established in 1854, Leavenworth was the childhood home of Buffalo Bill Cody.

224.4   *Toll plaza.*

224.6   *West Mission Creek.*

225.0–225.8   *Tonganoxie Sandstone Member* of the Stranger Formation.

226.0   *East Mission Creek.*

411.0   At this point the numbering system for mileposts changes. To the west, Kansas Turnpike numbers are used to measure the distance along the turnpike from its southern terminus at the Oklahoma state line. Westbound I-70 is part of the Kansas Turnpike from here to Topeka. To the east, the numbering system for I-70 resumes at this point. This is the same system that marks I-70 from the Colorado border to Topeka, and the mileposts along eastbound I-70 mark the distance east of the Colorado state line.

411.5   South of I-70, the top of the *Plattsburg Limestone* is overlain by the *Vilas Shale,* which is overlain by the *Captain Creek limestone, Eudora shale,* and *Stoner limestone* members of the Stanton Limestone.

411.7 Interchange with *I-435.* This is also the valley of Little Turkey Creek.

411.8 *Farley Limestone Member* of the Wyandotte Limestone Formation.

412.1 *Farley limestone.*

412.3 *Plattsburg Limestone.*

412.5 *Vilas Shale,* overlain by the *Captain Creek limestone, Eudora shale,* and *Stoner limestone* members of the Stanton Limestone.

412.7 *Stanton Limestone.*

413.5 To the south 1 mile is the *Edwardsville Northeast oil field.* Discovered in 1983, this is one of only two active oil fields in Wyandotte County. During 1984, the three wells in this field produced 702 barrels of oil. Bonner Springs Shale is north of the road.

414.0 Two miles south is the site where the first ferry operated across the Kansas River. Moses Grinter built the ferry in 1831, providing a vital link in the military road from Fort Leavenworth to Fort Gibson in Oklahoma. This road is followed closely today by U.S. 69. A two-story brick farmhouse, built in 1867 near the site of the ferry, is open to the public; it is located along K-32.

414.5 *Cross-bedded limestone* in the Farley Member of the Wyandotte Limestone. This limestone is oolitic, composed of small spheres of calcium carbonate. These spheres are arranged in angled lines in the limestone, similarly to the grains of sand in the Tonganoxie sandstone to the west.

415.1 Along the westbound lane is a *tourist information center,* which provides information on Kansas towns, rodeos, fairs, festivals, and other points of interest. North and east of the information center are outcrops of the two limestone layers that make up the Farley limestone, the uppermost member of the Wyandotte Limestone. The lower limestone is oolitic and cross-bedded.

415.3 *Argentine Limestone Member* of the Wyandotte Limestone, on north side of the highway.

415.7 *Mill Creek.*

415.8 Interchange with *K-132.*

416.0 *Loess exposure.*

416.6 *Muncie Creek.*

416.9   Thick *loess deposits* on both sides of the highway.

417.3   *Brenner Heights Creek.*

417.5   *Loess* on north side of highway. While loess is common throughout Kansas, especially thick deposits are found in the northeastern corner of the state. Many of the bluffs overlooking the Missouri River in Doniphan County are composed of loess.

418.0   *Lane Shale* overlain by the *Argentine Limestone Member* of the Wyandotte Limestone. The Argentine is one of the limestones that has been mined extensively in the Kansas City area. A by-product of this mining is underground space; more than 120 million square feet of underground space exist within 25 miles of Kansas City. The constant temperature and humidity of these man-made caves make them ideal for storing all types of items, including government records. The cool temperature allows economical refrigeration and the cold storage of large amounts of food. The underground space has also been used for offices and factories.

418.1   *Kansas River* to the south. Between 1854 and 1866, 34 steamboats paddled up the Kaw River; one made it as far as Fort Riley. But shallow water and sand bars eventually halted steamboat traffic. As one Lawrence newspaper editor wrote, the Kansas River was "a hard road to travel."

418.2   *Lane Shale,* overlain by *Argentine limestone.*

418.4   North of the highway is a *quarry* in the Argentine limestone.

418.7   *Argentine limestone* north of the highway.

419.0   Interchange with *I-635,* which loops around the western edge of Kansas City. Connecting with I-29, it provides a quick route to Kansas City International Airport, north of Kansas City. Iola Limestone is exposed north of the highway. The Iola, named for a town in Allen County, is thickest in southeastern Kansas.

419.4   *Mattoon Creek.*

419.6–420.4   *Westerville Limestone Member* of the Cherryvale Shale, overlain by *Drum Limestone.* Oolites are common near the top of the Westerville limestone, which is sometimes called the Kansas City Oolite. The Westerville reaches a thickness of 20 feet around Kansas City, but it is replaced by a shale from Miami County southward. Cherryvale Shale is named for a town in Montgomery County. Sometimes rock formations have different names in different states. In Oklahoma, this formation is called the Nellie Bly Shale.

420.5   To the south, in the Kansas River valley, is an area of

A view inside the Consolidated Kansas City Smelting and Refining Company's plant in the Argentine during the 1890s (courtesy of the Kansas State Historical Society)

Kansas City, Kansas, called the *Argentine*. The name comes from the Latin word *argentum*, which means silver. A smelter that once operated in this area actually refined small amounts of silver.

421.5   At this point, I-70 drops down to the *flood plain of the Kansas River*. The fact that this is a flood plain would have been all too obvious on Friday, 13 July 1951—a day known in Kansas City as Black Friday. On that day, the swollen Kansas River left its banks, spilled onto its flood plain, and inundated low-lying neighborhoods and industrial districts on both sides of the state line. At the height of the flood, brown, silt-laden water stretched from bluff to bluff, reaching depths of 30 feet.

An unusually wet spring and early summer set the stage for the 1951 flood. A stalled frontal system provided the climax when it dropped up to 16 inches of rain on the saturated soil of the Kansas River basin. Flooding occurred up and down the Kansas River, along many of its tributaries, and on streams south of the Kansas basin. Forty-one people died in the flood, which was one of the most destructive in the nation's history, causing $900 million worth of damage.

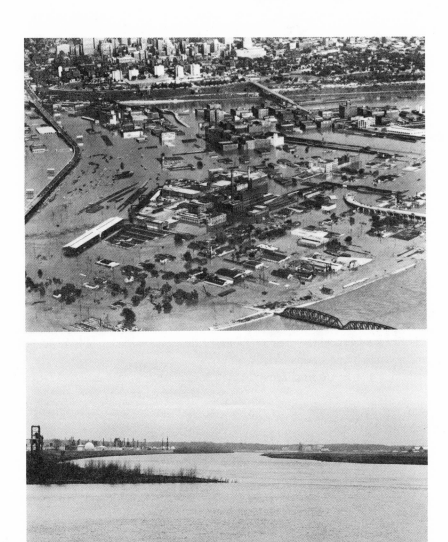

The 1951 flood covered the Kansas River valley and inundated the west bottoms area of Kansas City (above; courtesy of the Kansas State Historical Society). The confluence of the Kansas and Missouri rivers north of the Lewis and Clark Viaduct at Kansas City (below).

Also along this stretch of highway, I–70 reaches its lowest point in Kansas, approximately 760 feet above sea level. This is 3,150 feet lower than the highest point in Sherman County.

422.5   Thick *loess deposits* on the west side of the highway. Downtown Kansas City, Missouri, sits on the bluffs to the east.

423.4   To the west is a portion of Kansas City, Kansas, called *Strawberry Hill.* This area was home to many immigrants from eastern and southeastern Europe, who came west in the 1800s to work in the huge packing plants that operated in the Kansas River bottoms to the east. Many descendants of these immigrants still live in this neighborhood.

423.7   At this point, I–70 curves to the southeast and crosses the Kansas River by means of the Lewis and Clark Viaduct. Just to the northeast is the *confluence of the Kansas and Missouri rivers.* The intersection of the midlines of these two rivers marks the beginning point of the Missouri–Kansas boundary south of the Missouri River. The Lewis and Clark expedition camped at the mouth of the Kansas River during its exploration of the Louisiana Purchase. While in northeast Kansas, Lewis and Clark described the area as "some of the most charming bottom lands and uplands by no means bad."

424.2   *Kansas/Missouri state line.*

36·U.S. HIGHWAY·36

From St. Francis to St. Joseph

0.0 *Kansas/Colorado state line.* At one time U.S. 36 was one of the busiest roads in Kansas, providing a link between northern Kansas and Denver, about 160 miles west of here. With the completion of I–70 to the south and I–80 to the north, traffic along U.S. 36 dropped considerably. Across Kansas, the highway parallels the Kansas–Nebraska border for much of its route, passing through the northern tier of Kansas counties.

This is also the boundary between the Mountain and Central time zones. Parts of western Kansas, including Sherman County to the south, fall into the Mountain zone, while all of Cheyenne County is in the Central zone. Cheyenne County is named for the Cheyenne Indians, who ranged throughout western Kansas during the 1800s. In 1867 the tribe agreed to move to Indian Territory, but many of the Cheyennes continued to hunt in northwestern Kansas. In 1869 the Cheyennes were routed in a battle at Beecher Island on the Arikaree Fork of the Republican River, northwest of here. The loss effectively ended Indian resistance in western Kansas.

9.5 The elevation is 3,500 feet above sea level. The highest point along U.S. 36 in Kansas is at the Colorado border, where the elevation is 3,735 feet. The same geologic forces that lifted the Rocky Mountains also elevated western Kansas and tilted the entire state toward the east. The elevation here, for example, is nearly 3,000 feet above the elevation at St. Joseph, on the Missouri River. This period of mountain building caused steeper stream gradients and exposed more rocks, beginning a period of mountain erosion and stream deposition that continues today. The result is a vast sheet of alluvial material that covers much of the Great Plains from

The Arikaree Breaks, eroded canyons in the loess hills of Cheyenne County north of U.S. 36

South Dakota to Texas. This alluvium, called the Ogallala Formation, conceals most outcrops of underlying, older bedrock in western Kansas. These outcrops can only be found in the valleys of deeply incised streams. The first exposure of older rocks along eastbound U.S. 36 occurs 173 miles east of here, in Smith County. Windblown loess covers much of the uplands in northern and western Kansas, concealing many of the rocks near the highway, including much of the Ogallala.

The higher altitude of northwestern Kansas also means that the air is thinner. That, and the higher latitude, makes this corner of the state cooler than the rest of Kansas. The average January temperature in Cheyenne County is about 28°F; the July average is 77°F. Saint Francis reported the coldest temperature ever recorded in Kansas in the month of August—33°F in 1910.

11.0 *Loess* exposures north of the highway. Loess is a finely ground dust, a sort of rock flour that forms a mantle over much of the High Plains of western Kansas. It was deposited by the wind during the Ice Ages of the past million years. Streams have since eroded away the loess, carving draws and canyons that have exceptionally steep sides; loess can maintain a nearly vertical face without caving in or slumping. These loess canyons form

A pineapple-sized boulder of igneous rock found embedded in sedimentary deposits along a creek in Cheyenne County (left); the South Fork of the Republican River at milepost 12 along U.S. 36 in Cheyenne County (right)

a rugged landscape a few miles north of here, along the Arikaree River, where loess accumulations are up to 180 feet thick.

11.9 *K-27* runs north from here to the town of Haigler, Nebraska. South of Haigler, barely inside the Kansas border, geologists have discovered pineapple-size igneous rocks embedded in alluvial sediments at the base of loess hills. Nearly all rocks found at the surface in Kansas are sedimentary; these igneous rocks may have washed into Kansas from the Rocky Mountains to the west. Geologists also speculate that they may have been embedded in glacial ice that floated downstream, melted, and then dropped its load of rock far from the Ice Age glaciers of the Rockies.

12.4 *South Fork of the Republican River,* which begins on the plains of eastern Colorado, then cuts across the corner of northwestern Kansas before joining the Republican River in southwestern Nebraska. The Republican then makes a sweeping bend across southern Nebraska, before reentering Kansas in Jewell County. U.S. 36 crosses the Republican River near the town of Scandia at milepost 213. From Scandia the river goes southeast, where it is impounded to form Milford Reservoir, shortly before joining the Smoky Hill to create the Kansas River.

13.0–14.0 *Saint Francis* is the seat of Cheyenne County. The county courthouse, which was completed in 1925, is located at the east end of the central business district. Saint Francis isn't named for the Catholic saint, but for Frances Emerson, wife of one of the town's founders. In 1980 the population of St. Francis was 1,612; Cheyenne County's population numbered just over 3,500, the smallest of any Kansas county traversed by U.S. 36. According to the 1980 census, thirteen other Kansas counties have fewer residents than Cheyenne County. The least populated is Greeley County, about 100 miles south of here, which has a population of 1,845.

16.6 *West Fork of Sand Creek.* Many hills in this area are covered by sand sage, a close relative to the Great Sagebrush that is common in the dry valleys of the Rockies and the Great Basin. In Kansas, sand sage is usually found in the sand hills of the western half of the state. As the name Sand Creek suggests, windblown sand covers parts of this area, particularly along the southeast side of the valley of the South Fork of the Republican.

17.6 *Ogallala Formation.* In the western third of Kansas, the Ogallala is the most common, and in many places the only, rock formation at the surface. The Ogallala is composed of sands, gravels, clays, and other material that has eroded off the face of the Rocky Mountains. Some outcrops of the Ogallala are naturally cemented together to form a dense, hard rock called mortarbeds. While it is occasionally used as a source of building materials, the Ogallala is probably best known as an aquifer, a subsurface water-bearing formation.

The Ogallala underlies most of the western third of the state. In places it is up to 600 feet thick, although it is generally thicker in southwestern Kansas than in northwestern Kansas. Much of the Ogallala is saturated with ground water. Because it consists of coarse rock material with plenty of pore space between the particles, the Ogallala absorbs water like a sponge. Across western Kansas, the average saturated thickness of the Ogallala is about 100 feet, although large-scale irrigation has led to substantial declines in the water levels in some areas. The Ogallala is found throughout the Great Plains, underlying parts of seven other states in addition to Kansas. The formation was named for the town of Ogallala in southwestern Nebraska where it was originally described by geologist Nelson Darton in 1899.

18.6 The small town of *Wheeler* is 0.5 miles north on K–217.

22.0 *High Plains depression.* This isn't a state of mind; it is a place where the road crosses a topographic depression. The horizon on all sides is slightly above the highway. These dips in the landscape are common

Where the Ogallala Formation comes to the surface in western Kansas, it often forms a dense hard rock called mortarbeds.

throughout the High Plains, although their origin remains a topic of debate among geologists. The depressions may have been created by buffaloes wallowing in mudholes during wet weather, followed by wind erosion during dry spells. Solution of underground layers of salt and gypsum may also be responsible. Even the impact of meteorites has been suggested as a possible cause. Since the size and shape of these depressions is far from uniform, there is probably more than one cause for them.

28.0   *Bird City,* founded in 1885 by Irish Catholics, was originally called Bird Town. It was named for Benjamin Bird, the editor of the local newspaper. The town was sometimes called "Lindbergh's Playground," because Charles Lindbergh often barnstormed from here. He stopped to visit after his flight across the Atlantic in 1927.

34.8   *Cheyenne/Rawlins county line.* Like many western Kansas counties, Rawlins was named for a Civil War general. John A. Rawlins was appointed secretary of war in Grant's administration.

37.0   *McDonald.* Western Kansas is farming country. Wheat, silage, corn, milo, and alfalfa are major crops in northwestern Kansas, along with smaller amounts of beans, sugar beets, and sunflowers. Much of the corn and milo grown in Rawlins County is watered with center-pivot irrigation sprinkler systems that can be seen from the highway during the spring and summer. Most of these irrigation wells rely on the Ogallala Formation for water. Extensive pumping has lowered the water table throughout western Kansas; since 1940 the water level has dropped about four inches per year in Rawlins County wells. That decline is ominous, because very little of the scant precipitation in western Kansas makes its way back into the Ogallala. Geologists believe that the recharge to the aquifer is probably less than one inch per year in much of western Kansas. Most of the precipitation evaporates, runs off the ground's surface, or is captured by plant roots before it can soak through to recharge the Ogallala.

45.9   *Little Beaver Creek.* The beaver gave its name to several Kansas creeks and towns. Beavers, the state's largest living rodents, are found throughout Kansas, particularly in the eastern end of the state, where streams and trees are abundant. One Kansas specimen weighed more than 96 pounds.

46.5   *Ogallala Formation* north of the road.

51.0   For the next 3 miles, the eastbound highway runs along the *drainage divide between Little Beaver Creek and Beaver Creek.* Runoff north of the road flows into the Little Beaver, and on the south it joins the Beaver. The two streams converge east of Atwood.

53.0   The *elevation is 3,000 feet,* more than 700 feet below the elevation at the Kansas/Colorado border.

54.4   *Beaver Creek.* In 1867, eleven cavalry troops were killed by Indians along the banks of Beaver Creek in northeastern Sherman County. Hostilities between the Indians and the army were particularly severe in 1867 and 1868. Lt. Col. George Armstrong Custer and Gen. Winfield S. Hancock were sent to northwestern Kansas during the summer of 1867 to attack the Indians, but they met with little success.

54.6–55.1 *Atwood*, the seat of Rawlins County, was founded in 1879 by James Matheney and a group of Irish Catholics. The county courthouse, completed in 1907, is a red-brick building at the east end of the business district. Northeast of the intersection of U.S. 36 and K–25 is Atwood Lake, formed by a man-made dam on Beaver Creek. Most impoundments of surface water in northwestern Kansas have fallen victim to decreased stream flow. Atwood Lake is often dry.

61.3 *Ogallala Formation.*

69.2 Intersection with *K–117* to Herndon. In 1878 a group of German-Russians settled along Beaver Creek in northeastern Rawlins County. They first named their settlement Pesth, then Lincoln, and, finally, Herndon, in honor of William H. Herndon, Abraham Lincoln's law partner.

71.2 *Rawlins/Decatur county line.* Decatur County is named for Stephen Decatur, a naval hero in the War of 1812 and in the battle against pirates off the coast of Tripoli in 1815. Decatur is remembered for his pronouncement "My country! . . . May she always be in the right; but our country, right or wrong."

72.0–75.0 Northwestern Kansas has a *distinctive drainage pattern.* The major streams—such as the South Fork of the Republican, Beaver Creek, Sappa Creek, Prairie Dog Creek, and their larger tributaries—all flow to the northeast. The smaller tributaries, such as the ones that the highway passes over here, are aligned in a northwest-southeast direction. Though difficult to discern from the ground, these patterns are apparent on topographic maps and satellite photographs. The drainage patterns of the larger rivers are probably controlled by the uplift of the Rocky Mountains, which causes the rivers to flow down and away from the mountains to the northeast.

76.0 *Ogallala Formation* in the draw south of the highway, and loess exposed along the fence line.

77.3 County road to *Traer*. Near this town there is a highly eroded exposure of the Ogallala Formation, which forms a natural bridge outcrop called *Elephant Rock*.

78.0–81.0 To the south is the *Pollnow oil field*. Northwestern Kansas is not known for oil production. In 1984, wells in Decatur County pumped 542,000 barrels of oil, which ranked it forty-first among the 91 oil-producing counties in Kansas. The Pollnow field has produced nearly 1.2 million barrels of oil since it was discovered in 1953, making it the fourth-best producing field in Decatur County. That production comes from rocks about 3,700 feet underground, deposited during the Pennsylvanian Period of geologic history. Pennsylvanian formations are visible at the surface in eastern Kansas, along a band that runs from Kansas City southwest to Independence and Coffeyville.

82.1 *Intersection with U.S. 83* and the town of *Oberlin,* the seat of Decatur County. The geology along U.S. 83 is described in chapter 4 of this book.

83.2 *Oberlin cemetery* and *monument to the victims of the last Indian raid in Kansas.* In 1878 a band of Cheyenne Indians fled from their reservation in Indian Territory and headed north through their traditional hunting grounds. The Indians killed forty settlers in northwestern Kansas, marking the last Indian raid in the state's history. Several victims of the raid are buried in this cemetery, which also contains a monument commemorating the event.

83.9 *Sappa Creek.* The elevation here is 2,530 feet, or nearly 1,200 feet below the elevation at the Kansas/Colorado border.

84.7 South of the road is the *Jorn oil field.* The oil fields of Decatur and Norton counties are located on an ancient geologic uplift called the Cambridge Arch, a geologic structure that begins in southern Norton County and extends northwest across Nebraska to the vicinity of the Black Hills of South Dakota. Oil fields along this trend are also found in southwestern Nebraska.

During the early 1980s, geologists recorded over 300 small earthquakes, which may have been associated with oil production from the Sleepy Hollow oil field in Red Willow County, Nebraska, directly north of Decatur County. Most of the tremors were too small to be felt by people; they measured less than 2.0 on the Richter Scale. In the Sleepy Hollow field, oil is produced by water flooding, or pumping water underground to force additional oil out of the pores of subsurface rock formations. Geologists believe that water flooding may force apart faults in the granite basement rock, thus producing the earthquakes.

87.8 *Upper end of Cotton Creek.* Kansas is not usually associated with cotton, but agricultural reports show that farmers grew small quantities of cotton throughout the state, especially during the late 1800s. A small amount was grown in the early days of Decatur County, possibly providing a name for this creek. Today, cotton is raised near Sterling, in Rice County.

90.1 The town of *Kanona* is about two miles south, on a county road. Kanona lies on the path of the old Chicago, Burlington, and Quincy Railroad. As the railroads worked their way west during the 1800s, they established towns every 7 to 10 miles, to provide stops for water, fuel, freight, and passengers. Kanona is about 7 miles east of Oberlin and 10 miles west of Norcator. The railroads have abandoned many of the lines in western Kansas during recent years, severely affecting the economy of many old railroad towns. Today, virtually all the buildings in Kanona belong to a single owner.

90.5   *Ogallala Formation*, which is exposed in draws on both sides of the highway.

94.0   *Upper end of Long Branch Creek.*

100.4   *Norcatur*, situated near the county line. Its name combines the names of Norton and Decatur counties.

101.0   *Decatur/Norton county line.* Norton County is named after Capt. Orloff Norton, who was killed in a Civil War battle at Cane Hill, Arkansas. Lewis R. Jewell was a cavalryman killed in the same battle; Jewell County, three counties east of here, was named after him.

105.5   The town of *Reager.*

106.4 and 106.7   *Ogallala Formation.*

106.9   *Spring Draw*, which drains to the southeast into Prairie Dog Creek.

110.5   *K–383.*

112.5   Two miles south is *Keith Sebelius Lake*, a reservoir constructed by the federal Bureau of Reclamation and named after a long-time congressman from western Kansas. On the north shore of the lake is Prairie Dog State Park, a camping area that includes historic displays in a preserved adobe house. The park also harbors small colonies of black-tailed prairie dogs, which are common in the western half of Kansas. Prairie-dog towns, which can be identified by small dirt mounds dotting the landscape, are visible in several locations along U.S. 36 and other Kansas highways. The towns may shelter hundreds or thousands of the animals. The mounds serve as entrances to underground tunnels that reach depths of 5 to 15 feet.

Prairie dogs have many enemies, including ranchers, who have sought to exterminate the animals because their holes can injure cattle and horses. Natural predators include hawks, snakes, badgers, and coyotes. To protect themselves, prairie dogs keep the grass and vegetation around their towns closely cropped. That, combined with the vantage point at the top of their mounds, usually allows them to detect predators, to whistle an alarm, and to scurry to the safety of their burrows.

Prairie-dog burrows also served as home to black-footed ferrets, a member of the weasel family. These ferrets once ranged over the western two-thirds of Kansas, usually living in the same short-grass prairies that were home to the black-tailed prairie dog. Black-footed ferrets are now considered an endangered species; the last verified sighting of one in Kansas was in Sheridan County in 1957. Their disappearance is probably related to extermination measures taken against prairie dogs.

116.0–117.2 *Norton,* the seat of Norton County, lies at the intersection of U.S. 36 and U.S. 283, the latter of which runs south to the Oklahoma border. Norton calls itself the Pheasant Capital of the World. Every autumn, pheasant hunters from throughout the Midwest move into northwestern Kansas. Eastern ring-necked pheasants are native to Asia; 1,500 pair of the birds were introduced into 84 Kansas counties in 1905 and 1906. Today, pheasants are found throughout most of Kansas, particularly where fields provide food and cover. The brightly colored males and drab brown females are often seen along Kansas roads, though only the males are legal game for hunters, and then only from mid November to mid January.

119.3 *Prairie Dog Creek,* a tributary of the Republican River, was named by John C. Frémont, who explored the area in 1843. From here the creek flows northeastward, joining the Republican at Harlan County Reservoir, just across the border in southern Nebraska. Sappa Creek and Beaver Creek, which also drain northwestern Kansas, join the Republican just west of Harlan County Reservoir.

120.0 *Ogallala Formation* on the south side of the road.

120.7 *Norton State Hospital.* For many years this was a tuberculosis sanitarium. In 1969 it became a hospital for the mentally retarded. The Ogallala Formation is exposed south of the highway.

120.8 *Walnut Creek.*

122.4 *K–383* runs northeastward to Calvert and Almena. Near Calvert there are deposits of volcanic ash that are thick enough to be mined. Volcanic ash consists of tiny fragments of glass or congealed lava, produced during volcanic eruptions. Ash was deposited in this area during the past few million years; it probably came from volcanoes that erupted in New Mexico, Wyoming, or California. That ash was thrown high into the air (much as it was during the eruption of Mount St. Helens in 1980), and prevailing winds carried it over Kansas, where it fell to earth and was concentrated into deposits in streams and lakes. In some places those deposits are up to 20 feet thick. Ash is found throughout western Kansas and as far east as Douglas, Nemaha, and Chautauqua counties. However, it is mined only in two locations, including the spot south of Calvert. In the past, ash was used

This mine near Calvert, north of U.S. 36, produces volcanic ash from deposits that are more than 15 feet thick.

in toothpaste, cleansing powder, and even bedding material for chinchillas.

124.1 *Wildcat Creek.* Bobcats, sometimes called wildcats, are common throughout the state, particularly in the wooded areas of eastern Kansas. With the extinction of the mountain lion in Kansas, they are the only native member of the feline family in the state. Bobcats feed on rabbits, rodents, birds, and deer. Their keen senses of sight and smell allow them to avoid humans, so they are seldom seen.

125.0 *Ogallala Formation* crops out along the draws on both sides of the highway.

126.7 *South Fork of Prairie Dog Creek.*

130.8 *Norton/Phillips county line.* Phillips County was named for William Phillips, a Leavenworth lawyer who protested the election that produced the Bogus Legislature. He was killed in 1856 by proslavery forces. Phillips County was settled by a variety of ethnic groups during the 1870s, including Germans, Danes, Poles, and Swedes. These immigrants often named their towns after cities in the old country, such as Stuttgart, which is several miles east of here.

132.6 *Ogallala Formation* in the creek bank on the south side of the road.

133.3 About 11.5 miles south on this road is the small town of *Logan*, the home of the Dane G. Hansen Memorial Plaza, located in the center of town. This plaza, dominated by a modern museum and gallery, is a monument to philanthropist Dane G. Hansen, who died in 1967, leaving $15 million to his hometown. Despite its small size and out-of-the way location, Logan attracts prominent traveling exhibitions and art shows, thanks to these facilities.

133.5 *Deer Creek.* Two species of deer populate Kansas: the mule deer, found mainly in the western two-thirds of the state, and the smaller white-tailed deer, which is common throughout Kansas, particularly in wooded areas. Both were common when settlers first moved into the state, but by the 1930s their populations were so reduced that white-tailed deer were thought to be nearly extinct in Kansas. Since then, restocking and protection have increased their numbers, so that Kansas now has rifle and bow-hunting seasons every fall.

133.8 The town of *Prairie View* is immediately north of the highway. Loess is exposed in a creek bank on the north side of the road.

135.0 At about this point the Ogallala Formation reaches a feather edge on the uplands. To the west it generally thickens, while to the east is has been eroded away, exposing the underlying Cretaceous rocks.

135.4 *Loess* exposure south of the highway.

135.9 County road to *Long Island*. A location south of Long Island is famous for fossils of a short-legged barrel-bodied rhinoceros that became extinct at the end of the Tertiary Period, about 2 million years ago. This species was a grass eater that lacked the familiar nasal horn of today's rhino. Also at this point the highway passes over the *Long Island Syncline*. A syncline is a downward fold of subsurface rock layers. This one is named after the small town that it passes under.

136.0 *Loess* exposure to the south of the highway in the banks of Deer Creek.

139.4 *Deer Creek,* elevation 2,000 feet.

140.2 *K-121.* This short highway runs north to Stuttgart, one of many German communities in Kansas. According to one estimate, more than 100,000 Germans came to Kansas during the second half of the nineteenth century; for many years, more Kansas immigrants came from Germany than from any other European country. At Stuttgart, Deer Creek is joined by *Starvation Creek,* from the north.

141.1 *Deer Creek,* which the highway parallels and crosses over for the next 7 miles to the west. Deer are among the few large native mammals still found in western Kansas. At one time the plains teemed with a variety of unusual mammals, including camels, rhinoceroses, peccaries, bone-crushing dogs, saber-toothed cats, and several dozen species of horses. Most of those mammals were extinct by the end of the Tertiary and Pleistocene periods of geologic history, but their remains are still occasionally found in the sediments of western Kansas.

142.2 *Stuttgart oil field,* one of several oil fields that follow a linear trend northeastward from southwestern Phillips County, to the Nebraska state line and the Huffstutter oil field. These fields produce oil from an upward fold of rocks called the Stuttgart–Huffstutter Anticline. The oil here comes from rocks of Pennsylvanian age.

143.2 *Boughton Creek.*

143.3 The abandoned farmhouse north of the highway is built of rock quarried from the Niobrara Chalk Formation, which crops out in the surrounding hills.

145.3 *Bissell Creek.*

147.0 An *oil refinery and tank farm* are visible northeast of the highway. Most Kansas refineries are concentrated in the south-central part of the state, around Wichita and El Dorado. This is the only one in northwestern Kansas.

147.5 *Plottner Creek.*

147.6–148.8 *Phillipsburg.* Phillipsburg and Phillips County are not named for the same person. The county is named after a Leavenworth abolitionist, whereas the town is named for William A. Phillips, a reporter for Horace Greeley's *New York Tribune,* who was the founder of the city of Salina. Phillipsburg annually hosts one of the state's largest rodeos during the first week of August.

150.5 *Spring Creek* and a *loess exposure* south of the highway.

153.8 *Big Creek.*

154.0 The town of *Gretna* is 0.5 miles north of here. About 12 miles north of Gretna, in 1915, fossil hunter George Sternberg discovered a meteorite weighing 80 pounds. Northwestern Kansas has produced many large meteorites, including one discovered in 1892 near Long Island, which was found in 4,050 fragments and weighed 1,275 pounds. The Long Island meteorite was stony, made of heavy minerals, and contained iron. The other type of meteorite consists mostly of iron and nickel.

Astronomers believe that most meteors are fragments formed by collisions of asteroids between Mars and Jupiter. When they are deflected into Earth's orbit, they are caught up in Earth's gravitational field and are pulled to the ground. Though many meteors burn up in the atmosphere and are called shooting stars, a number make it to the ground and are labeled meteorites. Kansas has long been a prime location for searching for meteorites, probably because of the lack of vegetation in western Kansas, the widespread cultivation, and the predominance of sedimentary rocks at the surface, which makes it easier to identify meteorites. They often have a burned or pitted appearance, common to some igneous rocks.

156.8   *Plum Creek.*

159.0   *Turner Creek.*

159.2   This county road runs south to the town of *Kirwin* and *Kirwin Reservoir,* a man-made lake on the North Fork of the Solomon River and Bow Creek. Water levels in Kirwin have dropped in recent years, as they have in many western Kansas lakes. The lower water levels have been caused by lower flow rates in the streams that feed these reservoirs. These low rates of stream flow have two primary causes. One is increased irrigation along the streams, which lowers the water table and prevents ground water from entering the streams through seeps, springs, and underground flow. The other cause, ironically, is water conservation, practiced by farmers and ranchers in the watersheds of the streams. Through contour farming, terracing, and the construction of farm ponds, farmers have been able to conserve water by keeping precipitation on the land and by reducing the runoff. Reduced runoff decreases stream flow and means that less water can feed into the reservoirs from larger streams.

159.5   *Agra,* which shares its name with the city in north-central India where the Taj Mahal is located.

162.0   *West Cedar Creek* joins Middle Cedar and East Cedar creeks shortly before they flow into the North Fork of the Solomon River, south of here, near the town of Cedar.

162.4   *Phillips/Smith county line.*

163.6   *Middle Cedar Creek.*

164.0   *Kensington* and Athol, six miles east of here, are said to have been named after the wife and daughter of a railroad builder. Both towns are located along the route of the Rock Island Railroad.

167.6   *East Cedar Creek.*

169.4   *K–8* runs to the north, the route to the one-room cabin on Beaver Creek where Dr. Brewster Higley wrote the lyrics for Kansas' state song, "Home on the Range," in the early 1870s. Higley moved from Ohio to Smith County in 1871, and in 1873 the lyrics for "Home on the Range"—then called "My Western Home"—were published in the local newspaper, the *Smith County Pioneer*. The music to "Home on the Range" was composed by Dan Kelley, who moved to Kansas in 1872. Both Higley and Kelley subsequently left Kansas, Higley in 1886 and Kelley in 1889.

170.2 *Athol.*

172.3 *West Beaver Creek,* which joins Middle Beaver Creek and East Beaver Creek southwest of Athol before flowing into the North Fork of the Solomon River.

174.0 *Middle Beaver Creek.*

174.9 *East Beaver Creek.*

177.1 *Roadside park.*

177.2–178.2 *Smith Center,* the seat of Smith County. In the Smith Center park there is an octagonal Dutch windmill that was constructed in the 1880s in Reamsville, about 13 miles northwest of here, where it was used to grind grain for flour. The mill was moved to Smith Center in 1938.

179.5 *Spring Creek.*

182.1 The scar in the ground north of the road is a *borrow pit,* where soil was removed to build up the roadbed in the railroad overpass at this point. The result is the westernmost exposure of Cretaceous rocks along U.S. 36. The light-colored rock visible in the excavation is Niobrara Chalk, named for the Niobrara River in northeastern Nebraska, where it was first described by geologists in 1862.

The Niobrara is composed of two members: Fort Hays limestone, overlain by Smoky Hill chalk. Today the Niobrara underlies the northwestern one-fifth of Kansas, although outcrops of the formation in Mitchell and Osborne counties show that the Niobrara once covered a far greater area but has since been eroded away by wind and water. Smoky Hill chalk is also responsible for the unusual chalk formations at Monument Rocks and Castle Rock. Smoky Hill chalk is up to 700 feet thick; it includes layers of shale and bentonite as well as chalk. Smoky Hill chalk also contains concretions and fossils of clams, shark's teeth, fish, and swimming reptiles. Many of those fossils are on display at the Sternberg Memorial Museum at Fort Hays State University.

182.9 *Monkey Run,* a draw that drains into West Oak Creek and eventually into Waconda Lake in Mitchell County.

183.7 Oil and gas seeps have been found throughout Smith County. Many of them were discovered in the 1880s, when wells and cisterns became contaminated. One such incident occurred around 1880 in the town of *Bellaire,* one mile north of here. Wells were so polluted with oil that they were no longer usable, and a town row erupted when one resident was accused of putting oil down the town well. Other examples occurred around the county in various wells and farm ponds; local residents collected enough oil to use it for lighting and heat. Most of this was "light" oil, similar to kerosene and gasoline. Geologists are not sure about the source of these seeps, especially since they no longer occur. However, they theorize that the oil may have come from a leaky tank that contaminated ground-water sources, or it may have resulted from oil accumulations in deeply buried Pennsylvanian rocks. The more volatile portions of that oil may have migrated 2,000 feet to the surface and seeped out, although that seems unlikely, particularly because no oil fields have yet been discovered in Smith County.

184.1 *West Oak Creek.*

185.5 *Frog Creek.*

187.0 *Middle Oak Creek.*

188.4 *Spring Creek.*

189.7 U.S. 281 runs north to the town of *Lebanon,* the site of the lowest recorded temperature in Kansas history: $-40°F$ in the winter of 1905. A stone marker a mile north of town pinpoints the center of the 48 contiguous states. The geographic center should not be confused with the geodetic center of North America, which is about 40 miles south of here, at Meades Ranch. Marked by a bronze disc, the geodetic center is the beginning point for land surveying in North America. It is located in north-central Kansas because it is near the center of the conterminous United States and because it is near two major survey arcs along the 98th meridian and the 39th parallel. This geographic center is a major point of reference for surveyors. Thus, when a surveyor checks the lines of your property and uses geodetic markers as a reference point, he is positioning your lot in relation to Meades Ranch.

190.6 *East Oak Creek.*

192.6 *Smith/Jewell county line.*

193.1 *Porcupine Creek.* Porcupines are found throughout Kansas, except perhaps in the southeastern part of the state. A member of the rodent family, porcupines are generally nocturnal, solitary animals that feed on tree bark.

196.1 *West Limestone Creek.* Despite the amount of limestone in Kansas, limestone isn't a particularly popular place-name. Only four Kansas creeks and one township bear the name.

196.6 K–112 leads 2.5 miles north to the town of *Esbon.*

198.8 *Roadside park.*

199.5 *Middle Limestone Creek.*

201.2 *K-128* runs 6 miles south to the town of Ionia and on to Waconda Lake, 10 miles farther south.

202.6 *Limestone Creek.*

203.2 *K-28* runs six miles north to Burr Oak and on to Nebraska. Burr oaks are common in the bottom lands and pastures of eastern Kansas; they are noted for their large lobate leaves and for their large acorns.

204.3 *Elm Creek.*

208.1 *Middle Buffalo Creek* and *roadside park.* An early settler in Kansas, William D. Street, described a buffalo herd in this area in the late 1860s. He said that one night, while camping on Buffalo Creek near Jewell, he awoke to the sound of buffalo. The next morning, from a bluff west of Jewell, Street saw buffalo covering the Solomon River valley. "To the northwest, toward the head of the Limestone, for about twelve or fifteen miles, west across that valley to Oak Creek, about the same distance, away to the southwest to the forks of the Solomon, past where Cawker City is now located, about 25 miles south to the Solomon river, and southeast toward where Beloit is now situated, say fifteen or twenty miles, and away across the Solomon river as far as the field-glasses would carry the vision, toward the Blue Hills, there was a moving, black mass of buffalo, all traveling slowly to the northwest at a rate of about one or two miles an hour."

208.0–209.0 *Mankato,* the seat of Jewell County, was named by settlers who moved here from Mankato, Minnesota. The Jewell County Courthouse, at the north end of the business district, was built out of native limestone and was completed in 1937.

210.6 *K-24* runs south to Jewell and Beloit.

211.2 *Fort Hays Limestone Member* of the Niobrara Chalk. This rock layer is composed of chalk and chalky limestone, which is light gray and up to 85 feet thick. The formation was named after the fort at Hays by Samuel Wendell Williston, a famous paleontologist and early member of the Kansas Geological Survey.

212.0 *Fort Hays limestone.*

Buffalo, the state animal of Kansas, were once common on the plains of central and western Kansas.

212.2   *East Buffalo Creek.*

213.1   *K-14* runs north from here to Superior, Nebraska. The elevation here is 1,800 feet; it marks the approximate eastern limit of the High Plains along U.S. 36. In northern Kansas the boundary of the High Plains is marked by the eastward-facing break, or escarpment, formed by the erosion-resistant Fort Hays limestone. To the east the highway descends over the escarpment and enters the Blue Hills. Heading west, the highway crosses over the escarpment and climbs up the slope of the High Plains.

214.0   *Fort Hays limestone.* Near the east end of the north side of this massive road cut, about 100 feet west of the milepost marker, is a distinct fault. By tracing the layers of rock across the fault, it is clear that

A fault in the Fort Hays limestone at milepost 214 along U.S. 36

they moved three or four feet when the fault slipped, probably producing a sizable tremor. The fault must have moved some time after the rocks were deposited during the Cretaceous Period, about 100 million years ago.

214.5   *Carlile Shale,* named by geologists working in the Arkansas River valley of eastern Colorado in 1898. The Carlile underlies the western third of Kansas, cropping out in a line from Republic County to Hamilton County. Up to 300 feet thick, the Carlile is particularly visible where it is dissected by major drainage systems, as in the Blue Hills region in north-central Kansas. This blue-gray shale often colors the sides of hills in this area; it was probably the source of the name Blue Hills.

214.8   *West Marsh Creek.*

215.0   *Montrose.*

216.5   *Carlile Shale* crops out in the hills north of the highway.

217.4   *East Marsh Creek* drains into a salt marsh to the southeast of the highway.

218.0   Seven miles north is *Lovewell Reservoir,* a man-made reservoir on White Rock Creek. Lovewell supplies water for an irrigation district in Jewell and Republic counties. Canals carry the water to fields, where it is distributed by siphons and pipes. The water is used to grow crops, such as corn, which would be difficult to raise without irrigation, and to increase yields of other crops, such as alfalfa. The elevation here is 1,500 feet, or more than 300 feet lower than the crest of the Fort Hays escarpment, which forms the skyline a few miles to the west.

220.3 This county road runs north to the town of *Lovewell* and south to *Formoso*. A few miles to the east are Norway and Cuba.

223.3 *Jewell/Republic county line.* Republic County takes its name from the Republican River, which cuts across the western edge of the county.

224.4 Here the road crosses an *irrigation canal*, which takes water from Lovewell Reservoir to fields around Courtland and southeast to the area near Norway, a distance of nearly 25 miles.

225.3 *K–199* runs south to Courtland. From there a county road goes seven miles south to the Jamestown State Waterfowl Management Area, a large salt marsh that provides habitat for birds, fish, and other animals. Because Kansas is on the central flyway route for migrating birds, this marsh is an excellent location for bird watching.

This salt marsh is one in a series of salt springs, seeps, and marshes that occur in a band from here southwestward along the outcrop of the upper part of the Dakota Formation. Along this band, the Dakota is overlain by impervious shales that prevent the infiltration of fresh water that falls as precipitation. Thus, the fresh water cannot dilute the salty ground water that issues from the upper Dakota, which includes some salt-bearing shales. This saline ground water forms the salty marshes, such as those at Jamestown, where evaporating water leaves a white, salty crust along the edge of the water. To the southeast, farther from the protection of overlying rocks, precipitation falling on the Dakota outcrops has diluted or flushed out the salty water, and the ground-water quality improves.

226.1 *Beaver Creek.*

227.4 K–266 runs seven miles north, to a museum marking the spot where a *Pawnee Indian village* was located in the 1820s and 1830s. The village consisted of thirty to forty lodges built out of sod, timber, and thatch; the village's population probably numbered about 1,000. One of the larger lodge sites is now protected by this museum, with exhibits showing many of the items found during excavation.

This site also has a marker commemorating Zebulon Pike's visit to a Pawnee Indian village in 1806. Pike was said to have convinced the Indians to raise an American flag over the village, marking the first time a United

An irrigation canal that takes water from Lovewell Reservoir to fields south of here, in some cases as much as 25 miles from the reservoir

States flag was raised west of the Mississippi. Today there is considerable doubt that this is the site where Pike stopped, although the village might have been visited by mountain man Jedediah Smith, when he was on his way west with a band of trappers in 1826.

   231.0   *Republican River,* which was not named after the political party, but for the Republican Pawnee, one of four bands of Pawnee Indians living in Kansas and Nebraska in the 1500s. French explorers thought one band had a republican form of government and therefore gave them the name. The Pawnee in this area were farmers and hunters who lived in villages; twice a year they went west to the High Plains for buffalo hunts. In 1875, white encroachment finally forced the removal of the Pawnees to Indian Territory in Oklahoma.

The Republican River also was the scene of a Kansas gold rush during the 1930s, when miners panned for gold along the river. Prof. E. E. Kinney of the University of Kansas tested the claims by panning the river himself; he found virtually no gold. "Desirable as this would be," he wrote, "it is extremely improbable that a gold mining industry will ever develop in the state."

Kinney's words are probably still accurate. Kansas has no known gold mines, and geologists believe that the only part of the state that might harbor gold would be the Arkansas River, which drains the Rocky Mountains and might have carried gold flakes onto the plains. But would-be gold seekers shouldn't get their hopes up. After studying a false-alarm gold rush in Ellis County in 1895, Erasmus Haworth, director of the Kansas Geological Survey, warned, "False hopes once raised blast true and proper ambitions and unfit man for the ordinary duties of life."

231.2 *Roadside park.*

231.3 *Scandia.* A famine in Scandinavia during the 1860s drove many Swedes, Norwegians, and Danes to the United States. In 1868 a group of Swedes moved here and formed the town of New Scandinavia, which was renamed Scandia in 1876.

235.7 The town of *Rydal* and a *roadside park.*

236.8 *West Creek* joins Riley Creek and Salt Creek, east of Belleville, before they drain into the Republican River.

239.0 Intersection with *U.S. 81,* which runs north and south across Kansas and is generally regarded as the line of demarcation between "eastern" and "western" Kansas. Its geology is described in chapter 5.

239.2 *Riley Creek.*

239.0–240.0 *Belleville,* the seat of Republic County.

241.9 *Salt Creek,* whose name probably reflects the salt content from springs that carry saline water to the surface from the Dakota Formation. Salt Creek runs into a salt marsh about 10 miles south of here. Early reports of this marsh included descriptions of a salt flat composed of crystallized salt. Named the Tuthill Marsh after a nearby resident, it was one of the principal sources of common salt in Kansas' early days. Kansas is still a major salt-producing state, with mines and dissolution wells producing

Fossils of the clam species Inoceramus, *common in exposures of Greenhorn Limestone*

salt in the central part of the state. However, the first salt manufactured in Kansas came from this marsh. It was produced by scraping up the crystallized salt on the salt flat and by evaporating the marsh's brine. According to J. G. Tuthill, the brine was three times as salty as sea water, and 130 gallons of brine would yield a bushel of salt after evaporation.

242.5 *Greenhorn Limestone Formation* on the north side of the road. The upper part of this formation contains a rock layer known as the Fencepost limestone bed, which is used in north-central Kansas for fence posts and other construction.

247.9 *South Fork of Mill Creek.*

248.7 *K-139* runs south to Cuba.

249.2 *South Fork of Mill Creek.*

250.0 and 252.3 *Greenhorn Limestone*, named for a creek in southern Colorado. This formation crops out from Washington County in the northeast to Hamilton County in the southwest. Composed mostly of limestone and chalky shale, the formation is profuse with remains of the Cretaceous clam species *Inoceramus*.

253.0 *Graneros Shale*, overlain by *Greenhorn Limestone.* The

Graneros is a medium-gray to black shale that underlies the northwestern two-thirds of Kansas. It is generally about 40 feet thick; oysters and shark's teeth are common fossils.

253.8 *Washington/Republic county line.* This is also the line of the Sixth Principal Meridian, which is used for mapping and legal descriptions of land in Kansas, Nebraska, Colorado, Wyoming, and South Dakota.

254.3 *Graneros Shale,* overlain by *Greenhorn Limestone.* Geologists use the term *contact* to describe the surface that separates one rock formation from another of a different type or age. At this location, two different formations come into contact—the older Graneros Shale meets the younger Greenhorn Limestone. A contact is usually marked by an abrupt change in the character or color of the rocks, as in this instance, where the contact is between a limestone and a shale. Contacts are even more notable when they involve rocks from two different geologic periods. For example, near milepost 281 is the approximate contact between rocks of Cretaceous and Permian age.

254.9 *Davis Creek.*

255.1 *Graneros Shale,* overlain by *Greenhorn Limestone.*

255.4 *Greenhorn Limestone.*

256.3 *Hay Valley.*

256.8 *Mulberry Creek.*

257.7 K–22 runs north to the small town of *Haddam.* Recently, several small earthquakes have occurred in a cluster around this small town. Only one tremor, which occurred in June 1979, was strong enough for residents to feel. It registered 3.3 on the Richter Scale and knocked items off walls.

Geologists believe the quakes are associated with a zone they call the Midcontinent Geophysical Anomaly (MGA). Precise measurements of magnetic and gravity levels indicate that the North American continent may have begun to pull apart around this location about a billion years ago, splitting along a line that runs from near here, through Nebraska and Iowa, and on into the Lake Superior region. As the rift developed, igneous rock welled up to the surface. For some reason, the splitting stopped, and the igneous rock was eventually covered by sediments during later geologic time. That rift zone is responsible for the higher magnetic and gravity levels in the area and may be related to the continuing earthquake activity.

This point on U.S. 36 is an area of low gravity that occurs on either side of a strip of high gravity centered over the MGA, which crosses

This well, drilled in 1984 in Washington County south of U.S. 36, went more than 11,000 feet deep in search of oil, making it the deepest in the history of the state.

ton County's deep well is unusual because it penetrated about 7,000 feet of Precambrian rocks. Nearly all Kansas oil wells stop when they reach rocks of Precambrian age; most Precambrian rocks are igneous or metamorphic, and they contain little oil. However, geologists believe that the Precambrian here includes several thousand feet of sediments, and these sediments might contain oil. Texaco reported "no commercial amounts of oil" from the well, but the area continues to attract geological attention.

     271.3 *Mill Creek* and *roadside park.* This point marks the approximate westward limit of the glacial advance during the Kansas Stage of the Ice Age, about 750,000 years ago. The Kansan Glacier reached south into Kansas, to near Manhattan, and beyond Topeka and Lawrence, roughly paralleling the present Kansas River. Glacial deposits are not widespread in Washington or western Marshall counties, although they blanket the countryside east of Marysville.

     272.8 and 274.0 *Dakota Formation.*

     275.8 *Dakota Formation sandstone* north of the road.

     276.0 *Dakota Formation,* which underlies most of the north-western third of Kansas. Its most recognizable form is reddish-brown sandstone. Though sandstone is only a small portion of the formation, it composes most of the Dakota's outcrops because it resists erosion better than the softer Dakota sediments, such as clay and shale. In places, distinct, angling lines cut through the sandstone, evidence that it was deposited in river channels. Geologists call this cross-bedding. Here the Dakota lines the road and crops out in the hills surrounding the highway.

276.6 *Dakota Formation sandstone.*

278.5 *Lane Branch.*

280.0 *Little Blue River,* which runs south, joins the Big Blue outside of Blue Rapids, and then empties into Tuttle Creek Reservoir.

280.8 *Wellington Formation.* Rocks in this formation were deposited during the Permian Period, about 250 million years ago, making them significantly older than the nearby Cretaceous formations.

281.2 K–15–E runs north to the small town of *Hanover,* another German community in northeastern Kansas. Nearby are Bremen and Frankfort. Two miles northeast of Hanover is the Hollenberg Station, the only original Pony Express station still standing in its original location. Now a state-owned museum, the station is named after Gerat Henry Hollenberg, one of the founders of Hanover.

282.3 *Dakota Formation* south of the highway. This is the easternmost exposure of the Dakota along U.S. 36.

284.2 *Washington/Marshall county line.* Marshall County is named for a member of the first territorial legislature in Kansas. Of all the Kansas counties that U.S. 36 passes through, Marshall has the largest population, with just over 12,000 residents in 1980. This point is also the approximate boundary between Cretaceous and Permian rocks here in north-central Kansas. To the west, the rocks at the surface were deposited during the Cretaceous Period, about 100 million years ago. To the east, the Permian rocks were deposited about 250 million years ago. Areas underlain by the Cretaceous are generally considered part of the Smoky Hills physiographic province, while the Permian area is part of the northern extension of the Flint Hills.

285.0 *Walnut Creek.*

287.8 *Hop Creek,* the source of whose name is not known, although it may be the hop plant, which is used to flavor beer. No record of hop production has been found for Marshall County, although the 1876 Kansas Board of Agriculture reported that 86,000 bushels of barley were raised in the county, and a brewery was operating at Marysville. Thus, hop production cannot be ruled out as a source for the name.

291.8 *Florence limestone* south of the highway. This limestone is named for a small town in the Flint Hills of Marion County.

292.0 *Blue River.* Col. Frank Marshall, after whom the county is named, established a tavern and ferry across the Blue River. Marysville is named for Marshall's wife. The Blue River is dammed 40 miles south of here to form Tuttle Creek Reservoir, one of several lakes built by the U.S. Army Corps of Engineers to prevent floods, such as the 1951 flood that devastated the Kansas River valley. Tuttle Creek provides space to store large amounts of floodwater, and in times of flood the dam backs water in the Blue River valley as far north as Marysville.

293.0 *Marysville,* the seat of Marshall County, was on the route of several trails west during the 1800s. Both the Oregon and Mormon trails crossed the Blue River here. The Oregon Trail was the longest pioneer trail in the country. It started in Independence, Missouri, headed west to Topeka, and then north to Marysville, 174 miles from the trail's beginning. Marysville was also along the Pony Express route during its sixteen months of operation in 1860 and 1861. Today the Pony Express museum is in Marysville. Finally, the town was near the path of the Holladay Overland Mail and Express Company route from Atchison to Salt Lake City. Ben Holladay's stage line covered 3,300 miles, but it crossed the Blue River at Oketo, which is north of here, instead of at Marysville. Mark Twain described a trip on this stage line in his book *Roughing It.*

South of Marysville, along U.S. 77, is the Georgia Pacific gypsum mine. The Blue Rapids area was the site of the first gypsum mine in Kansas, opened in 1862. The Georgia Pacific Mine takes rock from an eight-foot-thick layer of nearly pure gypsum in the Easly Creek Shale. Operating about 80 feet underground, the miners produce about 1,000 tons of gypsum per day, mostly for use in manufacturing wallboard and in industrial plaster. Like underground coal mines, the gypsum mine employs the room-and-pillar method of mining, leaving behind pillars of rock to support the weight of the rock above. Most of the mined-out area is east and north of the plant. From core drilling and other studies, mine officials estimate that they have at least a 100-year supply of gypsum left underground at the Blue Rapids plant.

Gypsum is a soft white material that was deposited during the Permian Period, when an arm of an inland sea was cut off from the rest of the Permian sea. That sea evaporated, leaving behind rocks—such as gypsum and salt—that geologists call evaporites. Soluble in water, gypsum erodes rather easily in humid climates, so it is not visible at the surface here but can be seen in white, sparkling layers throughout the Red Hills, which are sometimes called the Gyp Hills.

One final note about Marysville. The town advertises itself as the Black Squirrel Capital. Technically, these black squirrels are a darker variety of the gray squirrel, which is common in the heavily forested areas of eastern Kansas.

293.2 *The Fort Riley Limestone Member* of the Barneston Limestone Formation is exposed on the south side of the highway.

293.7 The *Fort Riley limestone, Oketo shale,* and *Florence limestone* members of the Barneston Limestone Formation are exposed along the shore of a small lake north of the highway and in a road cut east of the lake.

295.5 *Lily Creek.*

298.5 *Spring Creek* and Spring Creek park.

299.5 *Home,* which is also called Home City, got its name because its first postmaster kept the post office in his home.

302.4 *Threemile limestone,* overlain by *Havensville shale.* Most shale outcrops are less prominent than are exposures of limestone or sandstone. Shale erodes much more easily, often only coloring the soil along the side of the road, rather than standing out.

302.8 *Crouse Limestone,* overlain by *Blue Rapids Shale.*

303.0 *Funston Limestone,* overlain by *Speiser Shale* and *Threemile limestone.* Many of the formations visible here are the same as those exposed further south in the Flint Hills. These chert-bearing, erosion-resistant limestones crop out in a line that runs from here to the south.

303.8 *Robidoux Creek.* The Robidoux family (pronounced Ruby-do) was famous in commerce throughout the old west. In 1841 Michael Robidoux carved his name on a ledge of Cottonwood limestone along this river. Nearby is Robidoux Ford. Farther west, in Wallace County, Peter Robidoux operated a general store. But one day in 1893, when he didn't make a single sale, he closed up shop and left $20,000 in goods to rot on the shelves. In 1843, Joseph Roubidoux III founded St. Joseph, Missouri, 88 miles to the east and named it in honor of Roubidoux's patron saint.

304.4 *Cottonwood limestone.*

305.0 *K–99* runs south through the Flint Hills and the Chautauqua Hills, and north to the town of Beattie, the home of an annual milo festival. Milo is a type of sorghum grown primarily as grain for livestock feed. Nemaha and Marshall counties produce substantial amounts of milo—

over 3 million bushels each in 1980—although several counties in southwestern Kansas produce more, making Kansas a leader in milo production.

From this point east to the Missouri River, U.S. 36 passes over glacial deposits of clay, sand, gravel, and even large boulders that date back to the time when ice covered much of northeastern Kansas. Some of the boulders are composed of rock formed hundreds of miles from Kansas, including metamorphic and igneous rocks that are common in the Canadian shield area of Canada, northern Minnesota, and Wisconsin. These rocks were embedded in glacial ice and were carried hundreds of miles to Kansas, where they were dropped when the ice melted. Since rocks of this type are different from the rocks formed here originally, they are called erratics by geologists. The most common and recognizable glacial erratics in northeastern Kansas are boulders of pink quartzite—a metamorphic rock—which were carried from outcrops around Sioux Falls, South Dakota, nearly 300 miles north of here.

308.3   *Snipe Creek.*

313.5   K-110 leads north two miles to *Axtell.* There is a *roadside park* at the northeast corner of this intersection.

314.4   *Nemaha/Marshall county line.*

315.5   *North Fork of the Black Vermillion River.*

317.2   *Baileyville.*

319.2   *South Fork of Wildcat Creek.*

320.4   *K-187* leads south 8 miles to Centralia, the hometown of the National Football League running back John Riggins. Before joining the Washington Redskins, Riggins played football at the University of Kansas.

322.4   *K-178* to St. Benedict. On the south edge of St. Benedict is *St. Mary's Church,* a country cathedral in the late Roman style. The church was constructed from 1889 to 1894 out of rock from the Reading Limestone Member of the Emporia Formation, quarried about three miles north of St. Benedict. The church stands 172 feet high, and the interior is lavishly decorated with a series of symbolic paintings.

323.4–324.1   *Seneca,* the seat of Nemaha County, gets its water supply from wells drilled into the alluvium of the Nemaha River and from Maxwell Spring, about 1.5 miles to the southeast. In northeastern Kansas, reliable sources of ground water are often obtained from glacial deposits—old river channels that were buried by the glaciers that invaded northeastern Kansas during the Pleistocene Period. The silts, sands, and gravels in these old river channels are often a good source of ground water, and many towns and farms in northeastern Kansas depend on them for water.

324.5 *South Fork of the Big Nemaha River,* one of the few northward-flowing rivers in Kansas. Drainage patterns in this area are probably influenced by a subsurface geologic feature called the Nemaha Ridge, which is a highly faulted, buried range of igneous mountains that runs north and south across the eastern end of the state, from Nemaha and Marshall counties in the northeast to Cowley County in the south. Those igneous rocks are within a few hundred feet of the surface here in Nemaha County. If the sediments that lie on top of the Nemaha Ridge were stripped away, the mountains left behind would stand over 3,000 feet high.

The ridge is subtly expressed at the surface here. Rocks of Pennsylvanian age—such as those quarried for the stone in St. Mary's Church at St. Benedict—crop out over the trace of the ridge, while younger Permian rocks are at the surface to the east and west. The ridge, which is named for Nemaha County, is also the source of many of the small earthquakes that Kansans have felt since the 1860s.

330.4 The town of *Oneida* is 1.5 miles north on K–236.

335.3 *Gregg Creek.*

335.6 This area is part of the *Strahm oil field,* the oldest field in Nemaha County. Discovered in 1948, the field has produced over 300,000 barrels of oil; in 1984 only two pumps were still operating. Nemaha County is not a major oil producer, pumping only about 254,000 barrels of oil in 1984. However, interest in oil exploration has increased recently in northeastern Kansas, and a major new field was discovered south of here in 1982. Production in this area is associated with the Strahm Anticline, a small fold in the rocks that is located east of the Nemaha Ridge. The rocks that produce the oil were deposited during the Devonian and Silurian periods of geologic history. These rocks are about 400 million years old—older than any rocks exposed at the surface of the state—and are buried nearly 3,000 feet deep.

337.5 *Cedar Creek.*

338.4 *Brown/Nemaha county line* and a *roadside park.* U.S. 75 leads north to Sabetha and to Omaha, Nebraska. About six miles northeast of Sabetha is Sycamore Springs, one of the spring-water resorts that were popular in Kansas in the 1880s and 1890s. An early report by the Kansas Geological Survey describes the location: "There is a frame hotel 24 × 60, three stories high, with accommodations for twenty-five guests. In this hotel is the post-office, dining-room, bath-rooms, with facilities for giving hot and cold baths, sweat baths, etc. There have also been erected two cottages, a refreshment stand, and liverybarn, and the grounds have been cleared of underbrush, so that there is a very convenient space for those who prefer living in tents." Today, Sycamore Springs is a private camping area, with a roller-skating rink and a swimming pool that is filled with water from the original spring. Near to Sycamore Springs there is another old resort area called Sun Springs.

338.6 *Grasshopper Creek* drains southeast to form the Delaware River, which flows into Perry Reservoir and eventually into the Kansas River. At one time the Delaware was called Grasshopper River, and the town of Valley Falls, on the upper end of Perry Reservoir, was called Grasshopper Falls. The Kansas Legislature removed "grasshopper" from both place names, although this creek retains that designation. Also at this location is an exposure of Cottonwood limestone on the bank of the creek.

341.3 *U.S. 75* runs south to Topeka.

341.5 *Fairview.*

342.8 *Cottonwood limestone.*

344.0 *Spring Creek.*

346.5 This county road runs south to *Powhattan* and to the *Kickapoo Indian Reservation,* one of several reservations that remain in Kansas. Before Kansas Territory was formed, numerous eastern tribes were moved here. By 1846, at least nineteen different reservations were formed within the boundaries of what is today Kansas. The Kickapoos were moved to eastern Kansas in 1832 from their lands in Illinois and Wisconsin. Their original Kansas reservation included a large area in Doniphan, Brown, Jackson, and Atchison counties. Today the reservation is a parcel of less than 40 square miles, just south of Powhattan. Other Indian lands in Kansas include the Potawatomi Reservation, southwest of Holton in Jackson County, and the Iowa, Sac, and Fox Reservation, in northern Brown County.

346.9 *Walnut Creek.* The elevation here is 1,000 feet, 2,700 feet below the high spot of U.S. 36 at the Colorado border.

347.0 *Falls City Limestone.* Because northeastern Kansas is covered with windblown loess and glacial deposits, few outcrops are exposed at the surface. Along eastbound U.S.36, this is the last outcrop for more than 40 miles, until the highway reaches rocks exposed by the erosion along the Missouri River.

350.3 Here the highway has been routed around *Hiawatha.* The old highway runs due east through the town, while the new stretch loops south of Hiawatha, the seat of Brown County and, with a population of 3,700 in 1980, the largest city along the route of U.S. 36 in Kansas. In the

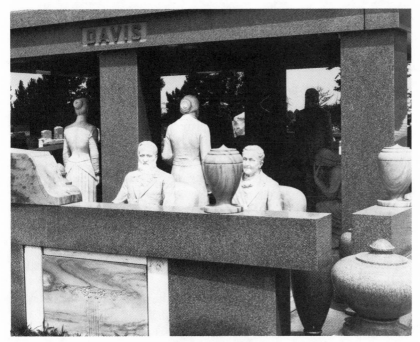

Marble statues at the Davis Memorial in Hiawatha

cemetery east of Hiawatha is the John Davis Memorial, a tomb that includes eleven life-size Italian marble statues of Davis and his wife at various times in their lives.

352.0   *North Fork of the Wolf River.*

358.6   County road to *Robinson,* 1.5 miles south of here.

360.6   Road to the *Brown County lake and state park.*

362.7   *Brown/Doniphan county line.* Doniphan County, named for Missourian Alexander Doniphan, was the first Kansas county to receive a name. Doniphan was a colonel in the Mexican–American War.

364.7   *Cedar Creek.*

365.8   *Roadside park.*

366.3   *Highland.* Immigration to Kansas is often associated with antislavery forces from East Coast states such as Massachusetts. Those abolitionists encouraged settlement in the state, particularly around Lawrence. But a great share of the Kansas immigrants came from midwestern states, such as Ohio, Indiana, Iowa, and Illinois. Highland is named after a town in Illionis, and not because it is located on the scenic loess hills of northeastern Kansas. Highland is also the home of Highland Junior College.

Established in 1858—three years before Kansas became a state—this is the oldest institution of higher learning in Kansas.

366.6 *K–120* leads south to Severance.

367.2 *Mission Creek.*

368.6 *K–136.* A quarter-mile north of here is the Iowa, Sac, and Fox Indian mission, built in 1837. Today it is a museum. The Iowa, Sac, and Fox Reservation covers about 30 square miles in northeastern Brown County, northern Doniphan County, and southeastern Richardson County, Nebraska.

370.5 The town of *Sparks* and the intersection with *K–7,* which runs south through Atchison and Bonner Springs, then along the eastern edge of Kansas to Cherokee County. To the north it passes through the old river towns of Iowa Point and White Cloud before terminating at the Nebraska border. This highway provides a trail along the banks of the Missouri River and through the vast loess hills that line the river along the northeastern corner of Kansas.

These loess hills are composed of a fine dust, more than a hundred feet thick in places, that was deposited by the wind during the last two glacial advances during the Ice Age. Geologists believe the loess was blown out of flood plains of large rivers that appeared at the close of the Ice Age, when the glaciers were melting and dropping their load of sediments. The difference in temperature from the snow-covered terrain to the north and the bare ground to the south may have created large differences in atmospheric pressure, resulting in strong winds capable of moving large amounts of windblown material.

In Doniphan County, loess composes some of the highest hills in the state. One of them, Lookout Mountain north of Sparks, is more than 350 feet above the flood plain of the Missouri River. Loess, by the way, is the same geologic material that blankets much of northwestern Kansas and has been carved by precipitation to form the steep hills and scenic canyons of Cheyenne County.

371.1 *Wolf River,* one of the major rivers in northeastern Kansas, flows north of here, emptying directly into the Missouri. The river was named by early French explorers. The elevation here is 850 feet, roughly the same as the flood plain of the Missouri River, which is three miles to the northeast, beyond the loess-covered hills north and east of the highway.

373.0 The town of *Fanning.*

373.7 *Coon Creek.*

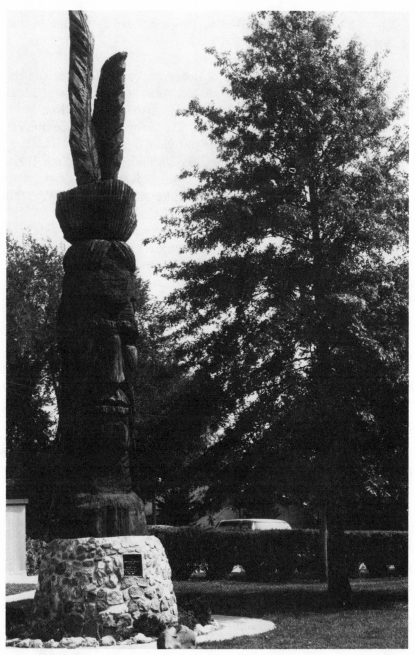

A wood carving by artist Peter Toth, on the lawn of the Doniphan County Courthouse in Troy

376.8   *Loess* exposed in the northeastern corner of the crossroads.

377.4   *Mosquito Creek.*

378.3   *Troy* is the seat of Doniphan County. The crests of the hills in Troy have an elevation of 1,150 feet. From here to the east, U.S. 36 descends through the valley of Peters Creek to the Missouri River, which has an elevation of 800 feet at St. Joseph. The Doniphan County Courthouse was erected in 1906; it is on the State Historical Register. On its front lawn is a wooden sculpture of an Indian, produced by the artist Peter Toth. Hewn from a burr oak, the sculpture stands 35 feet high and weighs eight to 10 tons. Toth hopes to produce similar statues for each state in the union; this is his twenty-ninth one.

South of Troy is the small town of Doniphan. Archaeologists have determined that the Kansa Indians, after whom the state is named, settled in a spot near Doniphan in the early 1700s. Here the tribe first came into contact with French traders and trappers. By the time of Lewis and Clark's expedition in 1804, part of the tribe lived near the Kansas River, while various Kansa villages were scattered throughout the Kansas River valley, including sites near present-day Manhattan, Wamego, and Topeka.

The town of Doniphan was established in the 1800s with dreams of becoming a major stop on the Missouri River. Doniphan hoped to rival Leavenworth and St. Joseph, but those dreams died when the river suddenly changed its course. Only a vestige of Doniphan remains today, more than a mile from the Missouri River.

380.2   *Roadside park.*

383.0   *Blair,* elevation 900 feet.

385.5   *Amazonia limestone* on east side of the highway. The Amazonia is named after a small Missouri town north of St. Joseph.

386.0–387.0   *Wathena.*

386.9   Thick exposure of shale from the *Lawrence Formation* in the bluff on the north side of the highway.

387.5   *Apple orchard* on the south side of the road. With its thick, rich soils, northeastern Kansas is an ideal location for growing fruit. In 1979, Doniphan County produced over 4 million pounds of apples, nearly one-third of all apples grown in the state that year. Just across the river, in northwest Missouri, farmers grow tobacco on the loess hills overlooking the Missouri River. Most of the crop is a light-brown burley tobacco that is marketed in the antebellum town of Weston, Missouri. Although tobacco was grown in

The change of course in the Missouri River at St. Joseph after floods in 1952

many parts of Kansas in the 1800s, production today is generally limited to the Missouri side of the river.

     388.6   *Oxbow lake.* Oxbow lakes get their name from their U-shape, similar to the bows in yokes used to harness oxen. These lakes result from loops where rivers, such as the nearby Missouri, have meandered and then have been cut off or abandoned by the main stream of the river. Many still contain water, because their bottoms extend below the shallow water table of the flood plain. One-fourth mile north of here, this oxbow lake connects with another one named Browning Lake, which forms the Kansas/Missouri border by making a large loop around St. Joseph's airport and an area called the French Bottom. Before the 1952 flood on the Missouri River, Browning Lake was part of the course of the Missouri River. During the flood the river carved a short cut, abandoning this meander loop and separating St. Joseph from its airport.

391.0 *Elwood* was the first Kansas station on the Pony Express and was the site of the first railway in Kansas, which was built between Elwood and Wathena in 1857. In 1866 the only paved road in Kansas also spanned the four miles from Wathena to Elwood, where there was a ferry to St. Joseph. There was a 25 cent toll for use of this road.

In November 1859, Abraham Lincoln made a speech and spent the night in Elwood during a week-long tour of northeastern Kansas. At the time he was a senator from Illinois, but he was already known as a leader of antislavery forces and was a potential candidate for president. After Elwood, Lincoln visited Troy, Doniphan, and Leavenworth. He later remarked, "If I went west, I think I would go to Kansas."

391.9 *Missouri River, Kansas/Missouri state line,* and the city limits of *St. Joseph, Missouri.* St. Joseph was the starting point for the Pony Express, which carried mail 2,000 miles to Sacramento, California, in ten days. The route of the Pony Express parallels U.S. 36 across northeastern Kansas to near Marysville, where it cuts northwest into Nebraska. Although it only operated in 1860 and 1861, it employed 120 riders, who traveled 650,000 miles with the loss of only one rider and one shipment of mail. The posters and handbills that were printed in order to enlist riders for the Pony Express minced few words; they would probably attract few recruits from today's job market. They read: "Wanted: Young skinny wiry fellows, not over 18. Must be expert riders willing to risk death daily. Orphans preferred. Wages $25 per week. Apply Central Overland Express."

83·U.S. HIGHWAY·83

The High Plains and the Chalk Beds

0.0 *Oklahoma/Kansas* state line. This is also the boundary between Seward County, Kansas, and Beaver County, Oklahoma. U.S. 83 runs north across the High Plains of western Kansas, bending to the east and leaving the state north of Oberlin. Along the way it passes through some of the state's most recent geological features and makes a brief cut through the Cretaceous chalk beds of Gove and Logan counties.

2.3 South edge of *Liberal*, the seat of Seward County and the regional trading center for much of southwestern Kansas.

2.9 Junction with *U.S. 54*, which runs southwest to Guymon, Oklahoma, and on into New Mexico. To the east it goes to Wichita and Fort Scott, Kansas. In Liberal, this road is called Pancake Boulevard, in recognition of the international pancake race that is held here every Shrove Tuesday. Women from Liberal run a quarter-mile course flipping a pancake, and then compare the winner's time with the winning time from a similar race held in Olney, England, on the same day.

Nine miles to the northeast, on U.S. 54, is the Sampson of the Cimarron, a railroad bridge 1,200 feet long and 110 feet high that crosses the Cimarron River. Four miles west of here is the Boles No. 1; on 20 July 1920, natural gas was discovered in this well, opening up the Hugoton Field, one of the largest natural gas fields in the world. The Hugoton underlies much of southwestern Kansas, extending into the Oklahoma and Texas panhandles as far south as Amarillo, Texas. Wells in Seward County

The Samson of the Cimarron — a railroad bridge crossing the Cimarron River northeast of Liberal in Seward County

produced over 26 billion cubic feet of natural gas in 1984, enough to rank the county seventh among the state's leaders in the production of natural gas.

4.0   The *Liberal airport,* an Army Air Force Base during World War II, is 2 miles to the west.

4.7   The *Southwest Medical Center* is west of the highway.

4.8   *Seward County Community College* is east of the highway. Two-year colleges are important educational institutions in western Kansas, where four-year colleges are rare. The nearest four-year college in Kansas is a private school, St. Mary of the Plains, about 80 miles away, in Dodge City. The nearest public four-year university in Kansas is Fort Hays State University.

5.5   The north edge of *Liberal.*

10.0   An area of *sand hills* west of the highway.

12.5   Junction with K–51 and westbound U.S. 270. The town of Woods is 10 miles west and Hugoton is 23 miles west.

17.6   Sand and gravel pits east of the highway.

18.0   *Cimarron River,* with a riverbed elevation of 2,650 feet. The Cimarron's path provides a rare break in the even topography of southwestern Kansas. In this area the Cimarron has a gradient of about 10 feet per mile, steeper than most large streams in Kansas. The Cimarron enters Kansas from Colorado in a broad, shallow valley that is progressively deeper downstream. Southeast of here, in southeastern Seward County, the Cimarron has cut a valley 250 feet into the young High Plains sediments.

19.3   The well to the west is in the *Lemert oil field*. This field, discovered in 1970, produces oil from wells more than a mile deep. These deep rocks are part of the Hugoton Embayment, a northern extension of the Anadarko Basin of western Oklahoma and the Texas Panhandle. The Anadarko is a deep downwarp in the earth's crust that has been filled with sediment over the course of geologic time.

All of Kansas is underlain by igneous and metamorphic rocks that were formed during the Precambrian Era of geologic history, more than 600 million years ago. Those Precambrian rocks have since been covered by layers of sedimentary rock. In Kansas, those sediments are deepest here in the southwest corner of the state, where they are up to 9,500 feet thick.

21.3   Junction with *U.S. 160*. U.S. 160 and U.S. 83 between milepost 21.3 and 41.4 are described on pp. 29–30.

41.4   Junction with westbound U.S. 160 and eastbound K–144. Ulysses is 27 miles to the west. For a description of the geology along southbound U.S. 83 for the next 20 miles see pp. 29–30.

46.9   Ivanhoe cemetery. This is all that remains of *Ivanhoe*, a town that was established in the 1800s and was named after Sir Walter Scott's novel. The Ivanhoe post office closed in 1905, although the township immediately to the north in Finney County is still named Ivanhoe.

53.5   *Haskell/Finney county line*. Finney County, the second-largest county in the state, is named after David Wesley Finney, lieutenant governor of Kansas when this county was organized in 1883. At that time, Finney County was even larger, including all or parts of nine other Kansas counties. Its size was reduced in 1887, though Finney still includes a panhandle, which is partially responsible for its unusual shape. Today the county covers just over 1,300 square miles, an area larger than the state of Rhode Island.

58.5   The town of *Plymell*.

61.0   South edge of *sand hills*. Sand hills cover the landscape south of the Arkansas River from southeastern Colorado to near Hutchinson, Kansas. Only a few isolated areas of sand hills cover the area north of the river in this same stretch, apparently because the prevailing winds were from the north when the sand was deposited. This was during the Ice Age, about a million years ago, when a huge ice sheet sitting to the north may have reversed the patterns of air flow from those we are familiar with today.

A sand dune south of the Arkansas River in Kearny County

Current wind patterns in Kansas come mainly from the south, and the name Kansas means "people of the south wind."

67.2   Junction with *Alternate U.S. 83,* which loops east around Garden City.

68.5   The *Finney County State Game Refuge* is immediately west. This 3,600-acre game preserve, which is home to the largest herd of buffalo in Kansas, also occupies the northeast corner of a vast tract of land that was once the Kansas National Forest. Stretching from here westward to the Colorado line, the Kansas National Forest covered more than 300,000 acres of sand hills south of the Arkansas River. But it was a forest in name only. The popularity of forestry around the turn of the century—and the continuing desire to transform the Great Plains into a well-watered, tree-studded land to match the easterner's ideal—provided the impetus for establishing a national forest in Kansas. Early efforts at forestry met with success in the sand hills of Nebraska, and these forests still exist today.

Efforts in Kansas began in 1906, with the first plantings in what was then known as the Garden City Forest Reserve. The following year a prairie fire scorched 200 square miles, killing most of the year-old plantings. In 1908 the Kansas National Forest was established, and 125,000 trees,

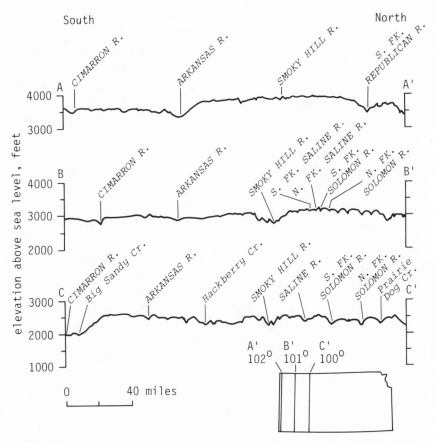

Topographic cross sections in western Kansas, showing the anomalous elevation of the Arkansas River relative to other large streams. Section B–B' follows U.S. Highway 83.

mostly pines, were planted. The next three years saw the planting of hardwoods such as locust and Osage orange, which are common in eastern Kansas. In 1911, severe drought wiped out nearly all of the hardwood plantings. After that, foresters concentrated on planting more conifers. However, the arid conditions of southwestern Kansas proved to be too much for the young trees, and on 14 October 1915, President Wilson abolished the Kansas National Forest.

69.0 *Arkansas River*. Like many of the rivers and streams in southwestern Kansas, the Arkansas is dry much of the time, carrying water primarily during times of heavy precipitation. In fact, in parts of southwestern Kansas the river stays dry so long that farmers cultivate the old riverbed.

Here the elevation of the Arkansas River stream bed is 2,830 feet. This elevation is anomalous in western Kansas when compared with that of other large streams in the north and south. This is particularly apparent along U.S. 83. The highway crosses the Cimarron River at an elevation of 2,650 feet. To the north, U.S. 83 crosses the Smoky Hill River at 2,625 feet in elevation. Yet here the Arkansas has an elevation roughly 200 feet higher than other streams, in spite of the fact that the Arkansas enters Kansas at a lower elevation than do the other two rivers.

Thus, the Arkansas River has a much lower gradient, or less steep course, than other streams in western Kansas. The reason for this may be found in the central Colorado Rockies, where the Arkansas has its source and is supplied with runoff, snow melt, and the rock debris that weathers from the mountains. As the Arkansas heads out onto the High Plains, it receives little flow of additional water; in fact, it loses water to its sandy riverbed. The decreased flow means that the Arkansas is unable to carry as much sediment as it could in the mountains; it therefore begins to drop its load of sediment. In other words, the Arkansas changes from a degrading stream—one that is cutting downward in its channel—to an aggrading stream, which is building up its riverbed.

69.2 *Finnup Park,* to the east, includes the world's largest free concrete municipal swimming pool and one of the largest zoos in Kansas.

69.7 Junction with eastbound *U.S. 50.* Pierceville is 14 miles and Dodge City is 53 miles to the east. Garden City is the center of much recent growth in southwestern Kansas, including the construction of the world's largest packing house. Much of this growth is based on the availability of feed grain, which is produced via ground-water-based irrigation. New irrigation systems have allowed extensive cultivation of areas, such as the sand hills south of the Arkansas River, that were previously unproductive. That irrigation, in turn, has lowered ground-water levels throughout western Kansas, as farmers have taken water out of the Ogallala Formation aquifer that underlies the western third of the state.

Western Kansans, however, have taken steps to deal with water-related problems. They have formed governmental units, called ground-water management districts, to provide local control over the use of ground water. The Southwestern Kansas Groundwater Management District, for example, has its headquarters here in Garden City. The districts encourage water conservation, undertake research, and in some cases have placed limits on drilling new wells in areas where ground-water withdrawal has been particularly extensive. One district, headquartered in Scott City, even has an active weather-modification program, which attempts to increase precipitation and to suppress hail. While the results of such efforts won't be completely known for many years, the districts appear to be making headway on water-related problems.

70.4 Junction with eastbound *K–156.* The town of Jetmore is 56 miles to the east. Garden City is the seat of Finney County. The county courthouse, which occupies a full block west of the central business district, was completed in 1929. It has limestone walls and trim, and it includes touches of exterior detail in the art-deco style.

71.0 Junction with westbound *U.S. 50,* which leads 6 miles to Holcomb and 23 miles to Lakin. Near Holcomb is the farm where Richard Hickock and Perry Smith murdered Herb Clutter and his family in 1959. The murder and subsequent trial were made famous by Truman Capote in his book *In Cold Blood;* much of the movie version of the book was filmed on location here in Finney County and in other parts of Kansas.

71.5 *Irrigation canal.* The settlers' first large-scale irrigation projects in Kansas were begun around Garden City in 1880. Unlike today's irrigation, which generally uses ground water brought to the surface by irrigation wells, the early irrigators diverted water from the Arkansas River, which, in the 1800s, was hardly a dry bed of sand. It carried a dependable flow of water and was considered navigable as far upstream as Kinsley, 80 miles to the east.

The first irrigation experiment diverted water from the Arkansas River 4 miles west of Garden City by means of an earthen dam which was 5 feet high. Soon, other diversions and irrigating canals were constructed up and down the Arkansas River, sending its water onto the thirsty uplands. In 1892, troubles arose when the flow in the Arkansas began to dwindle. By 1896 the Arkansas seldom flowed east of the Colorado line. Most of its water was diverted by irrigation projects upstream in Colorado. This marked the beginning of conflict between Kansas and Colorado concerning the use of the water in the Arkansas River, a dispute that has flared up many times over the years and continues today. Some of the old irrigation ditches are still used in the Garden City area, but they are fed by water wells, rather than by the Arkansas River.

72.5 At about this point, U.S. 83 climbs out of the Arkansas River valley to the south and enters a broad north-south depression known as the *Scott–Finney Depression.* Measuring 8 miles east and west and more than 30 miles north and south, the Scott–Finney Depression is the "Great Basin" of Kansas—an area of internal drainage. Streams flow into but not out of it.

73.0 *Farmer's Ditch irrigation canal.* This canal was constructed to draw water from the Arkansas River near Deerfield, about 13 miles to the west.

77.3 South edge of *White Woman Bottoms.*

78.0 *Ackley and McCoy lakes,* two ephemeral lakes, are 4 miles to the west.

79.0 North edge of *White Woman Bottoms.*

84.0 The town of *Tennis* is 1.5 miles to the west.

84.5 *Corrigan Lake,* an ephermal lake, is a mile to the west.

86.0 The site of the town of *Gano,* now only an elevator, is 1.5 miles to the west.

88.5 *Sondreagger Lake* is 3.5 miles to the west. It is another ephemeral lake, which only holds water after heavy rains.

90.0 This is the north edge of the *Hugoton gas field,* part of a subsurface feature called the Hugoton Embayment, which, in turn, is a shelf on the edge of a larger, deeper feature called the Anadarko Basin. During the late 1970s and early 1980s, rising prices for natural gas inspired deep drilling throughout the Anadarko Basin, particularly in Oklahoma. Some of the wells went more than 25,000 feet deep in search of natural gas. Financing those deep wells led, at least in part, to the failure of the Penn Square Bank in Oklahoma City in the early 1980s.

90.7 The town of *Friend* is 1.5 miles to the west.

91.2 *Finney/Scott county line.*

92.5 *Pawnee Mound* is about 4.5 miles to the east.

94.0 *Dry Lake* is 7.5 miles to the east. This oblong-shaped lake is located in a High Plains depression, which allows water to drain in but not out. The water that collects here generally evaporates in the arid southwestern Kansas climate, leaving behind a white, salty crust 2.5 miles long and nearly 0.5 miles wide. Features such as this are called playas; they are common in the Great Basin of the southwest. Rogers Dry Lake, at Edwards Air Force Base in southern California, where the space shuttle lands, is a much larger playa.

99.0 The town of *Shallow Water,* immediately west of the highway, is also the south edge of White Woman Basin.

102.6 *White Woman Creek.* This is one of several east-flowing streams that enter the Scott–Finney Depression only to dry up and

Dry Lake, an ephemeral lake east of U.S. 83 in southeast Scott County

disappear. White Woman Creek vanishes about 1 mile to the northeast in a saucer-shaped depression known as White Woman Basin or Modoc Basin. In 1951, a year of record rainfall in much of Kansas and of disastrous floods in the eastern end of the state, White Woman Creek was swollen with runoff, which it dumped into White Woman Basin, creating a lake covering 12,000 acres. This is roughly the size of Milford Lake, the largest lake in Kansas. Even more than a year later, 4,000 acres were still covered by this ephemeral lake.

105.0   North edge of *White Woman Basin*. This also marks the north edge of the Scott–Finney Depression, an area of internal drainage extending south to Garden City.

105.5   The south edge of *Scott City*, the seat of Scott County.

106.5   Junction with *K–96*. Dighton is 23 miles to the east, and Leoti, the seat of Wichita County, is 25 miles to the west. The Scott County courthouse, completed in 1924, is one block northwest of here.

107.2   North edge of *Scott City*.

109.2   At this point the highway passes through a small High Plains depression.

111.5   A sandpit 3.5 miles west of here on Ladder Creek was the source of a fossilized bison skull, with nearly straight horns that measured 6

Outcrops of the Ogallala Formation form bluffs and canyons around Scott County. State Lake, west of U.S. 83

feet from tip to tip. These horns belonged to a super bison, a much larger ancestor of the modern-day bison. In the same gravel pit, teeth belonging to camels, elephants, and horses that lived during the late Tertiary Period were found.

112.0   A *High Plains depression* 0.5 miles east of here. This depression is larger than most, covering nearly 0.5 square miles and measuring 25 feet in depth.

114.5   Junction with eastbound *K-4*. Healy is 17 miles and Shields is 25 miles east of here.

115.0   The valley of *Ladder Creek*, also known as Beaver Creek, is to the west. Ladder Creek drains into Lake Scott.

116.2   Junction with K-95, which leads north through Ladder Creek valley to *Lake Scott* and *Lake Scott State Park*. K-95 rejoins U.S. 83 at milepost 121.3.

Lake Scott is particularly scenic because of the steep canyons and bluffs formed by the erosion of Ogallala Formation mortarbeds. The Ogallala is well known as an aquifer in western Kansas, but here it crops out at the

Abandoned house built out of rock from the Niobrara Chalk, visible near milepost 143 on U.S. Highway 36 in Phillips County

Road cut in the Ft. Hays limestone along U.S. Highway 36 in Jewell County

Rolling loess hills along U.S. Highway 36 in Doniphan County

St. Mary's Church in the town of St. Benedict, north of U.S. Highway 36 in Nemaha County

A scene from the Jamestown salt marsh, south of U.S. Highway 36 in Republic County

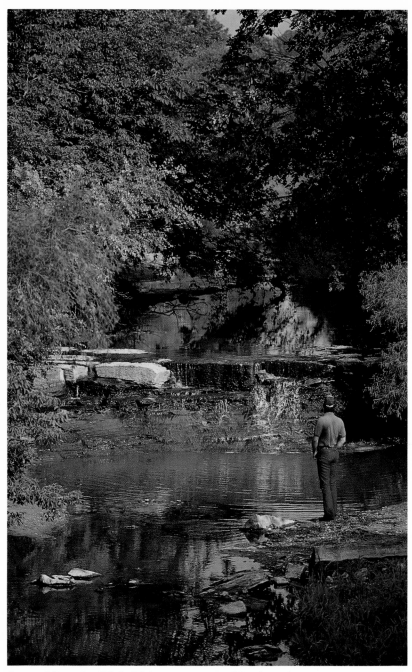
A waterfall south of U.S. Highway 36, south of Atchison

The Little Pyramids, highly eroded chalk outcrops west of U.S. Highway 83 in Logan County

The Bazaar cattle pens along the turnpike in Chase County

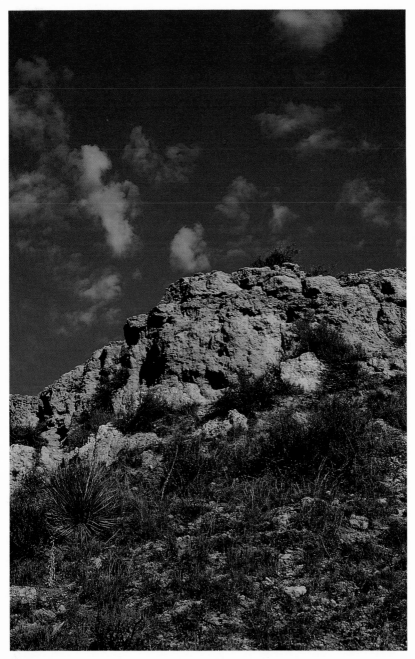

Point of Rocks, an outcrop along the Santa Fe Trail in Morton County, north of U.S. Highway 56

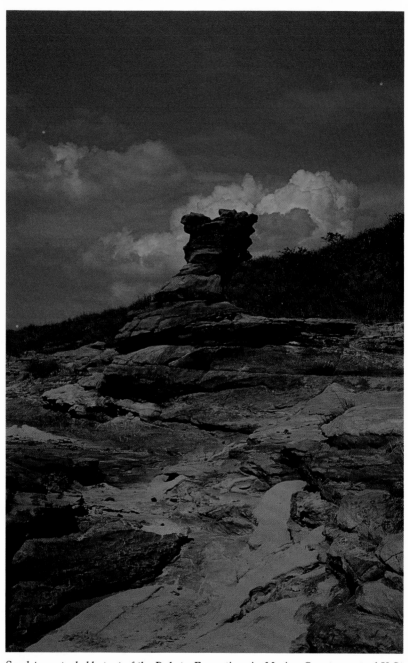

Sandstone, probably part of the Dakota Formation, in Marion County, west of U.S. Highway 56

surface and is naturally cemented together by calcium carbonate. The result is a hard, dense rock, which, at the south end of the park, forms a long ridge called Devils Backbone, as well as a number of canyons, bluffs, and draws.

Lake Scott is something of an oasis, one of the few large bodies of water in west-central Kansas. Originally called Lake McBride, it was one of the first areas set aside in the Kansas parks system. It is fed by Ladder Creek and a series of springs that continue to pour hundreds of gallons of water per minute into the lake. Among them is Big Spring, which, according to scientists at the Kansas Biological Survey, is home to a type of riffle beetle that is found nowhere else in the world. These springs occur at the contact between the very permeable water-bearing Ogallala Formation and the less permeable chalk of the Cretaceous Niobrara Formation.

Near the west edge of the lake was the site of El Quartelejo, the only Indian pueblo known to have been in Kansas. It was probably built in the 1600s by Taos Indians, who moved here to escape Spanish rule. The plentiful and dependable supplies of spring water attracted the Indians to the canyon of Ladder Creek, where they dug ditches to direct spring water to garden plots. In the process, the Indians created the first irrigation project in today's Kansas, beating the white man's efforts by 200 years or more. Some of the earliest white settlers along Ladder Creek in the late 1800s used these ancient ditches in constructing their own irrigation system.

The Indians later returned to New Mexico, and the site was occupied by Spanish and French explorers. The structure was eventually covered by blowing dirt; it was not excavated until 1898. A monument, the excavation site, and a nearby museum can still be visited. Two miles south of El Quartelejo is White Woman's Grave, the burial site of a Mrs. Combs, who died in 1883 and was the first white woman to be buried in Scott County.

117.1   *Christy Canyon.* Ogallala Formation mortarbeds east and west of the highway. About 1.5 miles west is Battle Canyon, where Col. William H. Lewis, commanding troops from Fort Dodge, was killed in 1878 in a battle with Chief Dull Knife and a band of Northern Cheyenne, who had fled their reservation in Indian Territory and were headed north when they were intercepted. Lewis was the last army officer killed by Indians in Kansas. The Cheyenne continued north and later killed nineteen settlers in Decatur County. In all, forty Kansans were killed during this last Indian raid in Kansas.

117.8   The view to the northwest is of the *Ladder Creek valley.*

118.0   The elevation is 3,000 feet.

The Little Pyramids, eroded remnants of Niobrara Formation chalk in southern Logan County

121.3   Junction with *K-95,* which runs west and south through Lake Scott State Park and past Shafer Canyon, Epler Canyon, Morgan Draw, and Sand Draw, rejoining U.S. 83 at milepost 116.2.

122.2   *Scott/Logan county line.*

123.5   *Hell Creek.*

123.8, 124.4, and 124.6   *Ogallala Formation* in the draw to the west at these points.

126.3   The Little Pyramids, chalk formations similar to but smaller than Monument Rocks, are about 5 miles to the west. Like most of the chalk monuments in this valley, they are composed of rock from the Niobrara Formation, deposited during the Cretaceous Period, about 80 million years ago.

126.8   The *Pierre Shale* and the *Niobrara Formation* are visible in the draw to the west, where selenite, a crystal form of gypsum, has weathered out of the Pierre Shale and sparkles in the hillsides. These formations have been faulted in this location, so that the younger Pierre Shale has been down-dropped next to the Niobrara. This fault is visible on the surface as a line separating the dark gray shale of the Pierre Formation (Pierre is pronounced like *pier*) from the yellowish chalk of the Niobrara.

128.3   This county road leads 4.7 miles west and 2 miles south to the *Little Pyramids*.

128.7   The town of *Elkader*. For a time in the 1870s, Elkader was a point of operation for fossil-hunting expeditions into the chalks of Gove and Logan counties. Today, it is a ghost town, with only a couple of deteriorating chalk buildings indicating its location.

129.0   The *Smoky Hill River*, streambed elevation 2,625 feet, has cut a deep gash across western Kansas, eroding the Tertiary-age Ogallala Formation and exposing underlying Cretaceous rocks that extend upstream almost to Colorado. The erosion of the Ogallala has also removed the major source of ground water for this area. Thus the Smoky Hill valley lacks the irrigation that is found to the north and south. The valley is used primarily as range land and for the raising of dry-land wheat.

The native vegetation and availability of surface water make the Smoky Hill River valley an ideal habitat for large herds of mule deer and pronghorn antelope, which were once widespread over the western two-thirds of Kansas. Pronghorns are not true antelope, but they have characteristics of deer and goats. They are but one of the native Plains animals that were misnamed by the pioneers. Prairie dogs are actually squirrels; jack rabbits are actually hares; buffalo are bison; elk are really wapiti; prairie chickens are grouse; and horned toads are actually lizards.

The Smoky Hill and its tributaries have been responsible for the erosion that has produced the chalk bluffs in Gove and Logan counties. A few miles to the west, along the river, an area of badlands covers several hundred acres. There the shales and chalks of the Niobrara Formation have been heavily sculpted to form an intricate network of mazelike stream channels, draws, and even a natural bridge. Outcrops of the Niobrara are a fossil hunter's paradise; the Niobrara has produced remains of thousands of vertebrate and invertebrate creatures from the Cretaceous, including huge fish, mosasaurs, pterosaurs, sharks, birds, and other animals.

Particularly common are oyster shells, some of which established themselves on the shells of a larger clam species called *Inoceramus*. Because of its size—as much as several feet in diameter in some spec-imens—these clams did not sink into the muck at the ocean floor. Instead, their weight was spread out, in much the way that a snowshoe distributes the weight of a person. Some paleontologists believe that oysters could colonize both sides of the clam, because the environment, even underneath,

A small natural bridge in the chalk badlands west of Monument Rocks in Logan County

was an aqueous sediment, watery enough for oysters to survive in and, apparently, flourish.

Fossils hunters should remember a couple of things before they wander into the chalk bluffs. First, nearly all of this area is private property, and they should secure the owner's permission before they enter a pasture. Second, this area is home to rattlesnakes, which are active during the warmer months of the year.

129.3 This is approximately the point at which the *Butterfield Overland Despatch trail* crosses U.S. 83. This trail, also known as the Smoky Hill route, was a stagecoach trail operated by David A. Butterfield. The route, which went from Denver to Atchison, operated for only a few months in 1865 before going out of business. Ruts from the old stagecoach line are visible near Monument Rocks, a few miles east of here, and particularly at Castle Rock, a chalk outlier in eastern Gove County.

129.3–129.5 *Smoky Hill Chalk Member* of the the Niobrara Chalk. The Smoky Hill crops out in a belt from Smith and Jewell counties, in north-central Kansas, to Finney and Logan counties here in west-central

Kansas. Probably the best-known exposures are here in Gove and Logan counties, where erosion has created scenic spires, bluffs, and monuments.

Chalk is actually a form of limestone that is composed of calcium carbonate, which precipitated out of sea water, and of the remains of tiny one-shelled animals called *Foraminifera*. These animals lived in the open Cretaceous sea, and their remains rained down on the ocean floor, forming a limy ooze that would later become chalk. At the same time, this soft muck claimed the remains of larger animals that sank to the bottom of the ocean.

Fossil collecting in the Niobrara in Kansas started in 1868, when the post surgeon at Fort Wallace, in Wallace County, found the remains of a mosasaur. By the early 1870s, fossil-collecting expeditions to the chalk beds were common. Some of the collectors—such as Benjamin Franklin Mudge, O. C. Marsh, and E. D. Cope—were supported by universities and museums, though the competition to claim new fossils spawned bombastic scientific rivalries. Other collectors operated independently. One of these, Marion Bonner, from nearby Leoti, developed a childhood interest in fossils into a lifelong passion. Though he lacked formal education, Bonner gained renown as an expert on fossils of the Niobrara, and specimens that he collected can be found in museums throughout the country. He and members of his family discovered new species of Cretaceous life that now bear the Bonner name.

These beds have yielded many scientifically significant fossils. The first birds with teeth were discovered here in western Kansas; the species was *Hesperornis,* a swimming penguinlike creature. Most of the finds from these beds made their way into various museums. Some, such as the Peabody Museum at Yale University, are on the East Coast, but others, such as the Museum of Natural History at the University of Kansas, are within the state. The Sternberg Memorial Museum at Fort Hays State University is named after one of the great fossil-collecting families in western Kansas.

129.8 *Smoky Hill Chalk Member* of the Niobrara Formation, west of the highway.

130.2 *Smoky Hill Chalk Member* of the Niobrara Formation, west of the highway. When these formations were deposited, the sea was several hundred feet deep and was slightly less saline than normal; the temperature was subtropical. The sea floor was almost perfectly flat, and the chalk was deposited at the rate of a fraction of an inch per year. The waters at the bottom of the ocean were low in oxygen; the ocean floor was dark, monotonous, and hostile to many groups of marine organisms.

Monument Rocks in Gove County, east of U.S. 83

Today the Smoky Hill chalk is interesting, not only for its fossils but also because it contains significant reserves of natural gas in parts of northwestern Kansas and adjoining states, where it is deeply buried. Much of that natural gas probably formed from the remains of the organisms that lived in the Cretaceous sea.

131.4   The road here turns 7.5 miles southeast to *Monument Rocks,* among the most famous of the chalk formations in western Kansas. Geologists were aware of the chalk here as early as 1858, but they were kept away by hostile Indians and the lack of transportation. In the first report of the Kansas Geological Survey, Benjamin Franklin Mudge wrote only that rocks of Cretaceous age occurred in Kansas, and "chalk is said to have been found in it." With the discovery of large vertebrate fossils, however, geologists were willing to brave the Indians, and during the 1870s they made regular forays to western Kansas.

These geologists often came to Monument Rocks, which is still a favorite stop for visitors who want to take photographs and look for fossils. Because of the number of visitors, vertebrate fossils around Monument Rocks are generally rare but are easier to find in the less-visited parts of

The Sphynx, a chalk pillar north of Monument Rocks, which toppled in 1986.

chalk. The main cluster of Monument Rocks is part of the Monument Rocks National Natural Landmark, although it remains on private land, which visitors should keep in mind.

Because of the rock falls that clutter the ground around the pyramids, it is easy to assume that erosion is quickly wearing them away. However, when recent photographs are compared to historic photographs of Monument Rocks and Castle Rock, it is apparent that the formations have changed relatively little. Geological Survey paleontologist S. W. Williston wrote in 1897: "My first knowledge of the Rock (Monument Rock) dates from October 1874, and since that time I have seen but little evidence of erosion. In various places throughout the chalk beds of the Smoky Hill River I have observed marks scratched by myself eighteen years previously that appeared as clear almost as when they were made. The erosion in general is not nearly so rapid as one would think. The smooth, worn surfaces made on the projecting angles of many low cliffs by the buffaloes are still to be seen nearly as smooth as they were twenty years ago." Probably the greatest danger of erosion today is not from the elements but from vandalism.

132.0   *Chalk badlands* east of the highway. While the Smoky Hill

A fault in an outcrop of the Niobrara Formation east of U.S. 83 in Logan County

chalk is best known for vertebrate fossils, invertebrates are much easier to find, though they are usually less spectacular. Among the most common fossils are the clam species *Inoceramus* and oyster shells, which can be found in nearly any outcrop of the chalk. Specimens of crinoids, or sea lillies, and spiral-shelled ammonites can also be found.

132.5   A fault in the Niobrara occurs in a creek bank about one-fourth mile east of the highway. Many faults have been mapped in the Cretaceous rocks of the Smoky Hill valley. However, their apparently random orientation makes their cause difficult to pinpoint, and geologists still debate their origin. Geologists suspect that the faults may have been caused by different rates of compaction in underground layers of rock. Or the faults may have been due to the solution of underlying evaporites, such as salt or gypsum.

134.1 and 134.8   *Niobrara Chalk Formation* to the east.

136.0   The view to the south is of the *Smoky Hill valley*. The elevation is 3,000 feet, while the valley floor is nearly 400 feet lower, dissected by the passage of the Smoky Hill over eons of geologic time.

137.3   *Plum Creek.*

140.7   *South Branch of Hackberry Creek.* The *Ogallala Forma-tion* is exposed in the bank east of the highway.

142.5   *Ogallala Formation* east of the highway.

143.0   *Ogallala Formation.*

143.2   *Middle Branch of Hackberry Creek.* On some old maps this creek is labeled Castle Rock Creek, because Castle Rock, a natural chalk monument, is located in this creek's valley, about 40 miles down-stream, in eastern Gove County.

146.2   Small area of *Ogallala Formation* on the south side of the creek east of the highway.

148.3   North Branch of *Hackberry Creek.*

150.5   The wells here are part of the *Oakley oil field,* discovered in 1956, which produced more than 600,000 barrels of oil, more than any other field in the history of Logan County.

151.4   South edge of *Oakley,* and the junction with U.S. 40. The elevation is 3,050 feet. Oakley is the seat of Logan County. The county seat was once located in Russell Springs, about in the center of the county, but was moved here in the 1960s in a bitterly contested election that Oakley won by one vote. Oakley is also home to the Fick Fossil Museum, located on the west edge of town. Although this museum contains exhibits related to local history, it focuses on the chalk beds and fossils from Gove and Logan counties. There are displays of large vertebrates, such as Cretaceous fishes and sharks, along with more recent Pleistocene mammals and huge numbers of invertebrates—clams, crinoids, oysters, and the like—that were found by the Fick family on a ranch near Monument Rocks in Gove County.

153.0   The north edge of the town of *Oakley* and the *Logan/Thomas county line.*

154.1   *South Fork of the Saline River.* This branch of the Saline joins the river's north fork 8 miles east of here, forming the Saline River.

156.5   *North Fork of the Saline River.*

156.9   Junction with *I-70.* I-70, described in chapter 2 of this book, is the major east-west transportation link across Kansas. Like the railroad during the 1800s, the Interstate's route has determined the fate of many small towns in western Kansas. Those that it passed through or near, such as Oakley, often continued to thrive because of the business that travelers brought to the towns. Places further away from the Interstate, however, had to fight to survive. I-70 attracted much of the east-west traffic that was once carried by such east-west highways as U.S. 36.

A sample of moss opal from Gove County

Two major destinations along I-70 are Denver, which is 248 miles to the west, and Kansas City, 357 miles to the east. Colby is 17 miles and Goodland is 52 miles to the west; Grainfield is 24 miles and WaKeeney is 51 miles to the east.

161.0   *Ohl oil field.*

163.4   *South Fork of the Solomon River.* This branch joins the north fork of the Solomon at Waconda Lake in Mitchell County, 130 miles to the east.

165.8   The elevation is 3,090 feet, the highest point along U.S. 83 in Kansas.

169.6   The town of *Halford* and a feedlot lie west of the highway.

170.6   *North Fork of the Solomon River.*

171.0   Junction with *U.S. 24.* Colby is 10 miles and Goodland is 47 miles west. Hoxie is 25 miles east.

173.9   *Gem* is 1.5 miles to the west. This town was probably named after the Gem Ranch, once located in Thomas County, not because of any precious stones. Gems are uncommon in most of Kansas. Probably the most common gem found here in west-central Kansas is moss opal, which is occasionally discovered in the Cretaceous chalk beds of Trego, Gove, and Wallace counties. The opal is stained by dark, branching deposits of manganese oxide, which gives the rock its name.

174.0–194.0 For the next 20 miles, northbound U.S. 83 follows the drainage divide between Prairie Dog Creek to the north and the North Fork of the Solomon River to the south.

176.7 The town of *Breton* and a roadside park.

179.5 The elevation is 3,000 feet.

181.1 *Rexford* is immediately north of the highway.

182.5 *Thomas/Sheridan county line.*

191.5 Roadside park on the west edge of Selden.

191.8 *Selden.*

193.7 Junction with eastbound K–23. Hoxie, the seat of Sheridan County, is 18 miles southeast.

194.0 Junction with K–383, running to the northeast. Dresden is 8 miles and Norton is 45 miles to the northeast. At about this point the highway leaves the drainage divide between the North Fork of the Solomon River to the south and Prairie Dog Creek to the north.

194.7 *Sheridan/Decatur county line.*

195.3 The gravel pit east of the highway is in the *Ogallala Formation.*

198.3 *Prairie Dog Creek,* elevation 2,650 feet.

199.8 *North Fork of Prairie Dog Creek,* which joins Prairie Dog Creek about a mile to the southeast.

202.0 *Loess deposit* east of the road in the creek bank.

208.0–209.8 Here the road is lined by exposures of Ogallala Formation mortarbeds.

211.6 *Sappa Creek.* This is also the south edge of *Oberlin.*

212.6 Junction with *U.S. 36,* described in chapter 3 of this book. Oberlin is the seat of Decatur County. The county courthouse, located in the town's central business district, was completed in 1927. The elevation here is 2,640 feet, or about 0.5 miles above sea level. From this point, Atwood is 28 miles west; Saint Francis is 69 miles west; Norton is 35 miles east; and Phillipsburg is 66 miles east.

214.4 Here the road passes through *Johnson Draw,* with Morrison Lake, now a dry lake bed, on the west.

217.6 About 7.5 miles west is a highly eroded remnant of the Ogallala Formation mortarbeds called *Elephant Rock.*

220.5 The elevation is 2,750 feet.

Elephant Rock, an eroded outcrop of the Ogallala Formation south of the town of Traer in Decatur County

223.0   *Ogallala Formation* in the draw to the west.

223.6   The town of *Cedar Bluffs* is west of the highway.

223.7   *Ogallala Formation* in the cliffs east of the highway.

223.8   *Beaver Creek.* The elevation is about 2,520 feet, the lowest point along U.S. 83 in Kansas.

225.0   *Kansas/Nebraska border.* The northbound highway enters Red Willow County, Nebraska. McCook, the birthplace of University of Kansas' Chancellor Gene Budig, is 14 miles to the north. North Platte is 82 miles north.

Red Willow County, Nebraska, is the site of one of the largest oil fields in Nebraska. It is also the center of a spate of seismic activity—small earthquakes that occur at the rate of one or two per month. Nearly all of these quakes are too small for people to feel. They are probably related to water flooding in the oil field, or the pumping of additional water into the oil reservoirs to force additional oil out of the rocks. Geologists believe the water may put pressure on faults and allow them to move, thus creating the small tremors.

INTERSTATE·135/U.S. HIGHWAY·81

Through the Heart of the State

0.0 I–135 begins in Wichita, then heads north to Salina, where it joins U.S. 81, which runs north to the Nebraska border. From Wichita to Salina, I–135 parallels the route of U.S. 81, historically one of the major north-south routes across the state and an informal line of demarcation between eastern and western Kansas, although it is east of the exact north-south dividing line in the state. Because the two highways are so closely related, the logs for I–135 and U.S. 81 are combined in this book.

From Wichita, I–135 goes north to Newton, then cuts northwest, paralleling the Little Arkansas River. About the only topographic relief encountered along the way comes between McPherson and Salina, where sandstone from the Dakota Formation crops out, forming steep buttes west of the highway. From Salina, U.S. 81 runs north to Concordia and on to Belleville, passing through a few outcrops of Cretaceous limestones, sandstones, and shales.

0.3 *Tollgate,* where I–135 diverges from I–35, which runs northeast to Emporia. I–135 continues north through Wichita. From here to its junction with I–235 at the north edge of Wichita, I–135 has been designated the Martin Luther King Memorial Highway by the 1985 Kansas Legislature.

0.4 Wells to the west are part of the *Robbins oil field,* which was discovered in 1929 and is one of the oldest fields in Sedgwick County. These wells produce oil from Mississippian formations that are about 3,000 feet underground.

0.5   *Forty-seventh Street.* Here southbound U.S. 81 exits I–135 and heads toward Wellington and Oklahoma. Southbound U.S. 81's path is parallel to but slightly west of the turnpike.

1.0   *I–235 interchange.* I–235 loops west around Wichita, passing Wichita's Mid-Continent Airport and the Sedgwick County zoo before rejoining I-135 north of the city.

1.2   Wells to the east are part of the *Robbins oil field.* Although Sedgwick County produced only about 560,000 barrels of oil in 1984, Wichita is a center for the oil industry in Kansas. Many oil-related companies are located here; they service wells drilled in south-central and southwestern Kansas. The oil business in south-central Kansas got its start when oil was discovered during the early 1900s in subsurface features called the Sedgwick Basin and the Nemaha Anticline. However, with the discovery of oil along the Central Kansas Uplift in Russell, Ellis, and Barton counties, much of the state's oil industry shifted to north-central Kansas.

1.7   *Arkansas River.* The riverbed elevation is 1,270 feet. About 5 miles north of here, near downtown Wichita, the Little Arkansas and the Arkansas rivers join and flow south to Tulsa, then on into the Mississippi River. Sand, which is common along the path of the Arkansas, is regularly dug and dredged from the river bottom and is put to a variety of uses. The production of sand and gravel is the most widespread nonfuel mineral industry in the state, and sandpits are common here in Wichita.

1.8   *Sand hills* north of the highway.

2.3   *Hydraulic Avenue.* Hydraulic is a north-south thoroughfare across Wichita, paralleling I–135 for part of its route. Three miles west of Hydraulic Avenue there is another major north-south route, called Meridian Avenue, which runs along the Sixth Principal Meridian. This line of latitude was used for surveying; it is the basis for legal descriptions in Kansas, Nebraska, Colorado, and Wyoming.

3.3   *K–15* joins northbound I–135 here. K–15 begins at the Kansas/Oklahoma border in Cowley County, joins I–135 here in Wichita, and then leaves I–135 near Newton, continuing north to Abilene and Washington. Also at about this point, Gypsum Creek joins the channelized course of Chisholm Creek. Gypsum, or hydrated calcium sulfate, occurs in the shales of the Wellington Formation, a Permian rock layer that crops out in the eastern part of Wichita, the area drained by Gypsum Creek. Chisholm Creek, named after Jesse Chisholm of Chisholm Trail fame, begins north of Wichita and flows through the heart of the city. For much of its course through Wichita, the creek has been channelized; it runs through a concrete-lined canal in the median of I–135 for the next five miles.

5.0   *Harry Street.* Linwood Park is west of the highway.

6.0   *Kellogg Street.* I–135 intersects here with Kellogg Street, or U.S. 54. Kellogg is a major east-west thoroughfare across Wichita, going west from here into the city's downtown area, where it passes near the confluence of the Little Arkansas and Arkansas rivers. On the west edge of Wichita, Kellogg Street passes near the Wichita Well Sample Library office of the Kansas Geological Survey. This library houses cutting samples from oil wells—chips of rock that are brought to the surface during drilling. These chips are used by geologists to determine the type of rock that the well has encountered. With these chips, geologists can examine subsurface samples from nearly every part of the state, which aids in the search for oil and other minerals. Also at about this point, K–96 joins northbound I–135.

7.5   The elevation is 1,300 feet.

8.5   Here Chisholm Creek is diverted into a concrete-lined canal that runs down the median of I–135. Because much of Wichita is built on the flood plain of the Arkansas River, drainage improvements were necessary in order to prevent urban flooding. This canal is designed to handle drainage in east-central Wichita; the Wichita–Valley Center flood way, along Big Slough Creek, handles drainage in the northern and western parts of the city.

9.0   *Twenty-first Street.* Immediately west of the highway is the Derby refinery; beyond it are the Wichita stockyards. East of the highway are Wichita State University and the Institute of Logopedics, which is internationally known for work in speech and hearing. Originally established as Fairmount College in 1892, Wichita State became the Municipal University of Wichita in 1926, and in 1964 it joined the Kansas Board of Regents system.

9.6   The wells to the east are part of the *Wichita oil field,* which has produced more than 2.9 million barrels of oil since it was discovered in 1957. Also, at this point the road crosses the *East Fork of Chisholm Creek.*

11.5   *Middle Fork of Chisholm Creek.*

11.8   The *junction of I–135 and I–235,* which loops west around Wichita. This is also the intersection with K–254, which runs 3.5 miles east to Kechi and on to the towns of Towanda and El Dorado.

12.1   *Boundary of Indian reservation.* This point marks the northern limit of lands given to the Osages in the days when Kansas was Indian territory. The land north of this line was the hunting grounds of the Kansa, Pawnee, and other Plains Indians. Indian activities are still an important part of Wichita. The Mid-America All-Indian Center in downtown

Wichita is designed to aid the cultural awareness of Indian people. Near the center there is a 44-foot sculpture called "Keeper of the Plains," by the Indian artist Blackbear Bosin.

12.3   This point marks a break in the landscape, between the Arkansas River Lowlands to the south and the Flint Hills Uplands to the north. The lowlands form a flat, alluvial plain, composed of sediments deposited by the Arkansas River during the past 10 million years of geologic history. The Flint Hills form gently rolling uplands; most of the rocks here were deposited during the Permian Period, about 250 million years ago. I-135 passes only briefly along the west edge of the Flint Hills, where the rocks are part of the Wellington Formation and are generally covered by much younger deposits of loess.

14.2   *Sixty-first Street.* Park City is immediately to the east, and Kechi is 2.5 miles east.

14.5   The creek bed here was part of the natural channel of Chisholm Creek before it was rerouted and channelized to prevent flooding.

14.7   *Man-made channel of Chisholm Creek.*

17.0   To the east is the *Kansas Coliseum,* an indoor arena that houses rodeos, concerts, and the Wichita Wings professional soccer team. To the west 0.5 miles is the town of Spasticville.

17.2   *K-164* exits to Valley Center, 2.5 miles to the west.

19.0–20.0   The *Goodrich oil field,* which was discovered in 1928, has produced more than 6.4 million barrels of oil. The wells here are about 3,000 feet deep, producing oil from rocks deposited during the Mississippian and Ordovician periods.

19.2   *101st Street.* Sunnydale is 1.5 miles to the east; Furley is 6.5 miles east.

22.3   *Sedgwick/Harvey county line.* The town of Sedgwick, about 5.5 miles to the west, lies mostly in Harvey County.

22.5   *Gooseberry Creek.*

23.6   *Rest Area.*

24.4   *Jester Creek.*

25.3   K-196 goes east 11 miles to Whitewater, 16.5 miles to Potwin, and 31 miles to El Dorado. About 3.5 miles northwest of Potwin, in Butler County, is the *Potwin sinkhole,* which opened up on 22 September 1937, creating an oblong void that measured 90 by 150 feet. The sinkhole, which soon filled with water, was probably caused by the solution and collapse of a limestone layer about 75 feet underground.

27.0   The elevation is 1,450 feet.

28.3 *McLains Road.* The town of McLains is 3.5 miles to the east of the highway.

29.3 The southern city limit of *Newton,* the seat of Harvey County. In 1871 the Santa Fe Railroad reached Newton, and the town replaced Abilene as the northern terminus of the Chisholm Trail. Newton quickly became a wild and rough cow town, which lasted until 1873, when the railroad reached Wichita.

30.0–30.5 *Westbound U.S. 50* goes 19 miles to Burrton and 35 miles to Hutchinson.

31.3 *First Street.* Newton's business district is about a mile to the west. For many years Newton was a switching point on the Santa Fe Railroad, so a series of tracks runs through the middle of town. Newton is still a stopping point for Amtrak passenger service.

32.5 *Eastbound U.S. 50* exits here, going 5.5 miles to Walton, l5 miles to Peabody, and 25 miles to Florence.

33.0 Shale from the *Wellington Formation* is exposed on the creek bank west of the highway. The Wellington is composed of shale, limestone, dolomite, siltstone, gypsum, and anhydrite, although it is probably best known as the unit that includes the Hutchinson Salt Member, the salt bed that is mined throughout central Kansas.

34.0 Here northbound I–135 leaves the Flint Hills and enters the McPherson Lowlands physiographic region. The McPherson Lowlands, part of a larger region called the Arkansas River Lowlands, are mostly alluvial deposits that form a flat, even landscape.

34.3 *Sand Creek.*

34.8 *K–15,* which exits here, runs 22 miles north to Lehigh and 64 miles north to Abilene. To the south 0.5 miles is North Newton and the campus of Bethel College. Established in 1887, it is the oldest Mennonite educational institution in the United States.

37.2 *East Emma Creek.*

37.4 *Zimmerdale* is 0.5 miles southwest of the highway.

40.1 *Middle Emma Creek.*

40.5 K–165 exits west to *Hesston,* which was named after Abraham Hess, one of the town's founders. Like many of the towns in south-central Kansas, Hesston was founded by Germans, who still make up a large percentage of the population.

41.4 *Hesston Corporation factory* is about 0.5 miles southwest of the highway. Hesston is the largest manufacturer of farm equipment in Kansas and is the tenth-largest in the United States. Hesston is particularly well known for its windrower, or swather, which cuts hay and piles it into long rows. Swathers are the Hesston High School's mascot.

43.1 *Harvey/McPherson county line.* McPherson County is named after the Civil War general James B. McPherson, who was killed in the Battle of Atlanta in 1864. Harvey County is named after James Madison Harvey, the fifth governor of Kansas, who served from 1869 to 1873.

44.0 Wells here are part of the *Graber oil and gas field,* discovered in 1934. It extends south into Harvey County, and produced about 24,000 barrels of oil in 1984 from 64 wells. In all, wells in McPherson County produced about 1.3 million barrels of oil in 1983, making it the eighteenth-ranking oil-producing county in Kansas.

44.3 *West Emma Creek,* which joins Middle and East Emma creeks before draining into the Little Arkansas River.

46.1 *K-260,* which exits here, runs 2 miles south to Moundridge, whose name comes from a low ridge that outlines a 1-by-2-mile elliptical basin west of town. Sand Creek drains this basin, which may have been formed by the solution of underlying Permian evaporites. The Hutchinson salt bed may have been present in the subsurface here at one time, but it has since been dissolved by circulating ground water. Today, the bed's eastern edge is several miles to the west. The dissolving salt created a broad trough from north of McPherson southward to Wichita; the ancestral Smoky Hill River once flowed through this trough, joining the Arkansas River near Wichita.

During the Ice Age, the ancestral Smoky Hill River and other streams gradually filled this solution basin with sand, silt, and gravel. Because fossils of Ice Age horses were found in these deposits, they are now called the Equus Beds (*equus* is Latin for horse). Today the Equus Beds contain ground water; they are a major source of water for many surrounding cities, including McPherson, Newton, and Wichita. The Equus Beds underlie I-135 from Newton north to McPherson, an area that is known as the McPherson Lowlands.

46.6 *Winsinger oil field,* some of whose wells produce oil from Ordovician, Silurian, and Devonian rocks, which were deposited between 500 and 360 million years ago and do not appear at the surface in Kansas. Here they are about 3,300 feet deep.

48.0 *Black Kettle Creek.* Black Kettle was a Cheyenne chief who was killed by U.S. Cavalry forces led by George Armstrong Custer at the Washita Massacre in western Oklahoma.

48.6 *K-260* runs 2 miles south to Moundridge.

The Equus Beds are unconsolidated sands and gravels deposited by rivers during the past few million years of geologic history.

49.5   *Sand Creek.*

50.0   The elevation is 1,500 feet.

54.2   *Elyria* is 2 miles west of here.

54.4   *Running Turkey Creek,* which joins Turkey Creek about a mile south of Elyria.

55.8   *Turkey Creek* drains south into the Little Arkansas River.

56.2–60.0   The wells here are part of the *Johnson oil and gas field,* discovered in 1932.

58.4   *K-61,* which exits here, runs 28 miles southwest to Hutchinson. An oil refinery is visible about 2.5 miles west of here.

60.5   *U.S. 56 interchange.* McPherson, the seat of McPherson County, is 2 miles west of here. The towns of Galva and Canton lie to the east along U.S. 56; the geology is described in chapter 9.

65.0   *Dry Turkey Creek* flows south through McPherson.

65.2   *New Gottland Church* east of the highway.

66.0 *New Gottland*, 0.5 miles east of the highway, was established by Swedish settlers in the 1800s. While the nearby town of Lindsborg is known for its Swedish heritage, Swedes settled throughout central Kansas in such towns as Marquette, Falun, and New Gottland.

67.0 This point marks the approximate division between the Smoky Hills and the McPherson Lowlands. To the north, the highway enters the Smoky Hills, which received their name from the thick haze that often fills their valleys in the early mornings. These steep-sided hills are formed by Cretaceous sandstones, remains of deltaic deposits and beach sands, left behind about 100 million years ago as rivers drained into the Cretaceous sea that covered western Kansas. To the south are the flat lowlands formed by alluvial deposits from streams draining south into the Arkansas River. These deposits make up the Equus Beds, which are Pleistocene and Pliocene in age.

Much of this alluvium was laid down by the Smoky Hill River during the Ice Age, when it flowed south to the Arkansas River. During the late stages of the Ice Age, streams in the Kansas River basin to the north were cutting down through their beds and flowing at lower levels than the ancestral Smoky Hill. These streams were also eroding headward, attacking the limits of their drainage basins. Eventually, in northern McPherson County, one of these headward-eroding tributaries of the Kansas River system cut into the Smoky Hill River and captured its flow. This process is called stream piracy, and it accounts for the U-turn in the present-day Smoky Hill River in northern McPherson County. Although it enters the county as a southeast flowing stream, it exits flowing to the north.

This break in the landscape also marks the drainage divide between the Arkansas River system to the south and the Missouri River system to the north. South of here, runoff from precipitation flows into southbound rivers that eventually reach the Arkansas. North of here, runoff flows into northbound tributaries of the Smoky Hill, which dumps into the Kansas River and eventually the Missouri. The dissected landscape and steep gradients of streams north of here indicate that headward erosion is still going on; the Kansas River basin is expanding southward through this area of easily eroded material. In the geologic future, additional stream piracy will occur, perhaps causing the Little Arkansas, or even the Arkansas River, to be diverted northward into the Kansas River basin.

The elevation here is 1,530 feet; for the next 1.8 miles, northbound I–135 follows West Kentucky Creek.

68.3   *Rest areas* are on either side of West Kentucky Creek.

69.6   Red shale on the east side of the northbound lane is part of the *Ninnescah Shale.* This unit is present as far north as Salina, but is more widely exposed to the southwest, in the Red Hills of south-central Kansas.

70.0   Six miles to the east there is a steep-sided draw called *Black Canyon,* which is 60 to 70 feet deep and is dissected by Gypsum Creek. The creek cuts through resistant sandstone in the Dakota Formation.

70.6   Visible 8 miles to the northwest are the *Smoky Hill Buttes,* a series of flat-topped hills, which are capped by outcrops of sandstone in the Dakota Formation. The southernmost hill is called Coronado Heights. Its elevation is just over 1,600 feet, about 300 feet above the surrounding countryside.

71.0   The elevation is 1,400 feet.

73.0   The county road to the east runs 10.5 miles to *Roxbury;* Alternate U.S. 81 leads 4 miles west and north to *Lindsborg,* a town that Swedish immigrants founded in 1868. It retains much of its Swedish flavor: every other year it holds a Svensk Hyllningfest Festival, celebrating that heritage. Lindsborg is home to Bethany College, a private Lutheran school that is the site of the Birger Sandzen Memorial Gallery, which displays works by that Swedish-American painter. Sandzen, who was born in Bildsberg, Sweden, in 1871, moved to Kansas at the age of twenty-three. He spent the rest of his life in Lindsborg, producing thousands of paintings, prints, and water colors, many of them featuring the light and color of the Kansas prairie landscape. Bethany College also hosts the Messiah Festival, which has been held annually since 1882.

76.5–77.0   These *meander loops* are abandoned channels of the Smoky Hill River; if they contained water, they would be called oxbow lakes. The Smoky Hill River cuts a tortuous path through this part of Kansas, with many meanders, abandoned channels, and oxbow lakes. It has been calculated that in the 65-mile-long valley from the McPherson/Saline county line to Junction City, the Smoky Hill actually travels 150 miles. Mark Twain once wrote that the Mississippi was the crookedest river in the country, but the Smoky Hill couldn't be far behind.

76.9   *McPherson/Saline county line.*

77.2   Here the highway crosses the channelized course of the *Smoky Hill River.* The river's natural channel is 0.1 miles north of here, but the Smoky Hill was straightened and channelized to prevent flooding and to

A geologic cross section of the southernmost of the Smoky Hill Buttes

stop the river from changing its course. From here the river heads north through Salina. Before the river's course was changed to loop around the city, the Smoky Hill caused flooding in eastern Salina.

77.5   The southernmost Smoky Hill Butte, *Coronado Heights,* is 4.5 miles west of here. The name comes from reports that Coronado climbed the hill during his exploration of Kansas in 1541. Historians are uncertain about how far Coronado traveled into Kansas, although they know that he followed the Arkansas River during part of his trek and that he crossed the Smoky Hill River somewhere here in central Kansas before abandoning his search for gold and returning to Mexico.

Atop Coronado Heights there is a picnic ground and a castle, built out of blocks of Dakota Formation sandstone by WPA workers. Much of the sandstone on Coronado Heights is cemented together by iron oxide, which has stained the rock black, instead of the characteristic red. This dark sandstone is sometimes referred to as ironstone. It is harder and more erosion resistant than are ordinary sandstone or the soft shales that form the slopes of the Smoky Hill Buttes.

77.9   *K–4* goes 0.5 miles northeast to the town of Bridgeport, and 4 miles southwest to Lindsborg, which was the seat of McPherson County from 1870 to 1873, when the county seat was moved to McPherson. Coronado Heights can be reached by taking this exit and heading west on a county road.

78.0   *Dry Creek.*

78.6   Gray shale in the *Wellington Formation* on both sides of the highway.

80.2   The *Smoky Hill Buttes* are visible four miles to the west. The tallest of these hills, near the north end of the line of buttes, is 1,632 feet in elevation, or about 340 feet above the elevation here at the highway.

The highest point along I-135/U.S. 81 comes about 6 miles south of the Nebraska border, where the highway's elevation is 1,652 feet, or about 20 feet above the tallest of the Smoky Hill Buttes. In other words, the highway must travel 85 miles to the north to reach the same altitude as the tallest hill to the west.

81.3   The county road here leads 1 mile east to *Assaria* and 7.5 miles west to *Falun*. Falun, named after a city in Sweden, was the birthplace of Hollis Hedberg, a petroleum geologist who was in charge of exploration for Gulf Oil Corporation. Gulf later named one of its ocean-going seismic-exploration ships the Hollis Hedberg, in his honor.

82.0   The wells to the west are in the *Olsson field*, discovered in 1929. Saline County lies above a subsurface feature called the Salina Basin, which has produced relatively little oil. In 1984, only about 1 percent of the exploration holes in Kansas were drilled into the area underlain by the Salina Basin. Geologists theorize that much of the oil that is produced today in Kansas actually matured far to the south, in Oklahoma. Because oil is a buoyant liquid, it may have moved up through layers of rock that dip upslope to the north in Kansas. Eventually, the theory goes, that oil was trapped along the Central Kansas Uplift, a subsurface geologic feature that prevented the oil from moving further north. Thus, oil is discovered today in central Kansas in a feature that prevented its movement into the Salina Basin. Any oil that is found in the Salina Basin must have matured there; it cannot have migrated in, the way oil has in other parts of the state.

82.3   *Salemsborg* is 3.5 miles to the west.

83.0   A hill called *Soldier Cap* is 9.5 miles to the west in a pasture that is part of the Smoky Hill Air Force Bombing Range. Because Dakota sandstone crops out in many of the pastures west of here, the ground has not been cultivated; instead, it has been left in native grass. Like the Flint Hills of eastern Kansas, the Smoky Hills constitute one of the prime cattle-grazing areas in the state.

83.3   The county road here joins K-4 and runs 10.5 miles east to the town of *Gypsum*, which is named for the gypsum deposits found in the Wellington Formation in central Kansas. During the late 1800s, several mines operated in Dickinson, Saline, and Clay counties; they produced gypsum that was used in plaster. Probably the best-known mines were near the towns of Solomon and Hope, east of here. For a time, gypsum was produced from a site called Henquenet Cave, about 7 miles southwest of Hope.

85.0   Wells here are in the *Salina field,* which was discovered in 1943. It has produced over 4.4 million barrels of oil, making it the second-most-productive field in the history of Saline County. This is also the last oil field along northbound I-135/U.S. 81 in Kansas. From here north, the highway passes through the Salina Basin. This broad depression in the earth's crust underlies north-central Kansas and much of Nebraska. It contains thick accumulations of Paleozoic sediments, but unlike the Cherokee Basin of southeast Kansas and the Sedgwick Basin that underlies the highway south of Salina, the Salina Basin appears to contain few deposits of oil or natural gas that would be economical to produce.

85.4   *K-104* runs 1 mile east to Mentor. To the west, a county road runs 3.5 miles to Smolan, the hometown of John Carlin, who was governor of Kansas from 1978 to 1986.

87.1   The elevation is 1,250 feet.

88.4   The road exits here to the southern part of *Salina.* The Salina Municipal Airport, formerly Schilling Air Force Base, is 1 mile west of the road. The former air base is now an industrial park and the site of the Kansas Technical Institute.

90.4   *Dry Creek.*

90.6   *Salina* is home to two colleges. Kansas Wesleyan, a Methodist school, is 1.1 miles east of here. Marymount College, a Catholic school, overlooks the city from a hill on the east edge of Salina.

91.7   *Crawford Street* exits here from I-135. To the east, Crawford is a major east-west thoroughfare across Salina. To the west, it intersects with K-140.

92.8   Intersection with *K-140.* Salina is 1 mile to the east. To the west are the towns of Bavaria, Brookville, Ellsworth, and Wilson. Concretions called barite roses are common on a hill outside of Bavaria. Barite is barium sulfate, which, in this location, cements together grains of sand to form unusual rose-shaped concretions that weather out of the Kiowa Shale Formation. In nearby Brookville, many of the town buildings are built out of native sandstone from the Dakota Formation.

93.8   Mulberry Creek.

95.2   This is the point of change in the numbering system between U.S. 81, to the north, and I-135 to the south. Along southbound I-135 the milepost numbers grow smaller; along northbound U.S. 81, the milepost numbers grow larger. Also at about this point, the highway intersects with I-70, which is described in chapter 2.

0   C M   5   I O   I 5

Barite roses, a concretion formed from sandstone cemented by barite, are found near Bavaria in Saline County.

155.6   *Trenton* is 0.2 miles to the west.

156.7   The pit to the west is now used as a fishing lake.

157.6   *Shipton* is 2 miles west.

158.0   *Saline River.* The elevation at the riverbed is 1,200 feet. The Saline flows east of here before joining the Smoky Hill River east of Salina. The river's name comes from its salt content; Saline County and the city of Salina were, in turn, named after the river.

158.5   *K-143.*

158.8   The view to the south is of the *Saline River flood plain.* To the north the highway passes over Cretaceous shales and sandstones that form the Smoky Hills.

159.7   *North Pole Mound,* elevation 1,455 feet, is 1.7 miles to the east.

160.0   *Wary Lake* is west of the highway.

160.6   *Saline/Ottawa county line.*

161.0   *Rest Area.*

162.0   The hills east of the highway are capped by sandstone from the Dakota Formation. In some locations, Dakota sandstone contains fossil remains of Cretaceous leaves and plants. During the mid 1800s, paleobotanists often traveled to central Kansas to study flora of the Cretaceous.

164.8 *Interchange with K–18,* which runs west 13 miles to Tescott and 29 miles to Lincoln, in the heart of post-rock limestone country. To the east it goes 4 miles to Bennington and 45 miles to Junction City.

165.3 *Antelope Creek.*

165.6 At this point, geologists have mapped a small fault, trending northwest to southeast and paralleling the Solomon River valley. This fault has down-dropped strata to the northeast by about 20 feet.

166.5 Sandstone in the *Dakota Formation* on the east side of the highway. While sandstone is the most visible part of the Dakota Formation, it is not the most abundant. The primary components of the Dakota are clay and shale units that do not stand out as prominently as sandstone, but tend to color the road cuts.

166.7 Sandstone from the *Dakota Formation* lies to the west of the highway.

167.2 *Solomon River.* The Solomon joins the Smoky Hill River about a mile south of the town of Solomon. Because the Solomon, Saline, and Smoky Hill rivers are so close together, this part of central Kansas is sometimes called the Tri-Rivers area. The riverbed elevation at this point is 1,200 feet.

168.0 *Battle Creek* is 1.5 miles to the west.

170.2 The town of Lindsey is 1.5 miles to the west. About 4.5 miles to the west is *Rock City,* where a group of 200 sandstone concretions have weathered out of the Dakota Formation. These concretions, some up to 27 feet in diameter, sit on top of the ground like a collection of stone bowling balls. The sandstone in these concretions is cemented together by calcium carbonate; over time, the surrounding rock has eroded away, leaving behind these rock spheres.

These concretions show evidence of cross-bedding, or angled lines that formed in the sand as it was deposited, probably by running water. While this is the largest and most dramatic collection of sandstone concretions in the state, similar forms are gradually appearing in other parts of the Smoky Hills and in a few other locations in Kansas.

170.8 *K–93* leads to Ottawa County State Park and Lake, about 4 miles east of here.

171.1 Along northbound U.S. 81, this marks the end of the

Huge sandstone spheres from the Dakota Formation dot the landscape at Rock City in Ottawa County, west of U.S. 81.

divided four-lane highway. For most of the remaining distance to the Nebraska border, U.S. 81 is a two-lane road.

172.2 *K–106* leads 2 miles west to Minneapolis, the seat of Ottawa County. South of Minneapolis about 2.5 miles, on K–106, is a county road that leads west 0.5 miles to Rock City.

172.8–173.1 Sandstone in the *Dakota Formation.* The sandstone in this road cut is highly cross-bedded.

173.4 *Lindsey Creek.*

176.9 Sandstone in the *Dakota Formation* on both sides of the highway. The sandstone here is stained brown or black because of its iron-oxide content.

177.8 *Pipe Creek.*

183.2 *K–41* leads west 5 miles to Delphos.

184.4 *Dry Creek.*

184.9 The *Greenhorn escarpment,* formed by limestone in the Greenhorn Formation, is visible 11 miles to the west, across the Solomon River valley.

185.2 *Ottawa/Cloud county line.*

185.4 and 188.0 Shale and sandstone in the *Dakota Formation.*

189.2 Junction with *U.S. 24.* To the west, this highway goes 9 miles to Glasco and 27 miles to Beloit, another town in the post-rock country of the Smoky Hills. To the east, U.S. 24 runs 13 miles to Miltonvale and 30 miles to Clay Center.

190.6 The elevation is 1,500 feet.

192.0–193.0 During this mile the northbound highway climbs out of the Dakota Formation, through Graneros Shale, and onto an escarpment capped by Greenhorn Limestone. The Greenhorn, a common limestone west of here, is the source of stone for many of the fence posts and buildings in north-central Kansas.

193.8 The uplands here are formed by Greenhorn Limestone, which is covered by loess soil. The elevation is 1,638 feet, or about the same elevation as the tallest of the Smoky Hill Buttes near Lindsborg.

195.0 *Wolf Creek.*

195.4 *Greenhorn Limestone* to the west of the highway in the creek bank.

196.2 A few miles east of here is the homestead of Boston Corbett, the man who killed John Wilkes Booth. In 1878, Corbett came west and lived in a dugout on his 80 acres. In 1886 he moved to Topeka and got a job at the Kansas Capitol as assistant doorkeeper in the House of Representatives. One day, Corbett pulled a gun on two other employees and was jailed. He was later pronounced insane and was sent to the state asylum in Topeka, but he escaped in 1888 and was never heard from again.

197.0–198.0 *Greenhorn Limestone.* In addition to its characteristic limestone layer, the Greenhorn includes layers of yellowish-gray and light-gray shale. The formation's total thickness ranges from 64 feet, in Ellsworth County, to 95 feet, in Ellis County, to 132 feet in Hamilton County. Outcrops of the Greenhorn often contain large accumulations of fossilized clams of the species *Inoceramus.*

200.3 *Greenhorn Limestone* is exposed in the road cut at the highway's intersection with the county road, which leads 3.5 miles east to the town of Huscher.

201.1 *Greenhorn Limestone.* The elevation is 1,550 feet. A quarry was once located west of the highway, but now the site has been smoothed, and the top of a limestone layer is used for a parking lot. Straight joints are visible throughout the limestone.

202.0 About 2.5 miles east are several clay pits and a brick plant that takes raw materials out of the Dakota Formation. Clays and shales make up about 80 percent of the Dakota. In places the clay is suitable for making bricks, tile, and other ceramic materials. Bricks are the major product of this plant. The different-colored clays that are taken from the Dakota are used alone or in combination to produce bricks of different colors. Bricks from this plant are shipped as far away as Chicago.

202.3   The southern edge of *Concordia*. About 0.3 miles to the west is the campus of Cloud County Community College.

203.0   The brick building 0.2 miles to the west is the *Nazareth Convent and Academy*.

203.3   Massive sandstone outcrop, part of the *Dakota Formation*, on both sides of the road.

203.5   *Concordia business district*. Three blocks to the west is the Brown Grand Theatre, an opera house that was built in 1907 by Napoleon Bonaparte Brown, one of Concordia's wealthiest citizens. The theater seated nearly 1,000 persons; it attracted plays and concerts for many years. In 1980 the opera house was restored and reopened; it is currently registered as a National Historical Monument.

This is also the intersection of Kansas highways 9 and 28. K–9 runs 34 miles west to Beloit and 12 miles east to Clyde. K–28 goes 12 miles west to Jamestown, skirting the southern edge of the Jamestown Waterfowl Management Area, a huge salt marsh.

203.6   The Catholic Church one block to the east is constructed out of native Greenhorn Limestone.

204.0   The north edge of *Concordia,* hometown of Frank Carlson, governor of Kansas from 1947 to 1950 and United States senator from 1950 to 1969. The Frank Carlson Library, at Seventh and Broadway in Concordia, houses memorabilia from his political service.

204.7   *Republican River.*

205.4   *Coal Creek.*

206.2   To the east 1 mile is the former site of a World War II German prisoner-of-war camp. Camp Concordia opened in 1943, and during its time, many of the German war prisoners were loaned to area farmers to provide help during the labor-short war years. Because of the vast distance back to Europe, escape attempts were rare. A restored guard house built of native limestone is visible along the road to the east.

208.2   The town of *Hollis* is 6 miles east.

208.3   *Coal Creek.*

209.2   *Cloud/Republic county line*. The area east of here was

Old spoils piles, such as this one in Cloud County, mark the locations of lignite mines that once operated in Cloud and Republic counties, east of U.S. 81.

extensively mined for lignite during the late 1800s and early 1900s. Lignite, also called brown coal, is a low-grade coal, which is relatively low in heating value, but was still valuable on the plains where fuel was rare. Lignite was formed from the remains of plants and other organic debris that accumulated along the shores of a Cretaceous swamp. Although mining was common in much of central Kansas—lignite mines are still visible in western Ellsworth County—this area was probably the largest lignite-producing region in Kansas. Of all the lignite produced in the state, 40 percent came from Cloud County before mining ended there in the 1940s. Lignite had drawbacks as a fuel—it produced large amounts of ash, and because of its moisture content, it was difficult to store. Therefore other fuels soon replaced it. The old mine dumps are still visible 2 miles to the east, which was also the site of an old mining town called *Minersville.*

    209.9   *Coal Creek.*

    212.2   Junction with *K–148.* The town of Norway is 5.5 miles to the west, Talmo is 4 miles east, and Wayne is 7 miles east. A mile southeast of Talmo, in the flood plain of East Salt Creek, is a large salt marsh called the *Tuthill Marsh*. Today, most of the marsh is dried up, although white, crusty patches of salt are visible on the surface in a few of the fields. During the 1860s, this was one of the largest salt marshes in Kansas, and for a time was the source of commercial salt production.

214.5   *Roadside park.*

214.7   *West Salt Creek.*

215.3–215.5   *Greenhorn Limestone.*

216.0   The uplands here are mantled by *Carlile Shale,* and the elevation is 1,530 feet. This point marks the division between the Smoky Hills to the south and the Blue Hills to the north. Both of these areas make up the eastern margin of the High Plains, which have been heavily dissected by streams. The dominant outcropping rock in the Smoky Hills is sandstone from the Dakota Formation, while Carlile Shale is the most common rock to the north, and its bluish-gray shales probably gave these hills their name.

217.3   This is the halfway point between the towns of Cuba to the northeast and Norway to the southwest.

220.3   Junction with *U.S. 36,* described in chapter 3 of this book. Belleville is immediately east of here, on the business route of U.S. 81.

220.5   *Riley Creek.*

221.2   The *Belleville* business district is 0.4 miles to the east. Belleville is the seat of Republic County, and the county courthouse, completed in 1940, sits in a square block in the middle of the Belleville business district.

222.4   Junction with U.S. 81 Business Route, which heads south into Belleville.

226.0   The uplands here are blanketed by a layer of alluvium that has been named the *Belleville Formation.* It has been correlated with the McPherson Formation, which makes up part of the Equus Beds that underlie I–135 in the McPherson area. Like the McPherson Formation, the Belleville Formation was deposited by an Ice Age river, which has since been captured by a headwardly eroding stream and has been redirected, leaving alluvial deposits behind in a buried valley. This buried valley has been named the Belleville Channel; it is crossed by the highway near the Nebraska state line. At this point the Belleville Formation is thin, averaging only about 20 feet in thickness. It represents the depositon of alluvium by streams that spread out on their uplands adjacent to the river valley. The Belleville Formation thickens to the north along U.S. 81 as the highway approaches the Belleville Channel. The Belleville Formation is overlain by loess in much of northern Republic County; that loess also thickens in the Belleville Channel.

227.5   The elevation is 1,650 feet, the highest point along the combination of U.S. 81 and I-135.

228.8   The county road here runs 12 miles west to the town of Republic, and 4 miles east and 1 south to the town of Munden.

230.8   *Rose Creek.* Narka is 10 miles to the east.

233.7   *Belleville Channel,* which was cut and filled by the ancestral Republican River during the Ice Age. At that time the Republican entered northwestern Republic County, as it does today, but it continued flowing easterly and veered toward the north, leaving Kansas near this point. A headwardly eroding stream in the present-day lower Republican River valley was extending its valley to the north, and eventually it captured the ancestral Republican near the town of Republic, about 12 miles west-southwest of here. This diverted the Republican to the south and accounts for the near-right-angle turn that the river makes in northwestern Republic County. The abandoned course of the ancestral Republican River, the Belleville Channel, was left high but not dry. Its 120 feet of alluvium and 100 feet of loess—the Belleville Formation—is today partially saturated with ground water and is an important source of water for municipal supplies and irrigation.

233.9   *Kansas/Nebraska state line.* Chester, Nebraska, is immediately north of here; Hubbell, Nebraska, is 6 miles to the east. The twin cities of Byron, Nebraska, and Harbine, Kansas, are 8 miles to the west.

69·U.S. HIGHWAY·69

From the Little Balkans to Kansas City

0.0 *Kansas/Oklahoma state line.* U.S. 69 runs along the eastern edge of Kansas, from Cherokee County north to Kansas City. During much of its route the highway follows the same path as an old military trail that connected Fort Leavenworth, Fort Scott, and Oklahoma's Fort Coffey. Today the highway carries coal trucks and travelers on their way to the Ozark Mountains of Missouri and Arkansas.

Northbound U.S. 69 enters Kansas in the midst of the Picher field of the Tri-State lead-and-zinc-mining district. Now inactive, these mines were once the world's largest producers of zinc. The first major discoveries of ore were made around Joplin, Missouri, in 1848 and 1849, but large-scale mining didn't begin until 1870, about the same time when ore was discovered in Galena, Kansas. Production gradually moved west into deeper mines in the Picher field, named for the town of Picher, Oklahoma, which is immediately south of the state line on U.S. 69. This field straddles the state line from Commerce, Oklahoma, east to Baxter Springs, Kansas. Although small mines began operating here in 1904, the richest part of the field, centered along U.S. 69, was not discovered until shortly before World War I.

One of the most important discoveries occurred less than a mile south of the state line in 1914. A driller was boring exploration holes without much success, when his company decided to abandon the project and ordered him back to Joplin. On the way, he got stuck in a mud hole. With nothing to do until help arrived, he decided to sink one more well. That hole struck high-grade zinc ore; additional holes located other rich ore bodies in the area, and mining soon followed. By 1917 the Oklahoma deposits were well defined, and mining moved into Kansas. The years from 1918 to 1941 saw the greatest production from the Tri-State district, but after World War II, production gradually declined until the last active mine in the district, two miles west of Baxter Springs, shut down in 1970 because of environmental and economic problems. During its 122-year history, the Tri-State district produced 2.8 million tons of lead and 11 million tons of zinc.

The principal lead and zinc ores in the district are sulfides: zinc sulfide, or sphalerite, and lead sulfide, or galena. These ores occurred in the limestones and cherts that were deposited during the Mississippian Period. Mississippian formations are exposed in southwestern Missouri, northeastern Oklahoma, and extreme southeastern Kansas, mostly east of the Spring River. Deposited more than 320 million years ago, these are the oldest rocks at the surface in Kansas. Ore-bearing formations appear at the surface in the Galena area; they dip gently to the west and are a few hundred feet below the surface along U.S. 69 in the Picher field, where they are buried beneath younger Pennsylvanian rocks of the Cherokee Group.

Ore bodies in the Tri-State district had irregular shapes: some were flat sheets; others were linear "runs"; others were circular. After they discovered an ore body, miners usually followed the ore, mining in a room-and-pillar fashion. Some of these rooms were quite large and had high ceilings. A room in one mine just east of U.S. 69 measured over 100 feet from floor to ceiling.

Each mine in the district had one or more vertical shafts, through which ore was brought to the surface and was then taken to nearby mills, where it was crushed and treated so as to separate the minerals from the gangue, or worthless portion of the rock. The gangue was mostly chert and limestone, which was ground into fine gravel and was piled into man-made hills over 300 feet high. This waste material, which locally is called chat, is no longer considered worthless. Many of the large chat piles have been quarried for use in concrete aggregate, road construction, ballast in railroad beds, and roofing material. The remnants of these large chat piles can be seen along U.S. 69 in Kansas and just to the south in Oklahoma.

Although the mines are now closed, many of the shafts remain open. Other mine shafts, and even the mines themselves, have collapsed over the years. Because of the size of the mine rooms, some of the resulting collapses have been large and dramatic. There is one such collapse a few

Underground lead and zinc mines along U.S. 69 in Cherokee County

hundred feet west of U.S. 69 just north of the state-line road. In fact, the first mile of U.S. 69 in Kansas is extensively undermined.

0.2   The road to the west leads 0.5 miles to *Treece,* a former mining camp and company town, which came into existence with lead and zinc mining in the area.

0.4   A few hundred feet east of the highway is a *mine collapse* that measures 150 feet wide, 300 feet long, and 50 feet deep—just one of many such collapses in the Tri-State district. A 1983 study by the state geological surveys of Kansas, Oklahoma, and Missouri located 307 mine collapses in Kansas, ranging from 10 to 500 feet in diameter and up to 100 feet deep. In addition, nearly 600 hazardous mine shafts were found and mapped.

0.5   The forested hill that rises 100 feet above the highway to the east is *Blue Mound,* which is capped by the Bluejacket Sandstone Member of the Krebs Formation. The Bluejacket is one of several sandstones that occur in the Krebs and Cabaniss formations of the Cherokee Group of rocks, which also includes shales and coal beds.

The Bluejacket is one of several sandstones that are important oil reservoirs in eastern Kansas, particularly in the small, shallow fields east of the Flint Hills. Drillers and geologists often give names to the subsurface sandstones which differ from the names applied to outcrops. In some cases, the names change from one area to the next because the sandstones are not widespread and are difficult to trace from one place to another.

To the west 1.5 miles is Tar Creek, which is named for seeps of thick oil that ooze to the surface along its course. This oil, which resembles tar or asphalt because the more volatile components have been lost to the atmosphere, is also encountered in the lead and zinc mines in the area. However, this thick, viscous oil—called heavy oil in the petroleum industry—is difficult to pump out of the ground. Thus, in spite of numerous indications of oil in Cherokee County, petroleum has never been produced in commercial quantities.

Today, Tar Creek is better known for a different kind of contamination: acid mine water. Many mines in this area took ore out of rock layers that were also aquifers, or water-bearing formations. Thus, water often entered the mines through these formations, and as many as 63 pumping plants operated around the clock to keep the water in check. Much of that water was dumped into Tar Creek, eventually making its way to the Neosho River. In 1947, more than 36 million gallons of water were pumped from the mines every day, enough to cover one acre of ground with water 110 feet deep.

When the mines closed, the pumping stopped, and the tunnels filled with water. By 1976, they had contained an estimated 100,000 acre-feet of water, which had become acidized through contact with iron sulfides (or pyrite) that remained in the mine walls or was left behind by miners. In 1979, water began to spill out onto the surface downstream in Oklahoma, creating bright red stains as it came into contact with the air and as the iron oxidized and precipitated out of solution. Mines in Kansas, especially those that have collapsed and have open shafts, provide openings for water to recharge the mines, thus continuing the seeping of mineralized water. Federal, state, and local agencies have grappled with the problem, but this remains one of the most environmentally blighted areas in the nation.

1.1   At this point, U.S. 69 passes over the *Foley Mine,* which is located along a southwest-northeast trending subsurface structure called the Miami trough. Named for the Oklahoma city (pronounced My-am-uh) along its trend, the trough is in some places a syncline, or downward fold in the rock layers. In other places, the trough is a graben. *Graben* is a German word meaning ditch, but in geology a graben is a long, crustal block that is bounded on either side by faults. This block has been down-dropped—that is, it is below its original position in relation to other rock layers.

Lead and zinc mines stretch for eight miles along the Miami trough, which runs through the heart of the Picher field in the Tri-State lead and zinc district. This trough may have played a major role in the deposition of lead and zinc, perhaps by providing an avenue for the upward movement of hot mineral-laden waters from deep in the earth's crust. However, geologists do not agree on the actual method by which ore accumulated in the Tri-State district. Because the ore-bearing rocks in the Miami trough have been down-dropped, the mines here are lower than those on either side. Directly below the highway, the Foley Mine is the deepest mine in the Tri-State district, 480 feet below the surface.

    2.2 Junction with *U.S. 166,* which runs west to the towns of Melrose and Chetopa. Between these two towns, parts of the highway are surrounded by groves of pecan trees, which thrive in the warmer weather of southern Kansas. The average temperature from 1941 to 1970 in southeastern Kansas was 57.6° Fahrenheit, compared to a statewide average of 55.2°. This is nearly five degrees above the average temperature in northwestern Kansas. As a result, the growing season in southeastern Kansas is nearly 200 days long, or 40 days longer than in the northwestern corner of the state.

    To the east, U.S. 166 runs to Baxter Springs. In 1863, an army garrison there was attacked by Quantrill's raiders, and eighty-seven men were killed, including many members of the garrison band. Some of those victims are buried at Baxter Springs National Cemetery, where a granite monument commemorates the event. Also at Baxter Springs, U.S. 166 intersects with U.S. 66, the famous highway that used to go from Chicago to Los Angeles. Route 66 cuts across about 12 miles of Kansas, running south out of Baxter Springs to Oklahoma, and northeast out of Baxter Springs to the towns of Riverton and Galena. Kansas is the only state traversed by U.S. 66 that isn't mentioned in the lyrics to the song "Route 66."

    Six miles to the east, U.S. 166 crosses the Spring River, which marks the western extent of the Ozark Plateau. Covering about 35 square miles in the southeasternmost corner of Cherokee County, the Ozark Plateau differs strikingly from the Cherokee Lowland, which U.S. 69 crosses in Cherokee County and most of Crawford County. The Cherokee Lowland is a flat erosional plain that developed on Pennsylvanian shales, silts, and sandstones of the Cherokee Group of rocks. This region is physiographically more subdued than is the Ozark Plateau; it has gentle slopes and shallow stream valleys. The only areas of topographic relief are isolated sandstone hills, such as Blue Mound, 1.5 miles to the south, which stands out as an erosional remnant above the surrounding terrain. The soft rocks of the Cherokee Group have weathered to form deep, fertile soils. Combined with the gentle, well-drained topography, this makes the area good for farming, except where the surface has been disturbed by mining.

An outcrop of Mississippian limestone in the Ozark Plateau region of Cherokee County

Timber is generally restricted to the slopes of hills, such as Blue Mound, to the banks of larger streams, and to abandoned mining areas.

On the other hand, the Ozark Plateau is an upland that forms the skyline to the east from here to Columbus, Kansas. As the name implies, this area resembles the Ozarks of Missouri and Arkansas. Being a plateau, the uplands are generally flat but slope slightly to the northwest, at about the same angle as the cherty Mississippian limestones of the region dip into the ground. The area receives more than 40 inches of precipitation in an average year, which percolates through the joints and fractures of these rocks, creating caverns and feeding springs and seeps, which in turn drain into clear-running streams that flow over gravel-lined beds in steep-walled valleys. These stream valleys produce topographic relief, which is much greater in the Ozark Plateau than it is in the Cherokee Lowlands. The highest point in Cherokee County, with an elevation of 1,040 feet, is located in the Ozark Plateau, just a few miles east of one of the lowest points, along the Spring River, where the elevation is 770 feet.

Much of the Ozark Plateau is mantled with cherty gravel, which was left behind by the weathering away of more soluble limestones in the bedrock. Thus, the soil here is often thin and rocky, which, together with the numerous steep slopes, makes much of the area unsuitable for cultivation. Deciduous hardwood forests cover most of the hillsides. Oaks and hickories dominate these woodlands, along with numerous other trees, shrubs, and vines. Some of these, such as sassafras trees and mistletoe, are usually associated with areas farther south and are not found anywhere else in the state.

The geology, climate, and vegetation of the Ozark Plateau produce a vastly different environment from the rest of Kansas, and this area is home to many animals that aren't found elsewhere in the state. At least six species of amphibians are limited to this part of Kansas, primarily because of the wet habitat provided by caves and springs. Those amphibians include the dark-sided salamander, the cave salamander, the graybelly salamander, the grotto salamander, the northern green frog, and the eastern nar-rowmouth toad.

Unfortunately, several of those salamanders are considered to be endangered species because of the recent destruction of their habitat. Endangered-species status is granted by the Kansas Fish and Game Commission. It means that the animals are protected, and a permit is required in order to collect them for scientific purposes. Biologists at the University of Kansas have proposed a nonrecreational natural sanctuary in Cherokee County so as to save some of the habitat of these animals and enhance their chances of survival.

    5.5 *Willow Creek.*

    6.3 *Bitter Creek.*

    7.0 About two miles to the east is the town of *Neutral*, which was once located in the Cherokee Neutral Lands, a tract that included much of Cherokee County and the area north of here. The Cherokee Indians, after whom this county is named, originally lived in Georgia, where they were one of the Five Civilized Tribes, with their own written language and a settled way of life as farmers. In 1827 they adopted a constitution and declared their independence as the Cherokee Nation, but eventually they signed a treaty, and most of them migrated to a 7-million-acre reservation in Oklahoma. Many Cherokees still live in northeastern Oklahoma, particularly in the area around Tallequah. In addition to the land in Indian Territory, an additional 800,000 acres in Kansas were given to the Cherokees in 1834. That area was called the Cherokee Neutral Lands, although the tribe never actually lived here.

The southern edge of Cherokee County belonged to another Indian tribe, the Quapaws, who moved there in 1833. Once Kansas became a territory in 1854, however, Indian lands came under pressure from squatters and railroads. By 1866 the Cherokees had ceded away the entire Neutral Lands to the government, and a year later the Quapaws surrendered claims to their lands in Kansas.

7.2 *Bluejacket sandstone.* U.S. 69 climbs onto a small escarpment, formed by this sandstone. In several places between U.S. 166 and Columbus, Kansas, this sandstone is poorly exposed in shallow road cuts such as this one. The Bluejacket is named for a small Oklahoma town, about 20 miles southwest of here, that was named after Charles Bluejacket, who came to Kansas with the Shawnee Indians in 1832. His grandfather was actually a white man who was captured by the Shawnees at the age of seventeen and was eventually made a chief of the tribe. His name was taken from the coat that he was wearing when he was captured, though his real name was Marmaduke Van Swearingen. The younger Bluejacket was a Methodist minister who was also elected a Shawnee chief; in Johnson County, a street and a park are named for him. Charles Bluejacket moved to Oklahoma in 1871, where he died in 1897.

11.4–12.3 *Columbus.* The seat of Cherokee County, Columbus was the site of a well in the 1870s that produced mineralized water with the taste and odor of hydrogen sulfide, or rotten-egg gas. This well was one of several mineral-water springs and wells that were popular in Kansas during the late 1800s. Nearby Baxter Springs was particularly famous for a series of ten springs, and by 1883 a hotel and bathhouse had been built there, although the enterprise had been largely abandoned by the turn of the century.

12.3 Junction with *K-7 and K-96.* This is the southern end of

K-7, which runs north to the Nebraska border. K-96 goes east to the Missouri border and west to Wichita, Hutchinson, and the Colorado border.

12.8 *Brush Creek,* which drains southeast to join the Spring River. Coal was first mined in southeastern Kansas in the 1850s, but in the 1870s a small amount was stripped from the west edge of this creek in eastern Cherokee County. Southeast is the town of Quaker, which was established by members of the Quaker religion in the 1800s.

13.1 *Strip mine* to the north.

15.4 *Krebs Formation.* Though shale is the principal component of the Krebs Formation, this outcrop is composed of sandstone. The formation also includes layers of coal and underclay, a type of clay that is found beneath coal layers in southeastern Kansas. Some underclays can withstand firing at high temperatures, and here in southeastern Kansas some underclay is used in manufacturing brick.

18.2 *Bluejacket sandstone* caps this small hill and crops out alongside the highway.

19.4 Junction of *K-96 and K-26.* This point is near the northern extent of the Tri-State lead-and-zinc-mining district in Kansas. Mines were located two miles to the southeast. Four miles to the southeast, a few houses mark the site of Badger, a former lead-and-zinc-mining camp located along the Spring River. Mines once operated on both sides and under Spring River, which flows over Mississippian limestones. Six miles to the northeast, abandoned lead and zinc mines can also be found near the town of Lawton and along the Missouri state line.

Ten miles southeast on K-26 is the town of Galena, for many years the center of lead and zinc mining in Kansas. Ore was discovered around Galena in 1870, the year when lead and zinc mining began in earnest just across the state line in Missouri. Major discoveries in Kansas occurred along Shoal Creek, south of Galena, in 1876 and along Short Creek in 1877. News of the Short Creek discovery spread rapidly, and within thirty days at least 10,000 prospectors and fortune seekers had poured into the Short Creek valley, giving birth to the rival towns of Galena and Empire City. Additional rich deposits were found south of Short Creek and west and southwest of Galena. These shallow deposits were found in the cherty limestone, which appears at the surface in this area. In some cases the ore was very near the surface and was mined "from the grass roots down," to use an expression popular at the time. The mining lots around Galena were the size of city lots—100 feet by 210 feet. Because they were so small and easily affordable, the area became known as a "poor man's mining district."

Spoils piles, cast-off rock from the days of lead and zinc mining, dot the landscape around Galena in Cherokee County.

The shafts and mines here were dug by small crews, often only two men using hand tools and simple hoisting devices that were powered either by man or by animal. The miners began by sinking a shaft; if they found ore, they drifted out horizontally, leaving rock pillars for support. However, if these pillars contained ore, they were often robbed when the ore in the rest of the mine had been exhausted. Lead and zinc production peaked here at the turn of the century, when Galena and Empire City had a combined population of nearly 13,000—the largest population center in southeastern Kansas.

Empire City no longer exists, having been annexed by Galena, which now has a population of about 3,500 and is surrounded by a wasteland of open mine shafts and collapsed mines. The shallow mine workings and the unstable nature of the rock overlying these mines—together with questionable mining practices of the early miners—have caused many of the mines around Galena to collapse. This process continues today, threatening parts of Galena that sit over long-abandoned mines.

Lead and zinc mines at Galena produced the chat piles and mine shafts that are visible today (courtesy of the Kansas State Historical Society).

19.8 *Shawnee Creek* drains south into Spring River, one of the few rivers that both enter and leave Kansas. That is, the river cuts through Kansas, but its source and mouth are outside of the state. In this case, Spring River's source is in Missouri, and its mouth is in Oklahoma, where it empties into the Neosho River. Other rivers that both enter and exit Kansas are the Cimmaron in far southwestern Kansas, and the Arkansas, which cuts across central Kansas.

21.1 *Little Shawnee Creek.*

24.1 *Strip mines* to the east. The coal that is mined here in southeastern Kansas comes from a series of rock formations called the Cherokee Group. The outcrop of coals from the Cherokee Group extends from Columbus, Kansas, northeastward into Missouri and Iowa. Coal-bearing rocks of the same age, but with a different name, occur in the coal fields of Illinois. The Cherokee Group is close to the surface here in southeastern Kansas, and strip mining is common, although no longer as common as it once was. This same formation is found under nearly all of the eastern third of Kansas, although it is generally much deeper. For example, in northeastern Kansas the coal is around 700 feet deep, but the scarcity of fuel in the early days of Kansas made it economically feasible to dig deep shaft mines to produce coal. During the 1880s and 1890s, at least five mines around Leavenworth produced coal from the Cherokee Group, including one mine that was operated by the state of Kansas at the State Penitentiary in Lansing. Prisoners dug the coal; eventually they tunneled under the Missouri River, following a coal vein. Today those deep shaft mines have been abandoned, and all the coal produced in Kansas comes from surface mines like these.

The strip mines here take coal from a layer called the Rowe (pronounced to rhyme with cow) coal bed, which is generally 12 to 20 inches thick. Although it was not mined extensively in the past, it is dug today in southeastern Crawford County and in northeastern Cherokee County, as well as just across the border in Barton County, Missouri.

24.4 *Old tipple* to the west. A tipple—used at a coal mine to screen, sort, and load coal—is the site where mine cars were tipped and unloaded.

24.5 *Strip mine* to the east.

24.7 *Long Branch Creek.*

26.5 On the east side of the road there is an old gas station, now a residence, which is constructed from sandstone, possibly from the Bluejacket Sandstone Member of the Krebs Shale Formation.

Two miles to the east, near Cow Creek, is a sinkhole that formed in 1921. This is one of several sinkholes in Cherokee County, east of U.S. 69, that have formed in the recent past. Some of these sinkholes contain evidence of "fossil sinkholes," which formed in the geologic past at the same location as the recent sinkhole. This indicates both an ancient process of limestone dissolution or cave formation and the eventual collapse of overlying rocks. This process occurs in the Mississippian limestones that underlie this area beneath a thin, shallow cover of Pennsylvanian shales, silts, and coals. Before the Pennsylvanian Period, these Mississippian limestones were exposed to the atmosphere, in much the way that they are exposed today in the Ozark Plateau of southern Missouri, northeastern Oklahoma, northern Arkansas, and the southeastern corner of Cherokee County.

Underground cavities are formed when precipitation percolates through organic-rich soil and becomes so acidified that it is able to dissolve limestone. As the water moves through the limestone, the joints and fractures become larger, eventually becoming cavernous. When the roof rock of one of these caverns fails, a sinkhole develops, and overlying rock and soil begins to sink into the cavern, like sand through an hour glass. This process began more than 300 million years ago, and it continues today.

The county road here also runs 7 miles west to the town of *Scammon*, which was named for four brothers from Illinois who moved to Kansas in the 1800s and became famous for coal mining. The Scammon brothers pioneered new methods of mining coal from underground, rather than from strip pits. Erasmus Haworth, one of the first directors of the Kansas Geological Survey, described the Scammon brothers' operation: "In 1874 Scammon Brothers sank a shaft just north of the old town of Stilson, now Scammon, from which shaft large quantities of coal were obtained. There was considerable opposition to this enterprise by their friends who doubted the expediency of such a method of mining, fearing that the nearness to the surface would cause the roof to break and crumble to such an extent that mining by the room and pillar system could not be employed. But the firm of Scammon Brothers persevered and from the start succeeded admirably, both mechanically and financially. From a shaft which began with two or three car loads a day they soon had it developed to a capacity of forty cars a day."

The Scammon brothers were mining coal from the Weir–Pittsburg coal bed at depths as shallow as 30 feet. The Weir–Pittsburg can be as much as four feet thick, although thicknesses of 36 to 38 inches are more typical. With its development in southeastern Kansas, the Weir–Pittsburg became one of the most extensively mined coal beds in Kansas history, producing over 200 million tons. Although there are no mines in the Weir–Pittsburg today, small amounts of the bed still remain as reserves.

Today's mining is done entirely in strip pits, but until 1960 there were underground coal mines in southeastern Kansas. Miners used a room-and-pillar method similar to that used by the Scammon brothers, taking out about 55 percent of the coal and leaving the rest behind as pillars to support the ground above. According to one report, coal production from strip mines surpassed underground mining around 1930.

27.5–28.2 *Strip mine* in the Rowe coal bed, which crops out in this area and was mined by surface methods. The Rowe is older than the Weir–Pittsburg.

28.2 While the Rowe occurs at the surface here, it dips below the Weir–Pittsburg in locations to the west, where it is too deep to be mined economically.

28.4 *Brush Creek,* which is crossed by a sturdy cement bridge, called a Marsh Bridge after the engineer who designed it.

28.5 K–103 to *Weir,* about four miles to the west. Weir (pronounced to rhyme with beer and named after the coal miner T. M. Weir) was one of the first locations where coal was commercially mined in Kansas. In fact, it gave its name to the Weir–Pittsburg coal bed, which has since been heavily mined in southeastern Kansas. An early report by the Kansas Geological Survey says that in the 1850s, Missourians came into the

Marsh Bridge at milepost 28 along U.S. 69 in Cherokee County

area and mined coal for use by blacksmiths. In 1866 a Missouri blacksmith and a Cherokee County resident named W. H. Peters teamed up to mine the coal from the surface. "The blacksmith led the way to an outcropping of the Weir-Pittsburg Lower at a point near the eastern suburbs of Weir city," wrote Survey director Erasmus Haworth. "Here with almost no labor at all they stripped a little soil from the upper surface of the coal and loaded their wagons rapidly by the use of pick and shovel."

30.4   Junction with *U.S. 160* and *K-57.* U.S. 160 runs across the southern edge of Kansas, from Pittsburg, in the east, to Johnson in the west. This highway, which is described in chapter 1, cuts through some of the state's most scenic geology, including the Flint Hills and the Red Hills. K-57 runs five miles east to the small town of Opolis, which was once called State Line. It is bounded on the east by the Kansas/Missouri border.

This is also the *Cherokee/Crawford county line.* Crawford County is named after Samuel J. Crawford, who served as governor of Kansas from 1865 to 1868, when he resigned to lead a military expedition against the Indians. He was governor of the state when Crawford County was created out of segments of Bourbon and Cherokee counties.

33.3   *Cow Creek.*

33.5   At the south edge of *Pittsburg,* U.S. 69 loops west around the city while Alternate 69 becomes Broadway and runs through the downtown. Pittsburg was named after the city in Pennsylvania, which is also in a rich coal-mining region, although in Kansas the silent *h* on the end of the name was dropped. Today, Pittsburg is a regional trading center for southeastern Kansas, and its 18,000 population makes it the largest city in this corner of the state.

34.6 *Quincy Avenue* in Pittsburg. About a mile to the southeast is Pittsburg State University, one of the six state-supported universities in Kansas and the only one in southeastern Kansas.

35.2 *Strip mines* on both sides of the road. The Weir–Pittsburg coal bed is mined in the pits on the west and north sides of Pittsburg; like all coal mined in southeastern Kansas, it is bituminous, a soft coal that burns with a yellow flame and is a lower grade than anthracite, or hard coal, which is not found in Kansas. Coal is formed from the remains of plants that thrived in the swamps along the coastlines of Pennsylvanian seas. As those plants were buried under other sediments, the heat and pressure changed the organic material into coal. Had the Kansas coals formed under greater pressure and heat, they would have been anthracite coals instead of bituminous. A lesser grade of coal, lignite, is found further west in Kansas, in rocks of Cretaceous age.

36.0 *First Cow Creek* to the west.

36.5 *Lincoln Park* and the *Pittsburg golf course* to the east. The golf course is located on land that was once strip mined and has since been reclaimed. To get at the coal, miners removed the rock and soil above it, which is called overburden. Before widespread land reclamation in the 1960s, shovels removed the overburden and created trenches up to 100 feet wide and 50 feet deep. The land that they left behind looked like it had been cultivated with a giant plow; it was usually considered to be useless and was left to grow back to trees and brush while the trenches filled with water. In 1969 the Kansas Legislature required coal companies to level the newly stripped land and to plant it with trees and grass. Later, more stringent federal regulations were passed, so that today's strip mines are made into useful, productive land. One problem, though, is the amount of pyrite in the coal. This pyrite, which is removed as the coal is mined, can increase the acidity in water; some of the reclaimed ground may still have highly acid spots that are difficult to cultivate.

37.4 *Crawford County Historical Museum*, with an old coal-mining shovel outside, is located to the west. Commercial coal production in southeastern Kansas dates back to the 1860s, when coal was used in blacksmith shops and for domestic purposes. Steam shovels were first used to strip-mine coal near Pittsburg in 1876, but their size restricted their use to depths of 12 feet or less. In 1878 and 1879 the Joplin and Girard Railroad

A deep-shaft coal mine near Pittsburg around the turn of the century (courtesy of the Kansas State Historical Society)

was built through this area to carry coal from the Kansas mines, and the town of Pittsburg was founded. By the late 1800s the coal and natural gas in southeastern Kansas had fostered a growing zinc-smelting industry, and by 1900, half of all the country's zinc was smelted in Kansas, much of it in Pittsburg, where eight smelters operated for a time and made zinc smelting the largest nonfood industry in the state.

Coal production reached a peak of 7.2 million tons in 1917 and 1918. Much of that coal came from Crawford County, and most came from deep shaft mines, rather than from strip mines. Today, southeastern Kansas coal production totals about 1 million tons annually. With the zinc smelters gone and with trains running on diesel fuel, the most common uses for coal are electrical generation and providing fuel for cement plants. Much of the coal mined in this area is trucked to electrical power plants and to cement plants in eastern Kansas and western Missouri.

38.1 The north edge of *Pittsburg*. U.S. 69 loops west around Pittsburg, while Alternate 69 runs through the downtown.

39.1 Junction with *U.S. 160*, running east to Springfield, Missouri. Frontenac is immediately east of the highway. Although the town was named for a French explorer, its population is made up of several ethnic groups, including Italians, Germans, and Slovaks. Many of those nationalities moved into southeastern Kansas to work in the coal, lead, and

Underground coal mines along U.S. highways 69 and 160 north of Pittsburg

zinc mines. This part of the state became known as the Little Balkans, and it retains its ethnic flavor to this day.

40.1–41.1 *Crawford County State Park,* to the east, is located in land that was strip mined to remove coal from the Mineral coal bed, the next major coal bed above the Weir–Pittsburg. These old mines have been allowed to fill up with water, and here they provide excellent fishing, especially for bass. The Kansas Fish and Game Commission also maintains a small herd of buffalo here. While strip mines are the most visible reminders of the coal industry, northbound U.S. 69 travels over underground mines for the next 7 miles. The only visible reminders of these mines, dug in the Weir–Pittsburg coal bed, are conical waste piles near the mine shafts.

43.0   To the east is a waste pile that marks the old mine shaft for Jackson and Walker Coal Company's *Mine No. 17.*

43.2   *K–57* runs west to the town of Girard, the seat of Crawford County. The small town of Ringo is about 3 miles west of here. Area towns have picturesque and unusual names, including Red Onion, Scotts Camp, Camp 50, Capaldo, Coalvale, Gross, Croweburg, and Litchfield, the first coal camp in Crawford County. More than than 100 camps were established in southeastern Kansas, most of them between 1890 and 1910. Most of them were company towns; some included hotels, saloons, churches, schools, and a company store. They were often situated near mine shafts or at the junction of section roads. Names for the towns came from a variety of sources: Capaldo was named for the owners of land where a mine was sunk; Midway, because it was halfway between Fort Scott and Baxter Springs along a stage line; and Dogtown, for the number of dogs at the camp.

43.4   *Junction with Alternate Highway 69,* which goes north through the towns of Franklin and Arma.

43.8   To the west is a waste pile that marks the shaft of Western Coal and Mining Company's Mine No. 16.

45.3   *Arma,* immediately east of here, was named for coal miner Arma Post. East of Arma is an area that has been extensively mined. Strip mining continues there today. An old mining town called Camp 50 is two miles west of here, and the hill three miles to the east is called Breezy Hill.

46.2   Junction with *Alternate Highway 69,* which runs south through Arma and Franklin.

46.4   Two miles to the east is *Croweburg,* a small mining town that gave its name to the Croweburg coal bed. The Croweburg, which occurs above the Mineral coal bed, has been strip mined throughout this area, though it is generally less than 15 inches thick around here. The Croweburg is found throughout the United States, extending from central Oklahoma, through Kansas to Iowa, and eastward to eastern Pennsylvania and central West Virginia. The Croweburg coal bed and its eastern equivalent, the Colchester coal bed, represent one of the largest coal swamps that ever existed.

Strip mining coal east of U.S. 69 near Arma in Crawford County

46.5   *Missouri River/Arkansas River drainage divide.* The elevation is 1,000 feet. Runoff north of this point eventually reaches the Missouri River, while runoff to the south flows into the Arkansas River.

47.1   For the next eight miles to the south, from here to Frontenac, U.S. 69 passes over underground mines that have been dug in the Weir–Pittsburg coal. These mines are from 75 to more than 200 feet below the ground.

48.2   *Dry Branch of Cox Creek.*

49.0   *Englevale* is about a mile to the west.

50.0   At about this point the highway passes over the crest of *Bunker Hill,* which extends to the northeast. The elevation is 970 feet. Bunker Hill, as well as the hills along U.S. 69 to the north and south for four or five miles, is capped by the Fort Scott Limestone, which overlies the soft coal-bearing silts and shales of the Cherokee Group of formations.

50.4 and 50.6   *Fort Scott Limestone,* which is generally about 33 feet thick.

51.5   Just west of the highway is a grassy hillside that in 1981 was a *strip mine* where the Bevier coal was extracted. The area has since been reclaimed as a pasture.

51.7   *Fort Scott Limestone* on the east side of the road.

51.9   *Contour mine* in the Mulky coal bed. A contour mine is a strip mine that is common in hilly or mountainous terrain, such as West

Virginia. It follows a coal bed around a hill, much as contour plowing does, leaving behind a mined-out bench when the mine is completed.

52.2 and 52.4   A shale layer in the *Cabaniss Formation.* This formation includes many of the coal beds in southeastern Kansas, but the shale layers in the Cabaniss have not been named.

53.2   *Strip mines* in the *Bevier coal bed.*

53.5   The *Breezy Hill Limestone Member* of the Cabaniss Formation, which occurs just below the Mulky coal and is named for Breezy Hill just east of Arma.

53.7   *Strip mines* in the *Bevier coal bed.* The Bevier is a bright, hard, blocky coal that occasionally contains pyrite and calcite.

54.0   Two miles to the west, a few houses mark the site of *Cato.* Dating back to the 1850s, this is the oldest town in Crawford County.

54.4   *Crawford/Bourbon county line.* Bourbon is a French name, but this county was probably named after Bourbon County, Kentucky. Bourbon was one of the first thirty-five counties in the state to be named by the 1855 legislature. Elected primarily by Missourians who crossed over the state line to vote illegally, this legislature had pronounced Southern sympathies, which are reflected in many of the place-names it produced.

54.8   *West Fork Dry Wood Creek.*

About a mile southwest there used to be an old post office called Dry Wood.

55.4   *Strip mines* to the west of the highway are in the Bevier coal bed.

55.6   *Walnut Creek* empties into Dry Wood Creek just southeast of here. Dry Wood Creek joins the Marmaton River west of Nevada, Missouri, and the Marmaton flows into Harry Truman Reservoir.

57.2   *Strip mines* on both sides of the highway are in the Mulky coal bed, which is generally about 1 foot thick.

58.6   The south edge of an escarpment, or long ridge, formed by the *Pawnee Limestone Formation.*

59.3   The quarry to the east is in the *Pawnee Limestone Formation,* a layer of limestone and shale that ranges in thickness from 15 feet to 60 feet.

61.0   *K-7* runs southwest to Girard, the seat of Crawford County. Headed north, it briefly joins U.S. 69 before diverging again at Fort Scott. One-half mile to the southwest on K-7 is Godfrey, a town that once had a population of 1,200 and rivaled Fort Scott as a coal-shipping center. Today it is little more than a crossroads.

63.7 The north edge of an escarpment formed by the *Pawnee Limestone*.

64.5–66.2 *Buck Run* is an old creek that parallels U.S. 69 through the city of Fort Scott before emptying into the Marmaton River. This is also about the southern edge of Fort Scott. A half-mile to the east is a national cemetery, established in 1862 as a 10-acre burial place for United States soldiers.

65.0 *Fort Scott Limestone* on the west side of the highway.

65.2 *Fort Scott Limestone* on the east side of the highway. The Blackjack Creek Limestone Member of the Fort Scott Limestone is a natural cement that crops out in this area and has been extensively quarried, supplying cement for an industry that began in 1868. This impure limestone naturally contains all of the necessary ingredients for cement. It needs only to be heated in a kiln, ground up, and moistened to form a cement.

66.2 *Wall Street.* Immediately west of here are a tourist information center and the Fort Scott National Historic Site. Restored by the National Park Service, this is the site of the original fort, established in 1842 as an army base on the Indian frontier and situated on an old military trail that ran along the eastern edge of Kansas.

The fort was abandoned in 1855 as the frontier moved westward, but because of its proximity to the Missouri border, the town of Fort Scott continued to be the focus of political activity, primarily between Free Staters and proslavery forces in their fight over Kansas' entry into the Union as a free or slave state. According to some sources, the term *jayhawker* originated in this area. It was used to describe Free Staters who attacked proslavery Missouri farmers. Originally the term denoted those abolitionist fighters, but eventually it stuck as a name common to all Kansans. Today the jayhawk is probably best known as the mascot of the University of Kansas.

With the onset of the Civil War, Fort Scott was brought back into service and was used as a supply base for Union troops as far south as the Red River. With the end of the war, the fort, which was named for Gen. Winfield Scott, was again abandoned. Today the military barracks, parade grounds, and munitions-storage area are open to the public.

66.6 *The Marmaton River,* which marks the north edge of Fort Scott, the seat of Bourbon County and one of the oldest towns in southeastern Kansas. Although the river was named by French explorers, the meaning of the name isn't completely clear. From here it runs east into Harry Truman Reservoir. Truman spent much of his adult life in Independence, Missouri, but he was born in nearby Barton County, Missouri.

The town of Marmaton is 7 miles west of Fort Scott. For a time during the Civil War, Marmaton served as the seat of Bourbon County, so that county government would be farther away from the Missouri border. On 22 October 1864, Marmaton was attacked during the retreat of the Confederate army of Sterling Price. Price's troops had earlier been repulsed at the Battle of Mine Creek, and they were fleeing south when camp followers attacked Marmaton, killing six people, wounding several others, and burning a number of buildings in the town, which had been left largely defenseless when most of its men had joined the Union Army.

67.1 *U.S. 54* runs west from here through the Flint Hills and on to Wichita, Pratt, Liberal, and Tucumcari, New Mexico, before terminating in El Paso, Texas. To the east it leads to Nevada, Missouri, the Lake of the Ozarks, and on to Illinois.

67.8 Shale in the upper part of the *Cabaniss Formation* is exposed on both sides of the road in the walls of a former strip mine in the Mulky coal bed.

68.0 *Fort Scott Limestone* on both sides of the highway.

68.7 *Little Osage shale* overlain by the *Higginsville limestone,* the upper two members of the Fort Scott Limestone. The Little Osage shale is black and platy and contains small phosphatic concretions.

70.0, 70.5, and 71.3 *Pawnee Limestone.*

71.8 *Wolverine Creek.*

73.1 About a mile to the east is the small town of *Hammond.*

73.0–74.0 *Longhorn cattle* are raised on a ranch that borders the highway in this vicinity. Though Longhorns were once common on cattle drives in eastern and western Kansas, they are much rarer today, largely having been replaced by beefier cattle breeds. In recent years Longhorns have been making a comeback. When crossed with larger breeds, they produce smaller calves, which results in easier births with fewer losses. Longhorns ran wild in the dry scrubland of southern Texas in the years before the Civil War; they survived on food that was considered unpalatable by other cattle breeds. They do well on native grass pastures, even those of marginal quality.

73.1–73.4 An *unconformity* occurs at the irregular contact between the Perry Farm shale and the Idenbro limestone members of the Lenapah Limestone below and the brown Hepler Sandstone Member of the Seminole Formation above. This unconformity represents a gap in the geologic record during which sediment deposition ceased and erosion occurred. Sedimentation resumed with the deposition of the Hepler sandstone. In places, the erosional surface at the base of the Hepler sandstone cuts completely through the underlying Idenbro limestone. Because the units above and below this unconformity are parallel, it can be called a disconformity.

73.9 Sandstone in the Nowata Shale. This sandstone is probably the Walter Johnson Sandstone Member, which crops out in Montgomery County and is named after Walter Johnson School, east of Coffeyville. Walter (Big Train) Johnson, who was born in Humboldt in nearby Allen County, is probably the most famous Kansan in the history of major-league baseball. Pitching for Washington from 1907 to 1927, Johnson compiled a 414–281 record. He spent the off-season in Coffeyville.

Other Kansas natives, many from southeastern Kansas, also made their mark in baseball. Steve Renko, Bill Russell, Mike Torrez, and Ray Sadecki all came from Kansas, as well as managers Gene Mauch and Ralph Houk. By 1985 a total of 161 Kansans had reached the major leagues; many were from large cities such as Wichita, Topeka, and Kansas City, but an inordinate number came from small towns here in southeastern Kansas. La Cygne, Frontenac, La Harpe, and Galena all had representatives in the majors. Even the tiny town of Weir produced three major-league ballplayers. Many of these southeastern Kansans played during the late 1800s and early 1900s, as did Walter Johnson.

75.0 *Strip mine* to the east of the highway in the Mulberry coal bed, a layer in the Bandera Shale Formation. This coal averages slightly more than 2 feet in thickness, but in places it is as much as 4 feet thick. In terms of strippable reserves, it is one of the most important coal beds in Kansas. Along southbound U.S. 69, this is the first old coal mine that is easily visible from the highway. Much-larger strip mines are still visible northeast of here, near Prescott, and throughout the area around Fort Scott, Pittsburg, and Columbus.

77.1 K–31 runs west and north to the town of *Fulton*, which was named after Fulton County, Illinois.

77.4–77.8 *Labette Shale* overlain by *Pawnee Limestone*. The Englevale Sandstone Member of the lower part of Labette Shale occurs at the northern end of the road cut.

78.2 *Little Osage River*. The Little Osage shale was named for exposures along the banks of this stream.

79.0 *The Bandera Shale Formation* contains layers of shale and sandstone that range from 20 to 50 feet in thickness. Sandstone in the lower part of the formation sometimes contains plant fossils. The Bandera Quarry Sandstone Member of the Bandera Shale was quarried west of Fort Scott, near Redfield, for flagstone. Flagstone is thin-bedded rock, usually sandstone, that naturally separates along bedding planes to form large, flat slabs that can be used for sidewalks, patios, and house siding. Flagstones from the Bandera Quarry sandstone are known locally as the Fort Scott flags.

79.1   West of the highway is Freedom Township, and 5 miles to the west is the site of Freedom Colony, an industrial commune that grew up in the economic hard times of the 1890s and then disbanded shortly after the turn of the century. Members of the colony cooperatively operated a coal mine, a warehouse, and a farm; but they probably drew the most attention in 1900 when 1,000 people gathered to watch the laying of the cornerstone for a flying-machine factory. The factory was completed, but the business's prototype never got off the ground, and neither did the business.

80.1   *Linn/Bourbon county line.* Linn County was named for a Missouri senator. Like Bourbon County to the south, Linn was one of the original thirty-five Kansas counties named by the proslavery Bogus Legislature of 1855. The states of Virginia, Kentucky, and Missouri were the sources for many of those names, some of which were later changed by more pro-Union legislatures. Miami County to the north, for example, was originally named for a Southern sympathizer, but the name was later replaced.

80.3–80.6   All three members of the *Altamont Limestone* are exposed in this road cut: the Amoret limestone, overlain by the Lake Neosho shale, overlain by the Worland limestone.

81.9   The *quarry* east of the highway is in the Pawnee Limestone. The overlying Bandera Shale and Altamont Limestone are exposed in the quarry walls.

82.1   K–239 runs a mile east to the town of *Prescott.*

84.8   *The Ladore Shale Formation,* overlain by the Middle Creek limestone, Hushpuckney shale, and Bethany Falls limestone members of the *Swope Limestone Formation.* The Ladore Shale crops out in the Kansas City area, where it is a thin, marine shale. In southeastern Kansas the formation is thicker and sandier, probably deposited in the delta of a Pennsylvanian river. In Labette County, the Ladore includes thin beds of coal that contain numerous plant fossils.

85.1   *Swope Limestone.* This formation, common in the Kansas City area, ranges up to 35 feet in thickness.

85.2   Limestone in the *Tacket Formation.*

86.4 *Hepler Sandstone Member* of the Seminole Formation. The Hepler is a brown and grey sandstone that ranges from 2 to 25 feet in thickness.

87.0 *Mine Creek* drains to the northeast, through an area of strip mines in the Mulberry coal bed, before joining the Marais des Cygnes River. The creek got its name from a much older mine, located near its banks about two miles to the northeast. This mine produced lead and zinc long before these minerals were discovered in the now-famous Tri-State mining district to the south, but the exact date of this mine and the identity of the miners remain a mystery. Indians told early settlers about the presence of lead ore, and the settlers found evidence of an old mine dump. Growing on the dump were oak trees that were four to six inches thick, estimated to be at least 25 years old. The settlers estimated that at least 25 more years must have passed between the cessation of mining and the time the trees had established themselves on the waste piles. Early French explorers or Indians may have opened the mines, and several mining ventures have since operated at this site, which is called Big Jumbo. Shafts as deep as 250 feet are said to have been sunk. One operation in 1901 produced 15 tons of high-grade lead ore, which was shipped to a refinery in the Argentine district of Kansas City, Kansas.

Mine Creek is known today for the Civil War battle that took place along its banks west of here in 1864. About 2,500 Union soldiers were pursuing 7,000 Confederate troops, who were retreating south after a loss at the Battle of Westport, near Kansas City. The Confederates were hoping to attack Fort Scott when the Union forces caught up with them here. The Confederates were beaten—about 300 Confederates died, compared to only a few Union losses—and continued to flee southward. The engagement was important, because it protected Fort Scott and forced the Confederates out of Kansas. It was also one of the last significant Civil War engagements fought west of the Mississippi.

87.6 *K–52* runs about six miles west to Mound City. About a mile west of here is the *Mine Creek Battlefield,* a site that is operated by the Kansas Historical Society and commemorates the largest Civil War battle in Kansas.

89.0–91.0 *Pleasanton,* the seat of Linn County, is named after Gen. Alfred Pleasonton, one of the Union commanders at the Battle of Mine Creek. About a mile north of Pleasanton is an abandoned quarry, where asphalt rock was once mined from the Hepler Sandstone Member of the Seminole Formation. If the rock had been more deeply buried, it would have

The Marais des Cygnes waterfowl area in Linn County

contained oil. However, because erosion has brought it near the surface, the more volatile fraction of the oil has escaped to the atmosphere, leaving behind natural asphalt. Similar deposits can be found in southwestern Missouri. The primary use of asphalt rock is in paving roads. Since it is a sandstone impregnated with asphalt, this rock only has to be crushed to make a usable paving material.

92.1 *Muddy Creek.* East of the highway is a waste pile marking the site of a shaft for an underground coal mine that exploited the Mulberry coal bed of the Bandera Shale.

94.6 *Big Sugar Creek.* Many of the marshes and man-made lakes that surround the Marais des Cygnes River in this area are part of the Marais des Cygnes Waterfowl Area, which is managed by the Kansas Fish and Game Commission to provide habitat for many bird species.

95.6 *Marais des Cygnes River,* which was named by French explorers and means ''marsh of swans.'' After the Kansas and Neosho rivers, the Marais des Cygnes (pronounced mare-de-zeen') carries more water out of the state than does any other stream. Visible in the stream bed east of the river is a series of rapids that formed in the Lenapah Limestone Formation.

95.7 *Trading Post,* established in 1834 by a French trader, was used as a post for trading with Indians. It was also, for a time, the residence of John Brown, who was in Kansas for much of the time from 1855 to 1859.

96.3   The isolated hills in this broad valley are outliers capped by the Hertha Limestone, the lowermost layer in the Kansas City Group of formations. Erosion has separated these hills from a prominent escarpment, or ridge, that is north and west of here. That escarpment is capped by Hertha, Swope, and Dennis limestones, which are in the lower part of the Kansas City Group.

97.4   To the west, this county road runs along the edge of the *Marais des Cygnes Waterfowl Area* to the town of *Boicourt,* which was named for a local French settler. About four miles northeast of here is the site of the Marais des Cygnes massacre, perhaps the most famous event in the pre-Civil War dispute between slavery and abolition forces in Kansas. In May 1858, about thirty Missourians came into the area; they captured eleven Free Staters and then marched them into a ravine, where five were killed and five were wounded. One escaped by feigning death. Although this was just one in a series of altercations between the two sides—because of its proximity to the Missouri border, this area was particularly vulnerable to armed conflict—the Marais des Cygnes massacre was important because it received extensive coverage in Kansas newspapers and throughout the country. John Greenleaf Whittier wrote about the event in a poem that was published in the *Atlantic Monthly* in September 1858. Today the site is part of a state memorial park.

99.2   *North Sugar Creek.* The hill to the east is capped by the Hertha and Swope limestone formations.

102.4   K-152 leads west four miles to *La Cygne* and east to La Cygne Lake and nearby *La Cygne Generating Station.* The power plant, which uses the lake for cooling water, is owned by Kansas Gas and Electric Company in Wichita and by Kansas City Power and Light Company in Kansas City. In addition, a large strip mine, immediately southeast of the plant, produces coal for generating electricity. That mine is operated by the Pittsburg and Midway Coal Company, the largest coal-mining company in southeastern Kansas. In this mine, located on an 8,000-acre tract straddling the Kansas/Missouri border, up to 90 feet of overburden has been removed to reach the Mulberry coal bed. Because the coal is high in sulfur, this plant has been specially equipped with "scrubbers," which remove much of the sulfur from the emissions.

102.6   *Tacket Formation,* which contains layers of shale, bluish-gray limestone, and massive sandstone.

103.0   In this road cut, shale in the *Tacket Formation* is overlain by the Critzer limestone, Mound City shale, and Sniabar limestone members of the *Hertha Limestone Formation.*

103.2   *The Swope Limestone Formation* is composed of the Middle Creek limestone, the Hushpuckney shale, and the Bethany Falls limestone members. The Swope is named after Swope Park in Kansas City, Missouri; it is common from here north to Kansas City. The most prominent member of the Swope is Bethany Falls limestone, which crops out in the eastern part of Kansas City and dips underground beneath much of the city, where it has been heavily mined in both Missouri and Kansas. The abandoned mines are now being used for storage, for offices, and even for manufacturing.

104.0–105.2   *Dennis Limestone Formation,* which is named for the small town of Dennis in Labette County, where it is up to 60 feet thick.

105.4   *Miami/Linn county line.* Miami County was originally named after David Lykins, a missionary to the Indians who was a member of the Bogus Legislature and a representative to the Lecompton Constitutional Convention, which included slavery as part of the proposed constitution for Kansas. By 1861, opposition to slavery was so strong in Kansas that the county was renamed, this time after the Miami Indians who had been moved into Kansas from Ohio, where they had been settled after their original removal from New York. Miami is a French name for the tribe, which settled here in the Marais des Cygnes area before moving once again to Oklahoma.

106.4   The small town of *Jingo* is about 0.5 miles west of here.

107.2   *Swope Limestone.*

108.2–108.7   *Dennis Limestone.*

110.5   *New Lancaster* is about 2.5 miles west of here.

111.5   *Drexel Corner.* Drexel, Missouri, is 9.5 miles east on this county road. To the west is the town of Osawatomie—whose name was derived from the combination of Osage and Potawatomi. It is probably most famous as a residence of John Brown during much of his stay in Kansas. Five of Brown's sons lived near Osawatomie; he came here in October 1855, and quickly became a leader of antislavery forces. Brown left the territory in 1858, the same year that he led an attack on the federal arsenal at Harper's Ferry, in what is now West Virginia. During much of his stay in Osawatomie, Brown lived with his brother-in-law, Rev. Samuel Adair. That cabin is preserved today as part of John Brown Memorial State Park.

112.3   *Iola Limestone Formation.*

113.2   *Cherryvale Shale,* overlain by *Drum Limestone.*

113.3 *Middle Creek,* which gave its name to the Middle Creek limestone, the lowest layer in the Swope Limestone Formation.

116.4–116.6 *Wyandotte Limestone.* An apparent syncline is visible in the road cut on the west side of the highway. This downward fold in the rocks may have been caused by the solution of limestone below the surface, followed by the slumping of rocks into the cavity. The syncline formed in the subsurface and was subsequently exposed during highway construction. Evidence for the dissolution of limestone at this outcrop can be seen in the expanded joints, which have been enlarged by ground water and then filled with reddish-brown soil.

117.1–117.5 *Wyandotte Limestone,* another formation that is especially common in Kansas City, particularly in exposures along the Kansas River.

117.2 The elevation is 1,100 feet as the road passes through the *Louisburg oil field,* which is named for the town to the north. The Louisburg was discovered in 1927. Like many of the older oil fields in southeastern Kansas, it produces less oil than it once did. Although it covers much of eastern Miami County and includes 107 producing wells, it pumped only 131,000 barrels of oil during 1984, an average of 3.3 barrels of oil per day per well. Wells that produce less than 10 barrels per day are called "stripper" wells, because they strip the last oil out of a rock formation. The majority of wells in southeastern Kansas are stripper wells.

The wells in the Louisburg field are between 270 and 600 feet deep. The shallowest producing zone for these wells is the Knobtown sandstone, which occurs in the upper part of the Tacket Formation. The Tacket is exposed along U.S. 69 about 15 miles to the south. Oil wells in western Kansas are generally much deeper, although many of them produce oil out of the same formations as do these wells in Miami County. The same formations are much deeper in western Kansas, buried beneath additional accumulations of sediments that aren't present in eastern Kansas. That is why the pump jacks here are so small—the shallow oil doesn't need to be lifted as far as in western Kansas.

118.0 *Wyandotte Limestone.*

118.3 *Wyandotte Limestone.* The limestone exposed on the west side of the road shows an apparent dip to the south.

118.8–119.6 *Wyandotte Limestone.* The Wyandotte is one of twelve formations that make up the Kansas City Group of formations. Many of those formations crop out in the area around Kansas City, and some of them—like the Wyandotte and the Swope limestones—are named for locations in Kansas City.

A typical Pennsylvanian limestone along U.S. 69

119.8   A branch of *South Wea Creek*. Wea Creek is a notable name in the history of oil exploration in Kansas. Along its banks, about 0.5 miles east of Paola, a group of drillers put down the first wells in search of oil in Kansas. The wells were drilled in 1860, and the original investors included David Lykins, an early member of the Kansas Legislature. The wells produced mostly salt water, along with traces of oil, but the first Kansas oil-exploration company and its drilling venture were brought to an end by the Civil War. Before the war was over, two investors had been killed; two more, who had Southern sympathies, had left the company; and others, who lived in Lawrence, were financially ruined by Quantrill's raid in 1863. Exploration began again in 1865, when several wells were drilled near this area and in other parts of eastern Kansas. Many of those wells struck natural gas, and one field two miles west of here produced natural gas that was piped to Paola in 1884, making it the first city in Kansas to be supplied with natural gas.

The discovery of oil in this part of Miami County was no surprise. Oil and tar seeps here had long been known to Indians and early settlers, who occasionally used the oil for lubrication. The largest of these seeps, the Wea tar spring, was located about two miles west of here along South Wea Creek. In places the oil seeped so much that it impregnated the surrounding rock, creating formations that are popularly known as asphalt rock.

Geologists agree today that oil seeps are generally a good sign of places to drill for oil, and that's what those first drillers found. George C.

Swallow, the second director of the Kansas Geological Survey, said that the oil seeps were "sufficient to convince anyone familiar with the indications and developments of petroleum in the productive oil regions of the country, that it must exist in large quantities in this county." Probably because of those oil seeps and because of the county's location in eastern Kansas, Miami was the first county in the state to undergo detailed geologic study by early staff members of the Kansas Geological Survey.

120.4 *Lane Shale* overlain by *Wyandotte Limestone.*

120.4–121.8 *Wyandotte Limestone.*

121.6 *Louisburg* was originally named St. Louis, but the name was changed to avoid confusion with the larger city in Missouri.

121.8 *Lane Shale* overlain by *Wyandotte Limestone.*

122.1 *South Wea Creek.* The name Wea, pronounced We'-uh, comes from an Indian tribe that lived in the Ohio River Valley and was moved to a reservation here in eastern Kansas in 1833.

122.5 and 123.5–124.7 *Wyandotte Limestone.*

125.8 *Plattsburg Limestone.*

127.4 *Vilas Shale.*

127.8 At this point the road crosses a boundary line representing the south edge of the *Shawnee Reserve,* the reservation in the Kansas City area that was given to the Shawnee Indians when they moved into Kansas.

128.3 *Wea Creek.* Nearby is the small town of Wea, whose most prominent building is the Holy Rosary Catholic Church.

128.8 The exit from U.S. 69 to the towns of *Wea* and *Bucyrus* (pronounced byoo-sigh'-rus), which was named after a town in Ohio and is about two miles west of here.

129.7 *Stanton Limestone* on the east side of the road. This is also the *Johnson/Miami county line.* Johnson County was named after Rev. Thomas Johnson, a Virginian who moved to Kansas in 1829 and established the Methodist Shawnee Mission, just on the Kansas side of the border in what is today metropolitan Kansas City. Johnson was killed in 1865; he is buried in a cemetery in Fairway, Kansas. Johnson County is the second most heavily populated county in Kansas, behind only Sedgwick.

131.5–131.7 *Plattsburg Limestone.* The Plattsburg has an average thickness of about 25 feet, although it is as much as 115 feet thick in Wilson County.

131.7 The towns of *Aubry* and *Stilwell* are just east of here.

132.7 In an abandoned quarry east of the highway, the Wyandotte Limestone reaches a maximum thickness of 60 feet. The Wyandotte

Many of the limestones in eastern Kansas have natural joints, such as this formation along U.S. 69.

varies in thickness from place to place; 50 miles to the southwest, in Anderson County, it pinches out and disappears completely. Where present, it has been quarried for use in aggregate and in cement.

132.2   *Plattsburg Limestone.*

132.9–133.9   *Wyandotte Limestone.*

133.4   Here the highway passes through a mature stand of evergreens on a wooded hillside. These trees are commonly known as red cedars, because of their reddish tint in winter. However, they are actually junipers; they are the only coniferous (or cone-bearing) evergreen trees native to Kansas. They are found throughout much of Kansas, but they prefer rocky limestone hillsides, such as this one formed on the Wyandotte Limestone.

134.3–134.7   *Wyandotte Limestone.*

134.8   *The Blue River* begins 0.3 miles to the southwest, at the juncture of Wolf and Coffee creeks. The Blue then flows northeast through the southern and eastern parts of Kansas City, Missouri, before joining the Missouri River near Sugar Creek, Missouri. In 1879 a proposal was made to annex into Kansas all of the land in Missouri north and west of the Blue

River and south of the Missouri River, an area that included all of Kansas City at that time. The plan had support in the legislatures of both states, but it never came to fruition.

135.0–136.0   *Wyandotte Limestone.*

136.0   Limestone in the *quarry* to the east is taken from the Wyandotte Formation.

136.8   *Plattsburg Limestone* and the *city limits of Overland Park.* This is the formal southern boundary of Overland Park, although the city itself isn't immediately apparent. With a population of just over 80,000, Overland Park is the fourth-largest city in Kansas. It is the largest of the profusion of suburbs on the Kansas side of the Kansas/Missouri border. Overland Park is also one of the larger cities in Kansas in terms of the area that it covers. Its city limits stretch 14 miles, from here to the Wyandotte/Johnson county line.

137.4   *Negro Creek.*

137.7   The town of *Stanley,* now a part of Overland Park, is 0.5 miles east of here.

138.2 and 138.5   *Plattsburg Limestone,* which was deposited in the latter part of the Pennsylvanian Period, about 300 million years ago.

139.7   *K–150* runs about eight miles west to Olathe, the seat of Johnson County. Martin City, Missouri, is five miles east of here.

139.8   *Tomahawk Creek.*

140.4   *Lane Shale* on the east side of the highway.

140.7   *U.S. 169,* which becomes Metcalf Avenue through Overland Park, exits to the north.

141.6   *Plattsburg Limestone* on the east side of the highway.

142.6   *Indian Creek.*

143.1   *College Boulevard.*

143.6   *I–435 interchange.* This highway encircles much of the Kansas City metropolitan area.

144.1   *103rd Street* and exposures of *Lane Shale,* overlain by the base of the *Wyandotte Limestone.*

145.1   *95th Street* and an exposure of *Vilas Shale,* overlain by the base of the *Stanton Limestone* on the west side of the highway.

145.5 and 146.1   *Stanton Limestone* on the east side of the highway.

146.6   U.S. 69 joins I–35, and the milepost numbering system changes to that of I–35. The geology along I–35 and the remainder of northbound U.S. 69 is described in chapter 7.

35·INTERSTATE·35

Across Eastern Kansas

127.0 I–35 runs north and south across the middle of the country, from Laredo, Texas, to Duluth, Minnesota. It enters Kansas at the southern border, near Caldwell, as a toll road and part of the Kansas Turnpike. I–35 leaves the turnpike here at the *Emporia interchange* and cuts northeast to Ottawa and Kansas City. The description of I–35 from Emporia south to the Kansas border is included in chapter 8. This log covers only the portion of I–35 from Emporia northeast to Kansas City.

127.0–128.0 In this mile the road travels along the *contact between rocks of Pennsylvanian age and those of Permian age.* Contact is a geological term for a location where rocks of different ages or types come together; here, rocks deposited during the Permian Period are in contact with ones deposited during the Pennsylvanian Period. The Pennsylvanian rocks were deposited at the close of the Pennsylvanian Period; they were then covered by rocks deposited at the dawn of the Permian Period, about 280 million years ago.

While this is a point of contact between rocks of different ages, it is not particularly distinctive because it is a conformable contact. That is, the rock layers of different ages are parallel, the younger one on top of the older in layer-cake fashion, indicating that they were deposited during continuous conditions of sedimentation at the bottom of an ocean. An unconformity is a contact between rock layers that represents a break or interruption in the deposition process, which has created a gap in the geologic record. For example, an unconformity occurred when the sea floor was lifted above sea level and exposed to erosion. As the land subsided below sea level, sediment deposition resumed on the irregular surface created by erosion. Thus, an unconformity is often an eroded surface that separates different rocks above and below.

128.0 The top of this small hill is mantled with rounded pebbles of *brown and grey chert*. Though it resembles road gravel, this rock was probably deposited several million years ago. While disagreement exists about the time and method by which these gravels were deposited, their smooth and rounded appearance is evidence that they were dropped by water, perhaps at the bed of a stream. Yet they are found well above present-day streams. For example, the gravel here is 100 feet above the Neosho River, one mile to the northeast; farther east in Kansas the gravels are found at even higher levels above nearby rivers. In short, they remain something of a geologic mystery. Understanding their deposition would probably help scientists better understand the erosional history of Kansas since its emergence from the Cretaceous sea.

130.0 *K–99 interchange.* Here the highway moves onto the flood plain of the Neosho River. The elevation is 1,110 feet.

130.5 *Emporia State University* to the south. Founded in 1863 as the Kansas State Normal School, this university was later called Kansas State Teachers College of Emporia. Although the name has changed, Emporia State still emphasizes teacher training. It also has an extensive program for handicapped students, and it houses the Center for Great Plains Studies.

132.0 To the south, about 20 feet above the flood plain, is the *Emporia terrace*. In geology, terraces are broad benches of land that drop sharply on the side close to the river; they represent old flood plains that have since been dissected by more recent flow from the river. This terrace, for example, contains alluvium deposited during the time when the Kansas glacier occupied northeastern Kansas, indicating that it was then a flood plain. Since that time, the Neosho has deepened the valley and has deposited younger sediments, so that the Emporia terrace generally remains above the flood plain. This terrace also contains deposits of volcanic ash. For several years, beginning in 1910, a three-foot-thick bed of ash was mined in western Emporia.

133.0 *U.S. 50 interchange.* U.S. 50 runs west from here, through Emporia, and on to Newton, Hutchinson, Dodge City, and points west. To the south are several borrow pits, now filled with water, where alluvium was excavated to build up the I–35 road bed across the Neosho flood plain. The water level in these pits is generally the same as the water

A cross section of the Neosho River downstream from Emporia, showing its flood plain, terraces, and alluvial aquifer

table in the flood-plain deposits. Rivers are usually neighbored by alluvial aquifers, stream-deposited material that holds ground water. The water table in these aquifers is generally about the same as the rivers they surround. However, when streams are low, water may move out of the aquifer and into the river, thus helping to maintain a flow even during times of drought. During floods, water may move from the river back into the aquifer, thus helping to recharge the water table. In short, these aquifers help to stabilize flow rates in the rivers, and heavy pumping out of an alluvial aquifer may do more than simply lower the water table. It may also decrease the amount of water flowing in the stream.

134.0 *Neosho River.* From near its source in Morris County, the Neosho River has carved a linear valley 100 miles long to Iola, where it curves to the south. Where the Neosho follows an unusually straight line, the river's course might be influenced by subsurface geologic faults. Scientists have located a number of long, straight faults that run northwest to southeast across the eastern end of Kansas. These faults may influence the geology at the surface and may cause rivers to drain in patterns along the faults. Because faults may cause earthquakes, and because dams are located on rivers, geologists are particularly interested in the relationship between faults and drainage patterns.

About 10 miles downstream from here, near Burlington, the Neosho is dammed to form John Redmond Reservoir. The river leaves the state near Chetopa, enters Oklahoma, and is dammed to form the Grand Lake of the Cherokees and two other reservoirs before joining the Arkansas River near Muskogee.

135.2 *Thornburg interchange* and exposures of the three rock layers that make up the *Emporia Limestone Formation.* These layers are called the Reading limestone (at the bottom), the Harveyville shale, and (at the top) the Elmont limestone. These limestones contain a number of fossils, including remains of algae that flourished in shallow Pennsylvanian seas. The algae produced a carbonate sediment on the sea floor. Carbonate is composed of carbon and oxygen, and in combination with calcium it produces limestone, or calcium carbonate. In an attempt to understand the formation of these limestone layers, geologists often study areas where the conditions today are geologically similar to conditions in Kansas 300 million years ago. Many of the shallow carbonate platforms in the Bahamas, for example, resemble the environment that existed in Kansas when these rocks were deposited.

135.6 *Reading limestone.* The Reading (pronounced Red'-ing, after the nearby town) is a dense, hard limestone that includes layers of shale and, in southeastern Kansas, coal. In places the Reading contains a number of closely spaced vertical joints. Limestone may look impermeable, but water can move along these joints and pass through layers of seemingly solid rock.

137.2 On the north side of the highway, *Badger Creek* is joined by *Coon Creek* before flowing beneath the highway. Raccoons are common throughout Kansas, but a much rarer relative of the raccoon, the ringtail, is also found in this area. Ringtails are members of the same family as raccoons, and the first ringtail collected in Kansas was taken here in eastern Lyon County. Small and slender-bodied, these mammals have a tail with eight or nine alternating black and white rings. Because they are almost entirely nocturnal and spend much of their time in trees, ringtails are seldom seen in Kansas, which is probably the northern edge of their range in the Great Plains.

137.7 *Burlingame limestone,* named for a coal-mining town north of here.

139.4 *Wakarusa limestone* exposed on the south side of the highway. This is a dark blue-gray limestone that weathers to a brown color. Limestone often changes color after being exposed to the elements, so geologists must chip off fresh pieces of the rock to determine its true, unweathered color. That's one of the reasons that geologists carry the rock hammers that are so closely identified with the profession.

140.8 *Dry Creek.*

141.1 *K-130* exits south to the town of Neosho Rapids and the Flint Hills National Wildlife Refuge, in the upper reaches of John Redmond Reservoir.

142.5 *Plum Creek.*

143.1 *Lyon/Coffey county line.* Coffey County was named by the Bogus Legislature of 1855, honoring Col. Asbury Coffey, a native of Kentucky who served in that legislature and became an officer in the Confederate Army.

143.2 *Arkansas River/Missouri River drainage divide.* Runoff south and west of this line enters the Neosho River and, eventually, the Arkansas River. To the north, runoff enters the Marais des Cygnes River, which enters the Missouri River near Jefferson City, Missouri.

144.5 On the horizon to the north is *Bebb Hill,* which is capped by Bern limestone. The hill's elevation is 1,200 feet, about 100 feet higher than the surrounding countryside.

148.4 *K–131* goes 0.5 miles south to Lebo, which was named after Joe Lebo, a captain in the Tenth Kansas Cavalry. Melvern Reservoir, on the Marais des Cygnes River, is accessible from this exit.

149.7 To the north of the highway is a view of the *Marais des Cygnes River valley.*

151.5 The highway passes through the *Last Chance oil field,* which was discovered in 1977. It produces oil from rocks in the Cherokee Group at a depth of about 1,500 feet. These same rocks crop out in Crawford and Cherokee counties in southeastern Kansas. Most Kansas oil fields have somewhat less romantic names than Last Chance. Fields are often named after the owner of the property where a discovery well is drilled. In this instance, however, Last Chance is probably an appropriate name, since it is the only oil field in Osage County.

152.2 *Calhoun Shale,* which reaches a maximum thickness of about 50 feet in Shawnee County.

154.2 *Frog Creek.* Twenty-one species of frogs and toads live in Kansas. The largest and best-known member of that family is the bullfrog, which is found throughout Kansas. One specimen from Chase County, the largest ever taken in the state, had a body length of 7.25 inches. This species is generally restricted to rivers, lakes, and swamps, where the females lay up to 40,000 eggs during a single breeding period. According to biologists, bullfrogs will eat just about anything, including fish, squirrels, and small birds. In Kansas, bullfrog season begins in late summer, when they are hunted for the meat, which is used in the dish of frog legs.

154.4 *Ozawkie limestone,* named for a small town in Jefferson County that was displaced when Perry Lake was built. A new version of the town was constructed on a hill overlooking the lake, which is northeast of Topeka. Ozawkie was named for a Sac Indian chief.

155.5 *U.S. 75 interchange.* The intersection of these two highways is called Beto Junction. "Beto" is formed from the first letters of four towns that can be reached by these two highways: Burlington, Emporia, Topeka, and Ottawa. About 12 miles to the south, east of U.S. 75, is the Wolf Creek Nuclear Power Plant, the only nuclear generating plant in Kansas. It was constructed by Kansas City Power and Light Corporation and Kansas Gas and Electric Company; it went on-line in 1985.

156.5 *Coffey/Osage county line.* Osage County is named for the Indian tribe that occupied this part of eastern Kansas before being moved to a reservation along the southern border of Kansas in 1825.

157.0 At this point the highway passes over a *fault* that was discovered by a geologist from the Kansas Department of Transportation. A fault is not a crack in the ground; it is a location where rock layers have moved relative to each other. This fault has a displacement of 20 to 30 feet. That is, rocks that were in the same layer before the fault moved are now 20 to 30 feet apart. Because this fault is in the subsurface, it is not visible from the road. However, faults can be seen at the surface in several locations in Kansas, including one in the rocks that line the spillway at Tuttle Creek Reservoir, near Manhattan.

157.3 *Long Creek.*

157.6 *Beil limestone.* This rock layer is full of fossils, an indication of the abundant invertebrate life that populated the Pennsylvanian seas. Among the fossils found in the Beil are brachiopods, bryozoans, fusulinids, mollusks, and cornucopia-shaped horn coral. The hard body parts of the horn coral have often been chemically replaced by crystalline calcium carbonate, so that the fossils have a characteristic sparkle when broken.

158.2 *Coal Creek.* Coal is not mined today in Osage County, but during the 1880s and 1890s this was the leading coal-producing county in the state. In 1889, 118 mines in Osage County produced 400,000 tons of coal, mostly from shallow shaft mines. Old coal dumps from many of these mines are still visible at the surface.

159.5 *Beil limestone.* Pennsylvanian limestones contain a variety of fossils, which can be collected at Kansas road cuts (although it is illegal to stop along the Interstate for any reason other than an emergency). While fossils are usually found embedded in the limestone, they can often be collected by simply picking them up from the ground near an outcrop. Fossils often have a different composition than the rock in which they were embedded, so they may simply fall out as the surrounding rock weathers away. In some places, for example, fusulinid fossils have weathered out of the limestone and can be found covering the ground like spilled wheat.

160.6 *K-31* leads north to Melvern Lake and the town of Melvern, the source of whose name is thought to be a Welsh word meaning "bare hill." With Melvern Lake and Pomona Lake, Osage County is the only county in the nation with two federally constructed reservoirs.

160.9 *Willow Creek.*

162.6 *K-31* goes south to Waverly.

163.0 *Plattsmouth limestone,* which, at 15 to 30 feet in thickness, is the thickest member of the Oread Formation.

163.2 *Rock Creek.*

163.6 *Plattsmouth limestone* is exposed in a quarry north of the highway.

164.2 *Kereford limestone,* another member of the Oread Formation. In places it contains oolites, small spheres of calcium carbonate that have formed around some sort of nucleus, such as a grain of quartz. These spheres range up to .07 of an inch (or 2 millimeters) in diameter, about half the size of the lead shot used in shotgun shells. The word *oolite* comes from a Latin translation of the German word *Rogenstein,* or roe stone, so called because an oolitic limestone may have the bumpy appearance of fish eggs. These features supplied the name for Oolitic, Indiana, a town in the limestone-producing region of that state. When oolites grow larger than .07 of an inch, they are called *pisolites,* a word based on the Greek word *pisos,* meaning pea, which aptly describes their size and shape.

166.9 *Plattsmouth limestone* on the south side of the road.

167.3 *South Branch of Tequa Creek* and an exposure of shale in the *Lawrence Formation.*

167.8 *Plattsmouth limestone* on both sides of the highway.

168.0 *Osage/Franklin county line.* Franklin County was named for the colonial statesman Benjamin Franklin.

168.3 *Plattsmouth limestone.*

168.8 *Toronto limestone.* The Kereford, Plattsmouth, and Toronto limestones are all part of the Oread Formation, named for the hill that is the site of the University of Kansas.

169.0 *East Branch of Tequa Creek.*

169.2 About 2.5 miles to the south was the town of *Silkville,* an experimental utopian colony established by the Frenchman Ernest Valeton de Boissiere in 1869. Boissiere helped to bring in French immigrants, and he started a silk factory, with looms that produced as much as 300 yards of finished material a day. Later he gave up silk as unprofitable, and the community turned to dairy production. In 1894, Boissiere died in his native France. Even though his experiment failed, silk became a popular commodity, which was produced in forty Kansas counties from 1889 to 1900. At one time the State Board of Agriculture even organized a special silk station at Peabody.

169.4   Shale in the *Lawrence Formation* is exposed north of the highway.

170.2   Exposures of the *Plattsmouth limestone* on the south side of the road. K-273 goes south 0.5 miles to Williamsburg, which was named after William Scofield, a local farmer. This town provided the name for the Williamsburg coal bed, a part of the Lawrence Formation that crops out in this area.

171.3   *Coal Creek,* named after exposures of the Williamsburg coal bed. The Williamsburg was the source for most of the coal mined in Osage and Franklin counties; production from this layer totaled about 90,000 tons, most of which was mined in the late 1800s. Recent studies show that the Williamsburg is extremely low in sulfur—only about 1 or 2 percent—which makes it an environmentally attractive coal, particularly because many coal layers in Kansas are high in sulfur and require expensive pollution-control equipment if they are to be burned. In spite of the Williamsburg's advantages, it is not mined today because it is only 12 to 15 inches thick and because it is covered by the Toronto limestone, a thick layer of rock that must be removed to get to the coal. Still, the Williamsburg is considered a reserve that could be important some day.

171.7   *Plattsmouth limestone* on the north side of the highway.

172.3   *Plattsmouth limestone.*

173.0   Southeast of the highway is the small town of Ransomville, the site of a shaft mine that operated during the 1890s and early 1900s, producing coal from the Williamsburg bed. This mine used a horse-powered hoister and provided coal for trains along the now-abandoned Santa Fe line that operated between Ottawa and Emporia.

174.7   *Rest Area.*

175.8   *Mud Creek.*

176.4   County road south to *Homewood.*

176.7   Exposure of shale in the *Lawrence Formation* on the north side of the highway.

178.0   One mile to the northwest are the *Chippewa Hills,* which are capped by Oread Limestone. These hills were named after the Chippewa Indians, who lived on a small reservation in this area during the

1800s. The Chippewas were originally from Michigan, but in 1839 they moved to this 8,320-acre reservation. In 1859 the Chippewas were united with another tribe, the Munsies, who were making their second attempt to settle west of the area of white encroachment. The Munsies were originally from the upper Delaware valley of New York, New Jersey, and Pennsylvania; they had been moved to Ontario, Canada, before coming to Kansas. The Munsie Indian cemetery, established in 1857, lies in the hills north of the highway.

179.1   *Mud Creek.* The elevation here is 920 feet.

180.0–182.0   In these two miles the highway passes through the *Sand Hills.* Unlike the sand hills of central and western Kansas—which are composed of geologically young windblown sand dunes—these hills are made up of sandstone that was deposited in a large river valley about 300 million years ago, during the Pennsylvanian Period. Called Tonganoxie sandstone, this rock layer is more than 140 feet thick in places. The old river valley where the sand was deposited followed a southwestward course, 12 to 20 miles wide, from Leavenworth County to Greenwood County.

The Tonganoxie today is an important water-bearing formation in this part of Kansas. Most of the bedrock units here are limestones and shales that yield little or no water when wells are drilled into them. Because sandstone has abundant pore spaces between the particles of rock, it is conducive to the accumulation and pumping of water.

183.0   *U.S. 59 interchange.* North on U.S. 59 is the city of Ottawa, the seat of Franklin County. The town was named after the Indian tribe that was relocated to this area from its home near Lake Huron. A portion of the restored downtown and the Franklin County Courthouse, completed in 1893, are on the National Historic Register.

To the south, U.S. 59 passes a plant where Weston shale is mined and heated to form a lightweight aggregate material, which is used in construction and for decorative purposes. This expanded, or bloated, shale was used in the construction of the terminals at Kansas City International Airport.

183.5   *Rock Creek.* To the north is a quarry where rock is mined from the Stanton Limestone.

185.6   *Limestone quarries* in the Stanton Limestone on the west.

185.9   *Marais des Cygnes River.* This river, whose name means "marsh of swans," has it source in Lyon and Wabaunsee counties and flows into Missouri near Trading Post, Kansas. In Missouri it is called the Osage

River. In Kansas, the Marais des Cygnes is commonly associated with a massacre that took place in a ravine north of the town of Trading Post.

186.6 *Stanton Limestone* on the west side of the highway.

187.5 *K-68 interchange,* which leads west to Ottawa and east to Louisburg.

189.3 Though the highway sign calls this Ottawa Creek, the official U.S. Geological Survey name for it is *Tauy Creek.* It joins Wolf Creek and Walnut Creek about 1.5 miles southeast of here to form Ottawa Creek. John Tauy Jones, who was part Ottawa Indian, was instrumental in the founding of Ottawa University.

191.0 *South Bend limestone,* exposed on both sides of the highway, is generally about one to six feet thick, except in Montgomery County, where it may be up to 27 feet thick.

191.3 *Wolf Creek.* About 2.5 miles upstream is the town of Le Loup, French for "the wolf."

192.9 *Le Loup road interchange.*

195.3 Here the highway passes through the *Paola-Rantoul oil and gas field,* which extends eastward into Miami County. This is one of the largest fields in eastern Kansas, having produced over 9.4 million barrels of oil and a small amount of gas from Pennsylvanian rocks, which are only 400 to 700 feet deep. Some of the oil produced from this field comes from sandstone bodies that may be only a few hundred feet across but stretch for miles. These sandstones, such as the one 1.5 miles southeast of here, are called shoestring sands, because of their shapes when plotted on maps. Geologists believe these features are ancient sand-filled river channels that were formed during the Pennsylvanian Period, much like smaller versions of the river channel where the Tonganoxie sandstone was deposited. Because they are so narrow, shoestring sands provide a small drilling target, and they seem to "jump around" (a characteristic that produced another nickname— squirrel sandstone—because the drillers thought the rocks were "squirrelly"). Even wells that are close to existing production may miss the old river channels and produce no oil.

The Kansas oil business probably got its start by drilling into these shoestring sandstones. Miami County, southeast of here, was long known by the Indians for its oil seeps. The first pioneers in the state used the oil for lubricating their wagons. During the 1860s the first oil wells in the state were drilled around Paola, the seat of Miami County. They produced small amounts of oil from shallow sandstones. George C. Swallow, the second director of the Kansas Geological Survey, wrote in 1865 that there was "very strong evidence of large reservoirs near these localities (in Miami County)."

The years have proven Swallow right. Southeastern Kansas has produced about 22 million barrels of oil from these shallow sandstones, as well as from deeper formations. While that is a substantial amount, it isn't enough to rank Miami among the top ten oil-producing counties in the state. In Butler County, the state's all-time leading oil producer, more than 538 million barrels of oil had been pumped as of 1984. Today, southeastern Kansas is a mature oil-producing area. Many of its fields have been abandoned, and most of the wells that remain are stripper wells, producing fewer than 10 barrels of oil per day.

Nevertheless, drilling continues in southeastern Kansas, partly because oil is close to the surface, meaning that wells can be drilled more quickly and less expensively than in other parts of the state. Oil that is within a few hundred feet of the surface here, for example, may be several thousand feet underground in western Kansas, even though the oil is in the same formation.

196.9 *Spring Creek.*

197.6 *Walnut Creek.*

198.0 *K-33 interchange. Wellsville* is about a mile to the north.

199.2 *Rock Creek* and the *Franklin/Miami county line.* Miami County was originally named Lykins County by the first Territorial Legislature, honoring David Lykins, a missionary to the Indians and a member of the first legislature. The name was later changed to Miami, the name of the Indian tribe that relocated to this area from Ohio, which has its own Miami County.

202.0 *Miami/Johnson county line.* Johnson County, which was named for the Methodist missionary Thomas Johnson, is one of several Kansas counties that has the same name as a Kansas town that is not located within that county. For example, Johnson (also called Johnson City) is not in Johnson County; it is completely across the state in Stanton County. Similarly, Logan is in Phillips County, not Logan County; Mitchell, Crawford, and Chase are all counties in the state, but the towns of the same names are in Rice County. Other examples: Ottawa is in nearby Franklin County; Wilson is in Ellsworth County; Allen, in Lyon County; Jefferson, in Montgomery County; Franklin, in Crawford County; Reno, in Leavenworth County; Sherman, in Cherokee County; Wichita, in Sedgwick County; Linn, in Washington County; Lane, in Franklin County; Greeley, in Anderson County; and Shawnee, here in Johnson County. In addition, Cherokee is in

Crawford County, just across the line from Cherokee County, and Sedgwick is almost entirely in Harvey County. What's more, a couple of near misses add to the confusion. Lyons is in Rice County, not Lyon County; and Coffeyville is in Montgomery County, not Coffey County.

203.7 *South Bend limestone* on both sides of the road.

204.3 An exposure of the lower part of the *Stanton Limestone Formation,* from the Captain Creek limestone at the bottom, up through the Eudora shale and Stoner limestone.

204.5 *Bull Creek* is dammed a short distance downstream to form Hillsdale Lake.

204.9 *Stoner Limestone Member* of the Stanton Limestone Formation.

205.9 *Stoner limestone,* exposed on both sides of the highway, is one of several rock formations that make up the Lansing Group of formations, which was deposited late in the Pennsylvanian Period. The Lansing Group and the Kansas City Group, which is slightly older than the Lansing, are common sources of oil, both in eastern Kansas and in western Kansas, where they are much deeper. Petroleum geologists frequently lump the two groups together and talk about production from the "Lansing-Kansas City."

206.5 *Stoner limestone.*

207.0 *Stoner limestone* on the north side of the highway. The rock layer immediately above the Stoner limestone is called the Rock Lake shale. Though not present at this location, Rock Lake shale is notable because it contains fossils of a small, dinosaur-like reptile that is named after the rock layer—*Petrolacosaurus.*

207.7 Two miles to the north, by county roads, is the city of *Gardner,* named for Henry J. Gardner, who was elected governor of Massachusetts in 1854, the year that Kansas was opened to settlement. Near Gardner is one of the most important locations in the history of the American West: the spot where the trail leading from Independence and Westport split into two trails. One, the Oregon Trail, led to the northwest, while the other, the Santa Fe Trail, led to the southwest.

209.8 *Rest area.*

210.2 *Little Bull Creek.*

210.8 *East Gardner exit.* Just north of the highway, accessible by this exit, is the Johnson County Industrial Airport, formerly the Olathe Naval Air Station. The Kansas City area is served by several major airports including Kansas City International, north of the city, and the smaller

Downtown Airport. The airport here handles much of the small-plane traffic for Johnson County.

The city lake two miles north of Gardner was the source of the largest channel catfish ever caught on a hook and line in Kansas. Taken in 1962, the fish weighed 32 pounds and was 25 inches long. Channel catfish are pale, slender fish that often have dark spots and are distinguished by a V-notch in their tails. They are among the most popular game fish in Kansas; they are found in lakes and streams throughout the state.

211.4 At this point the highway crosses the *drainage divide* between the Kansas and Missouri rivers, to the north, and the Marais des Cygnes and Osage rivers to the south.

212.4 *Cedar Creek.*

214.0 *Olathe Lake* is visible north of the highway.

214.5 *The Stanton Limestone Formation* exposed, from the Stoner limestone, up to the Rock Lake shale, up to the South Bend limestone. In places the Stoner has a brecciated appearance, meaning that it looks broken up, or crushed.

215.0 *South Bend limestone* is exposed on the south side of the highway.

215.6 *K-7 and U.S. 169 interchange.* K-7 extends from Nebraska to Oklahoma through the eastern tier of Kansas counties. U.S. 169 exits to the south, leading to Oklahoma via Chanute and Coffeyville. At this point, U.S. 169 joins I-35 and continues northeast. Olathe is one mile north of here on K-7.

217.0 *U.S. 56 interchange.* Eastbound U.S. 56 briefly joins I-35, while westbound U.S. 56 exits. U.S. 56 is described in chapter 9 of this book.

218.1 *K-150 interchange* and *East Olathe exit.* Olathe is a Shawnee word meaning "beautiful." The town takes its name from a Shawnee Indian who was assisting the town's founder in a search for a town site. When they came across a small hill covered with wild flowers, the Indian shouted "O-la-the"; thus both a town site and a town name were chosen. Olathe is the seat of Johnson County, the second most populated county in Kansas—behind only Sedgwick, which includes the city of Wichita. Although Olathe is located in the approximate center of Johnson County, much of the county's population lives north and east of Olathe, in the suburbs of Kansas City. Johnson County includes three of the ten largest cities in Kansas: Overland Park, Olathe, and Shawnee.

222.7 *I-435 interchange.* This freeway loops around Kansas City, eventually intersecting with I-70 and again with I-35 in Missouri.

224.1 *95th Street exit* and the city of *Lenexa,* which was named after the wife of Blackhoof, a Shawnee chief.

225.0 At this point the elevation is about 1,070 feet, and eastbound I-35 enters the upper part of the *Turkey Creek valley* before descending down this valley toward Kansas City. From here to the point where it joins the Kansas River, Turkey Creek falls a vertical distance of nearly 350 feet over a 10-mile course. This is an average gradient of 35 feet per mile, which is comparable to the 40-feet-per-mile gradient of the Arkansas River as it flows through the Rockies in central Colorado. Much of Turkey Creek's watershed is composed of impermeable surfaces—such as roofs, streets, and parking lots—which do not allow water to be absorbed. Instead, nearly all precipitation in this area runs off, and heavy downpours can cause flash floods as violent as the ones that occur in the mountains. One such storm in September 1977 sent Turkey Creek and Brush Creek, the next stream to the east, out of their banks, causing millions of dollars in damage and killing twenty-five people.

225.3 *87th Street exit* and *Overland Park.* This city, which got its name from the Overland Trail, is the largest city in Johnson County and the fourth largest in Kansas, behind Wichita, Kansas City, and Topeka. Like much of Johnson County, Overland Park has grown dramatically since World War II. In 1940, Overland Park had a population of 572, and Johnson County's population was 33,000. By 1980, Overland Park had grown to 82,000, and Johnson County was up to 270,000, making it the second most populous county in the state.

225.9 Northbound *U.S. 69* joins I-35.

226.3 Southbound *U.S. 69* departs from I-35. U.S. 69 is described in chapter 6.

227.2 *75th Street exit.*

227.7 *Captain Creek Member* of the Stanton Limestone, named after a creek in Douglas and Johnson counties.

228.5 *Plattsburg Limestone Formation* on the east side of the highway. The Plattsburg averages about 25 feet in thickness in Kansas; however, in parts of Wilson County it is up to 115 feet thick. Geologists believe that this thickening was caused by reef-type material that was deposited on the floor of the Pennsylvanian sea in present-day Wilson County. The source of the reef was probably algae or other lime-secreting animals such as crinoids, bryozoans, or brachiopods. These animals built up a thick marine bank, which is today part of the Plattsburg Formation and is called the Plattsburg bank.

228.8   *63rd Street exit* and *Merriam.* Here, eastbound U.S. 56 leaves I-35; westbound U.S. 56 joins I-35 from here to Olathe. K-12 is accessible to the west, leading to Shawnee. Merriam was named for the secretary of the Kansas City, Fort Scott, and Gulf Railroad. In its early days, Merriam was famous for its amusement park and zoo, which were opened in 1880 with former President Ulysses S. Grant presiding. Gambling is an important part of Merriam's economy, because nearly all the dice used in Las Vegas are manufactured here.

228.9   In the bluff to the east is the standard section of the *Merriam Limestone Member* of the Plattsburg Limestone and adjacent beds. A standard section is an exposure of rock that shows, as completely as possible, all of the rock strata in a given area in the order in which they were deposited. This section is used as a standard by geologists in correlating rocks in different areas that are thought to be of the same age. A standard section is somewhat different from a type locality, which is the place where a rock unit is typically displayed and from which it derives its name. Generally, a type locality is where the rocks are first described. For example, Merriam limestone was first described in Merriam, from which it derives its name.

230.2   *Turkey Creek.* This stretch of I-35 is also called the Turkey Creek Expressway. To the east, I-35 briefly reenters Overland Park. The Iola Limestone is exposed in the banks of Turkey Creek east of the highway. Though less than 10 feet thick here, the Iola thickens to 40 feet in Allen County, where it is quarried and used to produce Portland cement.

231.0   *I-635 and U.S. 169 interchange.* I-635 exits to the north, toward Kansas City, Kansas, and I-70. U.S. 169 exits to the south, becoming Metcalf Avenue in Overland Park. At this interchange, westbound I-35 passes into Overland Park, and eastbound I-35 enters Mission.

Mission gets its name from the missions that were established in this area by religious groups to serve the Shawnee Indians. The Shawnees, who moved to Kansas from Ohio in 1825, were among the first Indian tribes to be resettled in the state. They were given a 2,500-square-mile reservation along the south bank of the Kansas River. The Shawnee Methodist Mission has been preserved and restored as a state park in Fairway, 2.5 miles to the east. This mission was established in 1830 by Rev. Thomas Johnson, the namesake of Johnson County. In addition to the Methodists, the Baptists and Quakers also operated missions in this area. Today the fourteen cities that make up northeastern Johnson County are often referred to as Shawnee Mission, both out of convenience and in recognition of those early settlements.

231.8 *Johnson/Wyandotte county line.* Eastbound I-35 also enters Kansas City, Kansas, here. Wyandotte County is named for the Wyandot Indians, who moved to this area from Ohio and organized Wyandotte City, which was consolidated with Quindaro and several neighboring towns in 1886 to form Kansas City, Kansas.

232.4 *Turkey Creek.*

232.6–233.3 *Drum Limestone* exposed on the south side of the highway. About 10 feet thick in this area, the Drum thickens to 60 feet near Independence, Kansas.

233.5 *Turkey Creek.*

234.1 The thick road cut north of the highway exposes several formations. At the bottom is the Cherryvale Shale, up through the Drum Limestone, the Chanute Shale, the Iola Limestone, and the Lane Shale. The hill is capped by the Wyandotte Limestone.

234.4 *Cherryvale Shale* and the overlying *Drum Limestone.*

234.8 *Drum Limestone* and *U.S. 169 interchange.* This part of Kansas City is known as Rosedale, which was a separate city until it was annexed by Kansas City in 1922. Rosedale was the site of the Geyser Mineral Bathhouse during the late 1800s. Built on the south side of Turkey Creek, the bathhouse was near a well that had been drilled for gas and was producing mineral water. In addition to bathing, the water was used for manufacturing soft drinks.

Also, the Rosedale Arch is visible on a high bluff to the south. This arch, which is a replica of the Arc de Triomphe in Paris, honors World War I veterans from the area. At the base of this bluff, leading out from this exit, is Rainbow Boulevard, which is named after the army's Forty-second Divison of World War I. That division, composed of National Guard units from Kansas, Missouri, and more than twenty other states, was known as the Rainbow Division. The University of Kansas Medical Center is located one mile south of here on Rainbow Boulevard.

235.0 Although the *Turkey Creek valley* continues on to the northeast from this point, Turkey Creek does not. It is diverted into the Kansas River by a 1,500-foot-long tunnel through the bluff north of the highway. This diversion was built to prevent flooding in the lower part of the Turkey Creek valley, which merges with the West Bottoms of Kansas City in the flood plain of the Kansas River, a short distance to the northeast.

The West Bottoms area straddles the Missouri/Kansas state line and is home to railroad switching yards, industries, warehouses, the Kemper Arena, and the famous Kansas City stockyards. The stockyards were once surrounded by huge meat-packing plants, but today the plants are gone, and the stockyards do only a fraction of their former business. The Ogallala aquifer of the High Plains of western Kansas and adjacent states is partly responsible for these changes. Irrigation from the aquifer in the last 30 years has allowed western-Kansas farmers to raise large, dependable yields of corn, alfalfa, milo, and other crops that can be fed to cattle. Vast feed lots have sprung up on the High Plains from Texas to Nebraska to be close to these areas of crop production. In turn, packing plants moved west to take advantage of the large concentrations of fattened cattle. Today the largest packing plant in the world operates near Holcomb, Kansas, nearly 400 miles southwest of here, and Kansas is the largest beef-producing state in the nation.

235.6   *Kansas/Missouri state line.* Northbound I–35 enters Kansas City, Missouri, at this point. A few miles to the north, I–35 intersects with I–70 before proceeding north to Des Moines, Iowa, and terminating at Duluth, Minnesota. Southbound I–35 enters Kansas City, Kansas, here.

THE KANSAS TURNPIKE

The Chisholm Trail and Beyond

0.0 *Kansas/Oklahoma state line.* The southern border of Kansas stretches 411 miles along the 37th parallel. The Kansas Turnpike enters the state here and runs north to Wichita, northeast to Topeka, and then east to Kansas City, connecting the three largest cities in the state. Opened in 1956 and built at a cost of $160 million, the turnpike covers more than 220 miles across Kansas. It is also part of the Interstate highway system. I-35 joins the turnpike from here to Emporia, and I-70 joins the turnpike between Topeka and Kansas City.

The southern border of Kansas played a role in Oklahoma history when settlers gathered here in 1893 to begin the last great land rush into Oklahoma. The settlers claimed pieces of the Cherokee Outlet, or Cherokee Strip, a parcel of land west of the Arkansas River along the southern edge of Kansas. Formerly Indian Territory, the strip was 59 miles wide and 150 miles long. On the night before the rush, about 70,000 settlers crowded into Arkansas City, where the run is documented at the Cherokee Strip Museum.

From here to Wichita, the turnpike follows about the same route as the old Chisholm Trail. Originally used by Indians, traders, and the army, the trail ran from the Red River area of Texas up to Jesse Chisholm's trading post on the banks of the Little Arkansas River at present-day Wichita. During the 1860s and 1870s, the trail was used for moving cattle from southern Texas to the railheads in such towns as Caldwell, Wichita, Newton, Abilene, and Ellsworth. More than 4 million head of cattle were driven north during those years. The trip, which covered more than 600 miles, took the cowboys 30 to 40 days.

Before the Chisholm Trail was established, cowboys moved cattle from Texas, through eastern Oklahoma, and into Missouri. However, drives were slowed by cattle quarantines, harassment from Missouri bushwhackers, and the rugged, forested terrain. When the railroads were extended into central and western Kansas, the era of cattle drives began in earnest. The gentle terrain and shallow stream crossings of the Great Plains—features that are apparent along the turnpike between Oklahoma and Wichita—made the drives much easier.

0.8 *Wellington Formation.* This formation, predominantely shale, is up to 700 feet thick. One of the most common formations in south-central Kansas, it was deposited during the Permian Period, about 250 million years ago, when an inland sea covered this area. Part of this formation is a rock layer called the Hutchinson Salt Member, a bed of salt, several hundred feet thick, that is mined throughout central Kansas. There are no outcrops of salt, because the mineral is easily dissolved by precipitation, unlike less-soluble formations such as limestone.

1.6 *Tourist information center.*

3.2 *Wellington Formation.* Where the Wellington is composed of shale, it may contain a variety of fossils, including corals, mollusks, shrimp, and insects.

4.0 *Exit to U.S. 166,* which runs east to Arkansas City and west to South Haven and Caldwell. This is the last free exit on the turnpike. To the north the turnpike is a toll road, with fifteen exits between here and its eastern terminal, near Kansas City, Kansas. In 1986 the toll was $7.00 to drive a car the entire distance.

4.3 *Wellington Formation.*

11.2 Visible three miles west is the elevator at *Rome, Kansas.* In Italy, the city of Milan is 300 miles northwest of Rome. In Kansas, Milan is northwest of Rome, although the distance is less than 25 miles and can include a stopover in Perth.

14.3 South side of the *Slate Creek valley.* Shale from the Wellington Formation is exposed along the roadside and in the hillsides to the east and west. Composed mostly of shale, the Wellington contains thin layers of limestone, which can be seen weathering out of the exposures.

14.9 The wells that border the turnpike pump oil from the *Rome Northeast oil field,* discovered in 1977. These wells, which are about 3,600 feet deep, pump oil from rocks that were deposited during the Mississippian

Period, about 330 million years ago. Mississippian rock formations appear at the surface in extreme southeastern Kansas.

15.1 *Slate Creek* drains southeast, joining the Arkansas River between the towns of Geuda Springs and Rainbow Bend. Near Geuda Springs there is a series of mineral springs; during the late 1800s, a bathhouse and hotel were built there so that guests could take advantage of the healing properties of the waters. The water also was used in a bottling plant that produced soft drinks. Such bathhouses were common throughout Kansas in the 1800s. They were built at locations such as Baxter Springs in Cherokee County, Sun Springs and Sycamore Springs in Brown County, and Chautauqua Springs in Chautauqua County. The best known of these was Mitchell County's Waconda Springs, which is now covered by the waters of Waconda Lake.

19.2 *Wellington toll station and U.S. 160 exit.* To the east is Winfield, the site of a bluegrass-music festival every September. To the west is Wellington, the seat of Sumner County. Kansas regularly leads the nation in wheat production, and Sumner County is often the leading wheat-producing county in the state. In 1981, for example, Sumner County produced 15.6 million bushels of wheat, almost twice as much as the next-highest producing county.

In 1887, Wellington also became the site of a salt mine when the mineral was discovered in a test hole at a depth of 240 feet. However, Wellington is near the eastern edge of the Hutchinson salt bed, and the salt was only about 50 feet thick. A small mine soon opened, but it was a financial failure. Today, salt mines operate in the central Kansas counties of Rice, Reno, and Ellsworth.

19.3 *Wellington Formation.* This formation is named after the nearby town. This area is rife with English names, including nearby Oxford, Cambridge, and Runnymede.

19.7 *Deer Creek.*

19.8 *Rusk oil field* is located immediately west of here. During 1984, wells in Sumner County produced 1.4 million barrels of oil.

25.1 *Wellington Formation.*

25.5 *Wellington service area.*

26.8 *Ninnescah River.* The Ninnescah reportedly received its name from the Indian word for "good water" or "clear water." The town of Clearwater is located on the Ninnescah, several miles northwest of here. From this point south, the turnpike traverses the Wellington Lowlands, a flat and fairly featureless area. Because the most common rock here is shale and because shale is easily eroded, rivers have created this even, low-lying plain. Headed north, the turnpike enters the Arkansas River Lowlands, where the topography is similar to the Wellington Lowlands but was created by the Arkansas River and its tributaries, such as the Ninnescah. From here until it crosses the Arkansas at Wichita, the turnpike passes through an area covered by sediments that were deposited by those rivers during the past million years or so.

27.2   Small area of inactive *sand dunes*. Now grassed over, this sand was carried in by the Ninnescah and was blown out of its bed by strong southerly winds during times of low flow in the river. Though these dunes no longer move, they give the surface a hummocky, rolling shape.

27.4   *Belle Plaine*, 2.5 miles east of here, is home of the Bartlett Arboretum, a 20-acre tract of trees, flowers, shrubs, and grasses, which is open from April to mid November. The arboretum was begun in 1910 by Dr. Walter E. Bartlett; it is especially well known for its tulip beds. Because there are only a few exits off the turnpike, in order to reach Belle Plaine you must leave the turnpike at Wellington or at Wichita.

30.5   To the west 3.5 miles is a small town called *Zyba*, the last word in any alphabetical listing of Kansas place names.

33.0   *K–53 exit* to Mulvane.

33.3   *Sumner/Sedgwick county line.* Sedgwick County was named for Maj. Gen. John Sedgwick, a former West Point cadet who was wounded at the Civil War Battle of Antietam and was killed at the Battle of Spotsylvania. According to one source, Sedgwick's last words were ''They couldn't hit an elephant at this distance.''

34.5   *Cowskin Creek*, which was originally called Crooked Creek. Its name was changed when cattle on the Chisholm Trail died from Texas fever, and their hides were left to dry on the creek bank.

36.1   Former channel of *Cowskin Creek*. Cowskin Creek has been channelized—its path has been straightened and levees have been built to try to prevent flooding—throughout much of southern Sedgwick County. Part of this new channel parallels the turnpike to the west, behind an earthen embankment.

38.0   The Wichita suburb of *Derby*, once called El Paso, is three miles east of here.

39.3   *Haysville* is 1.5 miles to the west. This town is not to be confused with Hays, a much larger city 150 miles to the northwest. Haysville has been called the Peach Capital of Kansas. Sedgwick County leads the state in peach production, accounting for nearly half of the commercial peach crop in Kansas.

40.0 *Wichita/Valley Center Flood Way.* This artificial channel extends from the north side of Wichita, around the city's western and southern edges, and eventually feeds into the Arkansas River near Derby. The channel was built to divert floodwater from the Arkansas and the Little Arkansas rivers around the city of Wichita.

42.0 *South Wichita interchange.* This city is named for the Wichita Indians. When Coronado came to Kansas in 1541, he found the Wichitas living in central Kansas. By the time of the Civil War they had migrated into the Red River region of Oklahoma, but they were driven back into Kansas by tribes who were sympathetic to the Confederacy. In 1863 the Wichitas settled at the confluence of the Little Arkansas and Arkansas rivers, where Wichita is today. In addition to this city, several locations in Oklahoma have been linked to the tribe's name, though they are spelled and pronounced differently. The Washita River, in western Oklahoma, and the Ouachita Mountains, in Oklahoma and Arkansas, may have been named for the same Wichita Indians.

Since its start in 1863, Wichita has grown into the largest city in Kansas, with a population of just under 280,000. Today the city is a regional trading center, known particularly for aircraft manufacturing, oil, beef, and wheat. While Sedgwick County does not produce much oil, Wichita is home to industries related to western Kansas drilling and production. The city bills itself as the "Air Capital," because it leads the world in the production of personal airplanes and because it is an important center for the manufacture of military aircraft.

Wichita's growth has been accompanied by problems, including the finding of sufficient water. The city's primary water sources are the Equus Beds, an aquifer northwest of the city, and Cheney Reservoir, located to the west. With the increasing demand for water, the city may face shortages in the years ahead. In fact, Wichita has begun to look at other water sources, which include piping water from reservoirs as far away as northeastern Kansas.

44.4 *Arkansas River.* The source of this river is high in the Rockies above Leadville, Colorado, along the Continental Divide. The Missouri is the only other Kansas river which has its source in the Rocky Mountains. The Arkansas River's name is spelled the same as the state of Arkansas, but when referring to the river and to Arkansas City, most

Kansans pronounce it ar-kan'-sas, with the emphasis on the second syllable so that it rhymes with Kansas.

The Arkansas River enters Kansas in Hamilton County; it exits into Oklahoma in Cowley County, near Arkansas City. Eventually it joins the Mississippi in southeastern Arkansas. The river's flow often appears to be meager at this point, considering the distance that it travels and the area that it drains. Through much of western and central Kansas, the Arkansas is what geologists call an influent, or "losing," stream. That is, the stream contributes water to the ground-water reservoir adjoining the stream bed. In addition, impoundments and irrigation in Colorado and Kansas have probably depleted the river even further. Only when the Arkansas reaches this part of the state does it begin to receive inflow of ground water and to become an effluent, or "gaining," stream.

44.5   The *Boeing manufacturing plant* to the southwest.

46.0   *McConnell Air Force Base* to the southeast. Traffic from this base is often visible in the skies of Wichita. In January 1965, a jet tanker from the base crashed into northeast Wichita, killing thirty people and destroying numerous homes.

46.8   *Cessna plant* to the southeast.

47.5   *Wellington Formation.* The elevation is 1,350 feet. Where the road enters Oklahoma, the elevation is 1,147, a drop of only about 200 feet over more than 50 miles. The highest point along the turnpike is in the Flint Hills, where the elevation is 1,575 feet.

Headed north, the turnpike leaves the Arkansas River Lowlands and enters the Flint Hills Uplands portion of the Osage Plains. Here it begins a gradual climb up the gentle backslope of the Flint Hills escarpment. This escarpment was formed by Permian limestones, some of which contain flint and thus resist erosion. These beds dip, or get deeper, to the west. On the east, where they crop out at the surface, they form a steep slope, or escarpment, that has been deeply dissected by stream erosion. As the turnpike climbs this escarpment, it gradually gains elevation until it reaches the crest of the Flint Hills, near milepost 111, and then descends toward Emporia.

49.5   *East Wichita interchange.* Three exits leave the turnpike at Wichita. This one joins U.S. 54, or Kellogg Street, a major east-west thoroughfare across the city.

50.0   The *Beech factory and landing field* are to the north.

51.7–52.1   *Wellington Formation.*

53.3   *Fourmile Creek.*

54.5 *Sedgwick/Butler county line.* The 1,442 square miles of Butler County make it the largest county in Kansas; it is even bigger than the state of Rhode Island. Butler County is named for Andrew Pickins Butler, a United States senator from South Carolina before the Civil War. The county was named during a session of the proslavery Bogus Legislature in Kansas, making it one of the few Kansas counties that are named for Southern sympathizers.

55.5 On the north side of the road are *deformed beds of Wellington shale.* The wavy pattern in the rock layers was probably caused when ground water dissolved underlying gypsum beds and caused the rocks above to collapse.

56.1 *Republican Creek.*

57.0 *Andover interchange.*

58.4 *Dry Creek.*

61.4 *Odell Shale.* The shale in this formation is generally colored red or green by iron compounds, although in places it is gray and yellow. It ranges from 20 to 40 feet in thickness and was deposited in the shallower part of a Permian sea about 250 million years ago.

62.2 *Winfield Limestone,* a formation deposited a few million years before the Odell Shale, which crops out at milepost 61.4. In Kansas, rock formations are generally older in the east than they are in the west. Also, deeper rock formations are generally older than shallower formations, because younger layers of rock were deposited over layers of existing formations.

62.5 *Badger Creek.*

63.5 *Whitewater River,* a major tributary of Walnut River, which joins the Arkansas in southern Cowley County. The town of Whitewater is located along the river in western Butler County. The artist Frederic Remington owned a sheep ranch near Whitewater during the early 1880s. The high school near Whitewater is named after him.

63.9 *Gage shale,* overlain by the lower part of the *Winfield Limestone.*

65.0 *Towanda service area.* The elevation is 1,350 feet.

66.9 and 67.2 *Cresswell Limestone Member* of the Winfield Limestone.

67.8 For several miles to the north the turnpike passes through the *El Dorado oil field.* It includes 747 wells, which have produced nearly 300 million barrels of oil since the field was discovered in 1915, making it the top producing oil field in Kansas history.

An oil field near El Dorado in the early days of the oil business in central Kansas

Oil from these wells comes from underground rock formations that trap the oil and allow it to accumulate. In this area the traps are created by subsurface domes of rock and by anticlines, or arching folds in the rock strata. These folds are located directly over the Nemaha Ridge, a buried granite mountain range that extends from southeastern Nebraska into northern Oklahoma. These domes and anticlines—and ultimately the oil itself—were discovered by the careful surface mapping of the angles at which the exposed rocks dipped into the ground. Although the folds may be only subtly expressed at the surface, they often increase in magnitude with depth, providing large traps for oil. One of the first geologists to study this structure was Erasmus Haworth, director of the Kansas Geological Survey from 1895 to 1915.

In 1918 the field produced 29 million barrels—64 percent of the oil produced in Kansas that year and at least 6 percent of all the oil produced in the United States, production that was especially welcomed by the Allies during the waning days of World War I. Production from the field continues. In 1984, it produced over a million barrels, helping make Butler County the fifth-largest oil-producing county in Kansas that year. In all, Butler County has produced more than 500 million barrels of oil, making it the leading oil-producing county in the state's history. The top ten all-time oil-producing counties are, in order, Butler, Barton, Russell, Ellis, Rice, Greenwood, Stafford, McPherson, Rooks, and Graham.

69.5   *Tank farms* to the southeast.

70.5   *El Dorado interchange.* Originally El Dorado was spelled as one word, but a newspaper editor accidentally made it into two words and refused to correct his mistake. The name is pronounced with a long *a,* so that Dorado rhymes with tornado. This exit connects the turnpike with U.S. 77 and U.S. 54.

70.8   *Oil refinery* about 1.5 miles to the southeast.

71.4   *Constant Creek,* the only creek with this name in Kansas.

72.0   At this point the turnpike passes over the crest of *Oil Hill dome,* one of the major producing areas in the El Dorado field. Stapleton No. 1, the well that opened the El Dorado field, was drilled in the Oil Hill dome in 1915, striking oil in sandstone formations at depths of 660 and 2,500 feet. The turnpike passes through the shallower oil-producing formation—a layer in the Wabaunsee Group—between Emporia and Topeka.

Also, the former company town of Oil Hill, located 0.3 miles north of here, was established following the discovery of the El Dorado field. It became a Cities Service company town, and by the late 1920s it had a population of nearly 3,000. With the depression and the subsequent decline in oil production, the town's population dwindled, and Oil Hill finally disappeared.

72.5   *Cresswell limestone.*

73.3   *Oil refinery* to the southeast.

73.8   *Winfield Limestone.*

74.0   *Winfield Limestone* on the northwest side of the road. This point is the northeast edge of the El Dorado oil field.

75.0   West branch of *Walnut River.*

75.4   *U.S. 77 overpass.* U.S. 77 runs south through El Dorado and on to Winfield and Arkansas City. South of El Dorado is one of the largest known caves in Kansas, extending at least a mile into a thick ledge of limestone, although it has not been completely explored and is on private property that is currently off-limits to the public. Most Kansas limestones are too thin to form large caverns, but the state does have more than 500 caves, according to the Kansas Speleological Society. Most of these are either limestone caves in eastern Kansas or gypsum caves in the Red Hills of south-central Kansas.

75.5   *Doyle Shale.*

Stapleton No. 1 (Butler County), the well that discovered the El Dorado field, the largest oil field in Kansas history (courtesy of the Kansas State Historical Society)

79.9   A small arm of *El Dorado Lake*.

80.8   *Cole Creek* arm of El Dorado Lake. This is where the creek ran before it was covered by the lake waters. Now under several feet of water, the old creek channel probably still exists. These channels often make good fishing spots.

81.5   *Walnut River* arm of El Dorado Lake.

83.0   A *quarry* in the Fort Riley Limestone Member of the Barneston Limestone, northwest of the highway.

84.3 and 85.3   *Fort Riley limestone*, a light gray or tan limestone that often forms a conspicuous outcrop, or rim rock, in the pastures of the Flint Hills.

87.1   *Towanda Limestone Member* of the Doyle Shale.

87.7   The elevation here is 1,450 feet.

88.5   *Towanda limestone*.

92.0   *The Walnut River* joins the Arkansas immediately south of Arkansas City, which was once called Walnut City.

92.4   *Cassoday interchange*. One-half mile south on K–177 is Cassoday, a town that bills itself as the Prairie Chicken Capital of the state. Kansas has one of the largest remaining populations of prairie chickens on the North American continent. During courtship the birds make a resounding booming noise, and the areas where they mate are called "booming grounds."

To the north, K–177 takes a scenic path through the Flint Hills to Cottonwood Falls, which is known for the Chase County Courthouse, the oldest county courthouse still in use in Kansas. Completed in 1873, the courthouse was designed in the French Renaissance architectural style and was built of Cottonwood limestone. Strong City is a mile north of Cottonwood Falls. Sometimes called the Twin Cities of Kansas, these two towns were connected by a horse-drawn trolley during the 1800s. Today, Strong City is the home of the Flint Hills Rodeo, held every June.

94.9   The elevation here is 1,500 feet.

95.9   The *quarry* to the northwest is in the Fort Riley Member of the Barneston Limestone, which also crops out along the turnpike here.

96.3   *Butler/Chase county line*. Chase County is named for Salmon P. Chase, secretary of the Treasury under Abraham Lincoln and later chief justice of the United States Supreme Court. Chase's portrait graces the $10,000 bill.

97.0   *Matfield Green service area*. Matfield Green, a Kansas town named for a London suburb, is located several miles north of here.

97.9 *Florence Limestone Member* of the Barneston Limestone. This limestone contains layers of chert, or flint as it is called locally. The Florence is one of the chert-bearing limestones whose rubbly outcrops give the Flint Hills their name.

While most of the rest of Kansas has been cultivated, these Flint Hills remain largely in native grass. Because much of the ground is too rocky for cultivation, ranchers use it for pasturing cattle, so that today it is one of the last great preserves of the tall-grass prairie that once stretched from Texas to the Dakotas and as far east as Ohio. Though there is disagreement about the extent of the Flint Hills in Kansas, most geologists believe they extend from Marshall County, in the north, to Cowley County, in the south, and on into Oklahoma, where they are called the Osage Hills. However, the heart of the Flint Hills seems to lie in these vast pastures of Chase and Morris counties.

The tall grasses in these prairies are predominately big and little bluestem, switch grass, and Indian grass. Shorter grasses, which are more common in western Kansas, can be found in drier parts of the Flint Hills. However, bluestem grass is the most apparent, providing the Flint Hills with their other name, the Bluestem Hills. Except near river bottoms, trees are rare in the Flint Hills, in part because ranchers burn off the hills every spring to eliminate undesirable weeds, grasses, and small trees from their pastures. The grasses then quickly regenerate from seeds or roots. Trees are generally restricted to river bottoms, where there is enough ground water and protection from wind and fires.

98.3 *Matfield Shale* and overlying *Florence limestone*. The Florence contains a number of fossils, including bryozoans, colonial animals that were common in Pennsylvanian and Permian seas. Though they resemble corals, bryozoans did not build reefs; instead, they attached themselves to coral reefs or to other animals living in the reefs.

98.5 *Kinney Limestone Member* of the Matfield Shale.

98.9 *Speiser Shale* and overlying *Threemile Limestone Member* of the Wreford Limestone. The Wreford (pronounced Ree'-ford) is another of the chert-bearing limestones that make up the Flint Hills.

100.0 *South Fork of the Cottonwood River.* Eight miles to the north, in a Flint Hills pasture, is the Knute Rockne Memorial Monument, marking the site where the Notre Dame football coach was killed in a plane crash in 1931.

Here in Chase County, on this branch of the Cottonwood, the

Kansas Fish and Game Commission is attempting to reintroduce river otters into Kansas. River otters were once found along the major rivers in the state, but their population was depleted during settlement, and the last native specimen was captured near Manhattan in 1904. However, the new otters, brought to Kansas from Minnesota, appear to be thriving in Chase County. Biologists say they travel throughout the oxbows and tributaries of the Cottonwood, feeding on crayfish and fish. The otters spend most of their time along the river in old beaver dens, but occasionally they cross open stretches of grassland on their way to other bodies of water.

East of here, in Lyon County, the Cottonwood joins the Neosho River. Emporia is situated near the confluence of these two rivers. Many Kansas cities are located near river junctions: Wichita, at the confluence of the Arkansas and Little Arkansas rivers; Salina, at the confluence of the Smoky Hill and Saline; Junction City, at the confluence of the Smoky Hill and Republican; Manhattan, at the confluence of the Kansas and Blue; and Kansas City, at the confluence of the Missouri and Kansas rivers.

100.2 *Crouse Limestone,* which is cherty in some locations, ranges in thickness from 6 to 18 feet.

100.5 *Funston Limestone* and overlying *Speiser Shale.* The Speiser is generally a red shale that is up to 35 feet thick.

100.8 *Wreford Limestone.*

101.3 *Little Cedar Creek.*

101.7 *Blue Rapids Shale,* which is named after a small town in northeastern Kansas where gypsum is mined.

101.9 *Little Shaw Creek.* Cherty gravel alluvium is exposed in the creek bed west of the highway. Alluvium is the unconsolidated material—usually sand, gravel, and clay—deposited by rivers. Often the alluvium is saturated by water, so that it may be a source of ground water.

103.2 *Funston Limestone,* overlain by the *Speiser Shale* on the west side of the highway.

104.1 *Schroyer limestone,* the youngest and uppermost member of the Wreford Limestone, on the east side of the road. Layers of flint in limestones such as the Schroyer have helped the Flint Hills to resist erosion, giving the area a rolling, undulating topography. The elevation here is 1,500 feet.

104.7 *Speiser Shale* and overlying *Threemile limestone.*

105.0 *Funston Limestone* on the west side of the road. This formation is named after a camp at Fort Riley, which is in turn named after Frederick Funston, who grew up near Iola and was a classmate of William

Allen White's at the University of Kansas before leading a Kansas regiment in the Spanish–American War.

105.1   *Sharpes Creek.*

106.0   *Speiser Shale* and overlying *Threemile limestone.*

107.0   *Kinney limestone.* For the next 4.5 miles to the north, the turnpike follows the drainage divide between the watershed of the Neosho River, to the northwest, and the Verdigris River to the southeast. Both streams join the Arkansas River near Muskogee, Oklahoma, and only then does runoff from one side of the highway mix with that of the other.

107.9   *Schroyer limestone.*

109.0   *Wymore shale* and overlying *Kinney limestone,* both members of the Matfield Shale.

109.9   *Havensville shale* and overlying *Schroyer limestone,* members of the Wreford Limestone.

110.1   *Threemile limestone.*

110.3   *Havensville shale.* This shale is named after a small town in northeastern Pottawatomie County. Many rock layers here are named after locations in northeastern Kansas, in part because it was the first area in Kansas to be carefully examined by geologists. Scientists were familiar with the geology of eastern Kansas by the mid 1860s, but they waited until the 1870s and the end of the Indian uprisings to explore much of the rest of the state.

110.6   *Havensville shale* and overlying *Schroyer limestone.*

110.8   *Bazaar cattle crossing and cattle-holding pens.* Ranching dominates the economy in this part of Kansas; it also plays a big role in the economy of the rest of the state. Kansas is the leading beef-packing state in the nation. In 1980 the sale of cattle and calves generated $2.8 billion, more than 50 percent of the income received by Kansas farmers.

These cattle pens along the turnpike provide a place for ranchers to collect and ship cattle, now that trucks have replaced railroads as the primary mode of transporting cattle. Cattle outnumber people in many Kansas counties, especially here in Chase County, where, in 1980, there were 56,200 cattle and calves, compared to a human population of just over 3,000. That's a ratio of about eighteen cattle for every person.

The elevation here is about 1,575 feet, the highest point along the turnpike.

111.0   *Havensville shale,* overlain by *Schroyer limestone* on the east side of the road.

111.5 *Speiser Shale,* overlain by the *Threemile Limestone Member* of the Wreford Limestone.

111.8 *Blue Rapids Shale,* upward through the *Funston Limestone* and the *Speiser Shale,* on the east side of the road.

112.3 *Bloody Creek* on the west side of the road.

112.8 *Bader Limestone.* Bader is pronounced to rhyme with raider. The elevation is 1,400 feet. Southeast of the road is a spring that flows even in dry times.

113.2 *Bader Limestone* and overlying *Easly Creek Shale.* The Easly Creek is composed mostly of red, green, and gray shale, although it contains layers of limestone in places. Shale occurs in many colors, due to the presence of iron-bearing minerals and organic material. Iron compounds may color it red or green, while organic content can cause shades of gray or black. In northern Kansas, this formation contains beds of gypsum up to 8 feet thick, which are mined near Blue Rapids, in Marshall County. This is also approximately halfway between Wichita and Topeka.

113.5 *Crouse Limestone.*

113.7 *Blue Rapids Shale* and overlying *Funston Limestone* on the west side of the road.

114.0 *Speiser Shale* and overlying *Threemile limestone.*

114.7 *Crouse limestone* and overlying *Blue Rapids Shale.*

115.2 *Cottonwood limestone.* This limestone was quarried to provide building material for the Chase County Courthouse at nearby Cottonwood Falls.

115.7 *Neva Limestone Member* of the Grenola Limestone. The Neva is a gray limestone that ranges in thickness from 9 to 28 feet. In places it contains fossilized fusulinids, one-celled organisms that populated the Permian seas. About the size and shape of a grain of wheat, fusulinids probably floated on the surface of the seas, feeding on smaller microorganisms.

116.0 *Johnson Shale,* up through the *Red Eagle Limestone,* the *Roca Shale,* and the *Grenola Limestone.*

116.2 *Jacob Creek* and the *Chase/Lyon county line.* The Bogus Legislature originally named Lyon County after John Breckinridge, vice-president under James Buchanan and later secretary of war for the Confederacy. After the Civil War, the Kansas Legislature thought it more appropriate to name the county after Gen. Nathaniel Lyon, a one-time commander at Fort Riley who fought the Confederates at the Battle of

Wilson's Creek, a few miles south of Springfield, Missouri.

116.8 *Sallyards limestone* and overlying *Legion shale*, both members of the Grenola Limestone.

116.9 *Burr Limestone Member* of the Grenola Limestone.

117.3 *Grenola Limestone* on southeast side of the turnpike. This formation is named for a small town in Elk County in southeastern Kansas.

118.0 *Johnson Shale.*

119.0 *Grenola Limestone.* Emporia is visible to the northeast.

121.7 The elevation here is about 1,200 feet as the road drops north into the valley of the Cottonwood and Neosho rivers.

122.5 Headed south, the turnpike passes into Permian rocks that make up the Flint Hills. Headed north, the road passes from the Permian into rocks of the preceding geologic period, the Pennsylvanian. This subtle boundary in geologic time occurred about 280 million years ago. Except for a small area northeast of Emporia, where the road climbs back into Permian rocks, Pennsylvanian formations underlie the road for the rest of its northeastward length in Kansas.

123.1 and 123.9 *Phenis Creek.*

124.7 *Cottonwood River.* The channel elevation of the river is about 1,100 feet. The Cottonwood and the Kansas–Smoky Hill rivers are the only streams that cut completely across the Flint Hills.

125.9 Here the turnpike passes over the main line of the *Atchison, Topeka, and Santa Fe Railroad.* Emporia is a major railroad center; its rail yards are visible to the east. This railroad follows the gentle grade of the Cottonwood River and its tributary Doyle Creek on its way through the Flint Hills and on to the former cow town of Newton. By following the Cottonwood, the Santa Fe avoids the steep climb up the rugged east slope of the Flint Hills escarpment, which the turnpike encounters between Emporia and Cassoday.

127.0 *Emporia interchange.* At this point, U.S. I–35 leaves the turnpike and heads toward Kansas City by way of Ottawa. The geology along I–35 from Emporia to Kansas City is described in chapter 7. Also, U.S. 50 goes west from here, passing through Strong City, Newton, and points west.

Emporia is perhaps best known as the home of William Allen White, the long-term editor of the *Emporia Gazette* and a nationally famous author and editorialist. White was born in Emporia but grew up in El Dorado. In addition to editing the *Gazette,* White was a progressive politician who once ran as an independent candidate for governor of Kansas. Today the School of Journalism at the University of Kansas is named for White, who attended KU for a time in the 1880s.

128.8 *Nebraska City Limestone Member* of the Wood Siding Formation.

129.2 *The Neosho River,* whose source is in Morris County, northwest of Council Grove. In Morris County the river is impounded to form Council Grove Lake. From there it flows straight southeast for 100 miles to Iola, where it bends south. Along the way it is impounded to form John Redmond Reservoir. The Flint Hills National Wildlife Refuge is located at the upper end of this lake, about 12 miles southeast of Emporia.

Because of its size, location, and dependable flow, the Neosho provides habitat for several unusual kinds of fish. One is the shortnose gar, which has a long jaw, lined with menacing needlelike teeth. Gar can survive in shallow, warm river waters that hold little oxygen in the summer, a characteristic that allows gar to survive longer out of water than most fish can. Gar feed on other fish and lay their eggs on rocky river bottoms, such as those found along the Neosho.

This river is also one of the few homes of the Neosho madtom, a member of the catfish family that grows only a few inches long. Because of impoundments and pollution in the river, populations of this fish have dwindled to the point that some scientists consider it to be the most endangered fish species in Kansas. The river also harbors some big fish. The largest flathead catfish ever landed in Kansas was taken from the Neosho in 1966; it weighed more than 86 pounds and was more than 55 inches long.

130.0 *Troublesome Creek.*

130.3 *Allen Creek.*

131.8 *Emporia service area.*

132.2 *Stillman Creek.* This is one of several area creeks that drain southeastward into the Neosho River.

135.8 *Dow Creek.*

139.0–140.0 For a mile to the northeast the turnpike passes over rocks of Permian age in the lower part of the *Admire Group.* The road then returns to rocks of Pennsylvanian age.

139.9 This location marks the crossing of one of the *major drainage divides* in the country. To the south is the watershed of the Arkansas River, and to the north is the drainage basin of the Missouri. One must travel to southern Oklahoma to reach the other side of the Arkansas River basin—the Arkansas/Red River divide. The Arkansas River basin extends westward to the Continental Divide in central Colorado. To the north, the Missouri River basin reaches to the southern prairie provinces of Canada and westward to the Continental Divide in western Montana. Runoff north of here reaches the Missouri River, which joins the Mississippi north of St. Louis. South of this divide, runoff enters the Arkansas River, which joins the Mississippi in southeastern Arkansas.

140.7   *Grayhorse limestone,* overlain by the *Brownville limestone,* both members of the Wood Siding Formation. Many of these Pennsylvanian formations extend into southeastern Nebraska and are named for small towns in that area, such as Brownville. Both the Grayhorse and the Brownville limestones contain numerous fossils. The Brownville occasionally produces trilobites, which are rare in Pennsylvanian rocks in Kansas. Trilobites were small, segmented animals—now extinct—that are related to horseshoe crabs, spiders, and other animals called arthropods. Trilobites burrowed into the soft sediments on the floors of Pennsylvanian seas.

141.4   *Nebraska City Limestone Member* of the Wood Siding Formation.

141.8   *Duck Creek.*

142.9   *Dry shale* and *Grandhaven limestone,* members of the Stotler Limestone. The regular sequence of limestones and shales is often apparent in these Pennsylvanian outcrops. Geologists believe that rocks were deposited in a specific sequence—sandstone, shale, limestone, shale, sandstone—as the level of the seas changed in eastern Kansas. Sandstone was deposited where the seas were shallowest. As the seas deepened, shale was laid down. Finally, when the seas were at their deepest, limestone was deposited. As the seas began to retreat, shale and then sandstone were left behind, creating a cyclic sequence of deposition. Such alternating limestones and shales, geologists believe, indicate slight fluctuations in the depth of the sea.

143.2   *Hill Creek.*

143.9   *Pillsbury Shale* and overlying *Dover Limestone Member* of the Stotler Limestone.

144.1   *One Hundred and Fortytwo Mile Creek.* Creeks named by distance are common in Kansas—from One Mile Creek, east of Fort Riley, all the way up to this creek, which received its name because it was about 142 miles from the beginning of the Santa Fe Trail in Independence, Missouri.

144.8   *Dover limestone.*

146.7   *Admire interchange* and *U.S. 56 exit.* Twenty-five miles west of here, U.S. 56 passes through Council Grove, a historic spot on the Santa Fe Trail. To the east it passes through the old coal-mining area of Burlingame and Scranton, on its way to Baldwin City and Kansas City. The geology along U.S. 56 is described in chapter 9 of this book.

147.9 *Pillsbury Shale* and overlying *Dover limestone.*

148.0 *Elm Creek.* Downstream about seven miles, Elm Creek and One Hundred and Fortytwo Mile Creek converge to form the Marais des Cygnes River, a French name meaning "marsh of the swans." One of the larger streams in eastern Kansas, the Marais des Cygnes is impounded to form Melvern Lake in southern Osage County. The Marais des Cygnes flows through Ottawa and leaves the state near Trading Post. Once in Missouri, it is called the Osage River; it forms Truman Reservoir and Lake of the Ozarks before flowing into the Missouri River near Jefferson City.

149.0 To the northwest 1.5 miles is *Log Chain Creek.*

151.4 *Salt Creek.*

153.3 *Pillsbury Shale* and overlying *Dover limestone.*

154.3 *Lyon/Wabaunsee county line.*

154.8 *Osage/Wabaunsee county line.*

155.3 *Soldier Creek,* which flows into Dragoon Creek.

155.7 Approximate crossing of the *Santa Fe Trail.* Although ruts from the trail are visible in many locations in Kansas, they are not apparent here.

156.7 *Dover limestone* on the east side of the road.

157.0 *Dragoon Creek.* Zebulon Pike crossed Dragoon Creek in his 1806 trek across Kansas. From here, Dragoon Creek flows into Hundred and Ten Mile Creek about 20 miles downstream. Together they are impounded to form Pomona Lake. Hundred and Ten Mile Creek then joins the Marais des Cygnes near Quenemo.

157.8 *Dover limestone.*

159.9 *Sandstone* layer in the Pillsbury Shale. Rock formations are named according to the type of rock that makes up most of a given layer. For example, shale is the most common type of rock found in the Pillsbury Shale, but the Pillsbury contains smaller portions of other rock types, including this sandstone bed.

160.5 *Switzler Creek.*

162.8 At about this point the northbound turnpike leaves the Marais des Cygnes/Osage River drainage basin and enters the Kansas River drainage basin.

163.4 *Tarkio Limestone Member* of the Zeandale Limestone.

164.6   *Willard Shale* and overlying *Zeandale Limestone.*

165.2   *Elmont Limestone Member* of the Emporia Limestone, overlain by *Willard Shale.*

165.5   *Osage/Shawnee county line.*

167.3   *Soldier Creek Shale,* which is overlain by the *Burlingame limestone.*

167.7   *Wakarusa River.* This river has its source just to the west, in eastern Wabaunsee County. It flows to the east and is impounded near Lawrence to form Clinton Lake, before joining the Kansas River near Eudora.

169.0   The elevation here is 1,100 feet.

170.8   *Sixmile Creek.*

172.3   *White Cloud Shale Member* of the Scranton Shale on the east side of the road. The White Cloud contains few fossils and ranges in thickness from 30 to 80 feet.

173.7   *Burlingame Limestone Member* of the Bern Limestone south of the road.

175.8   South branch of *Shunganunga Creek.*

176.0   The turnpike passes through a small area of partially obscured orange glacial deposits that contain pink Sioux quartzite boulders, which were carried from South Dakota and Minnesota into Kansas by glaciers. This location is near the southern boundary of the glaciers' movement into Kansas less than one million years ago. The ice sheets rearranged the landscape, leaving behind the rolling, hilly topography that characterizes northeastern Kansas.

To the northwest, partly obscured by hills, Burnett's Mound appears on the horizon. This 200-foot-high hill is capped by Bern Limestone. According to legend, it protected Topeka from tornadoes. On 8 June 1966, however, a funnel passed over the mound and cut a swath 0.5 miles wide through the heart of Topeka, killing thirteen people and destroying more than $100 million worth of property. Even the dome of the Statehouse was damaged.

177.0   *South Topeka interchange,* which provides access to U.S. 75, a north-south route through downtown Topeka. I-470 extends west to I-70, and U.S. 75 Bypass runs northwest to U.S. 75. From this point north, the turnpike follows I-470 to the east Topeka interchange, a stretch of road that has been designated the Martin Luther King Memorial Highway.

178.0   *Howard Limestone,* including an exposure of the Nodaway coal bed. Though coal deposits are common in the eastern third of Kansas, few outcrops are visible along the turnpike. This coal bed, which ranges from a few inches to 2 feet in thickness, has been mined extensively in the past, producing nearly 12 million tons from underground and surface diggings. More than 97 percent of the Nodaway coal mined in Kansas came from Osage County. Today the market for coal from this formation is hindered by the coal's high sulfur content—as much as 6 to 8 percent.

180.0–180.2   *Topeka Limestone.*

180.9   *Deer Creek.* The elevation here is 910 feet, more than 600 feet below the turnpike's highest point in the Flint Hills.

181.7   *East Topeka interchange.* Here the turnpike joins I–70. Eastbound I–70 heads toward Lawrence and Kansas City, while westbound I–70 passes through downtown Topeka before continuing west to Junction City, Salina, and Denver. For a guide to the geology of the remainder of the turnpike from this point to Kansas City, see chapter 2.

56·U.S. HIGHWAY·56

The Old Santa Fe Trail

0.0 *Kansas/Oklahoma state line.* U.S. 56 cuts diagonally across Kansas, from Elkhart, here in southwestern Kansas, to Kansas City, in the northeastern corner of the state. Like many Kansas highways, U.S. 56 follows a pioneer trail—in this case the old Santa Fe Trail, which went from Independence to Santa Fe.

Here in southwestern Kansas, on the sandy soils of the High Plains, U.S. 56 reaches its maximum elevation in Kansas, 3,610 feet. From here eastbound, it heads down the face of the High Plains, gradually losing elevation. This point also marks the southern edge of Elkhart.

West of here 0.2 miles is the southern terminus of K–27, which goes north through the western tier of Kansas counties. About 9.5 miles north of here, west of K–27, is *Point of Rocks,* an outcrop of Jurassic and Tertiary formations along the Cimarron River. The Jurassic Period occurred between 200 and 150 million years ago, immediately before the Cretaceous Period. During the Jurassic, sandstones and shales were deposited over the western fifth of Kansas. These Jurassic formations were subsequently covered during the Cretaceous; thus, they are found only in the subsurface in Kansas, except here in the southwestern corner of the state, where they jut out in a few locations. One of those locations is Point of Rocks, where a few exposures of bright red Jurassic shales are visible, covered by younger layers of the Ogallala Formation.

These Jurassic and Tertiary rocks played a role in the history of the West. The road to Santa Fe reached Gray County, northeast of here, before dividing near present-day Cimarron. One route continued west along

Point of Rocks, an outcrop of the Ogallala Formation and rocks of Jurassic age, in Morton County, north of U.S. 56

the Arkansas River, taking a longer but well-watered route to New Mexico. Another branch of the trail forded the Arkansas River at the Cimarron crossing and struck a more direct southwesterly course to Santa Fe, parallel to and a few miles north of present-day U.S. 56. This route was called the Cimarron Cut-off. Though shorter, this cutoff crossed an area of intermittent streams and widely spaced springs, which made it hazardous in dry weather, thereby earning it the additional nicknames of the Dry Route and La Jornada, "the journey."

Point of Rocks was a landmark on the Dry Route to Santa Fe. Because there was reliable water at nearby Middle Spring, pioneers used the area around Point of Rocks as a campground, and wagon ruts are still visible in the countryside east and west of Point of Rocks.

0.5 The *Elkhart* business district is northwest of here. Elkhart, established in 1913, is the southwestern-most city in Kansas, closer to the state captials of Wyoming, Colorado, New Mexico, and Oklahoma than it is to Topeka.

1.0 The north edge of *Elkhart*, seat of Morton County. Glenn Cunningham, probably the most famous middle-distance runner in Kansas history before Jim Ryun in the 1960s, represented Elkhart High School when he set a Kansas high-school mile record in 1929. He went on to the University of Kansas and finished second in the 1,500-meter race at the 1932 Olympics. For a time in the 1930s, Cunningham held the world record time for the mile: 4:06.7 minutes.

3.5–4.5   The wells here are in the *Taloga oil field,* which was discovered in 1955. The field produces oil from rocks that are 3,800 to 4,400 feet deep.

4.5–6.5   *Sand hills* are common throughout southwestern Kansas. They are the result of wind deposition during the past million years of geologic history. These sand hills provide some of the rare relief here in a physiographic area known as the High Plains.

5.5   The elevation is 3,500 feet, as the eastbound highway begins a gradual descent down the face of the High Plains.

8.0   The wells south of the highway are in the *Wilburton oil and gas field,* which is responsible for 10 million barrels of oil; it is the second-most-productive field in the history of Morton County. However, southwestern Kansas is probably better known for natural-gas production. Most of that gas comes from wells in the Hugoton field—which underlies much of southwestern Kansas—but substantial amounts also come from fields, such as the Wilburton, that produce gas from formations below the Hugoton.

9.2   The town of *Wilburton,* also the west edge of a tract of sand hills.

10.0   In the sand hills south of the highway there is a small blowout, an area where strong winds have blown away the soil and sand, denuding the landscape.

10.8   Morton County is near the center of a region that was called the dust bowl when it suffered severe dust storms during the drought of the 1930s. The dust bowl extended into Colorado, New Mexico, Texas, and Oklahoma. Overcultivation of marginal land, poor conservation practices, scanty precipitation, and crop failures all contributed to the dust storms. The silty and sandy soil of this region is highly susceptible to blowing when its natural vegetative cover has been removed, leaving it to the mercy of the strong winds of the Great Plains. The Cimarron National Grassland was formed when the federal government bought up some of the most severely wind-eroded land and took it out of cultivation, attempting to restore its natural vegetation. This is the eastern boundary of the *Cimarron National Grassland,* a checkerboard tract of 107,000 acres, which is managed and controlled by the United States Forest Service. These grasslands include much of the land along the Cimarron River valley in Morton County.

15.0 East edge of the sand hills.

17.4 The town of *Rolla*.

17.7 Roadside park and the junction with K–51, which leads north and west 18 miles to Richfield.

19.7 West edge of a tract of sand hills.

21.8 *Morton/Stevens county line*. Morton County is named for Oliver P. Morton, a senator from Indiana from 1867 to 1877. Stevens County is named after Thaddeus Stevens, an antislavery congressman from Pennsylvania.

22.0 The elevation is 3,250 feet.

25.0 East edge of sand hills.

26.2 The town of *Feterita* and the junction with *K–25*, which runs south, connecting with a highway to Guymon, Oklahoma.

27.7–31.4 These wells are in the *Gentzler oil and gas field*, which was discovered in 1971 and produces natural gas and oil from wells about 6,000 feet deep. Stevens County has long been the leading gas-producing county in the state. In 1984 the county produced over 92 billion cubic feet of natural gas, nearly one-sixth of all the natural gas produced in Kansas that year. Most of that Stevens County gas, about 74 billion cubic feet of it, came from wells in the Hugoton field, which underlies much of southwestern Kansas and extends into Oklahoma and Texas.

33.1 West edge of *Hugoton*, the seat of Stevens County. The town is named after Victor Hugo.

33.6 This is the intersection with northbound K–25, eastbound K–51, and eastbound U.S. 270.

34.9 North edge of *Hugoton*.

35.8 Junction with northbound K–25 and U.S. 270. Ulysses is 25.5 miles north of here. About 13 miles north, on the Cimarron River, is the site of Wagon Bed Springs. Although it is now dry due to lowered water tables brought about by irrigation, this was once an important watering spot on part of the Santa Fe Trail. In those days it was known as Lower Springs. Middle Springs is near Point of Rocks, north of Elkhart; and Upper Springs is farther up the Cimarron River, near the point where it crosses from Oklahoma into Colorado.

36.0 Southwest edge of a tract of *sand hills*.

39.5 Oil well in *Gentzler North field*. Beyond the well is a blowout. When the native vegetation in these sand hills is disturbed or destroyed, these stabilized dunes may become active. Blowouts are areas where the denuded sand is eroded and carried by the wind. A dune may form downwind from the blowout, burying additional vegetation and killing it. In this way, a blowout may grow into an active dune field. Active areas of dunes are scattered throughout the sand hills of southwestern Kansas, and landowners fight a constant battle to stabilize the shifting sand.

40.0 *Sand hills,* covered by sand sage.

40.5 A *feed lot* to the north.

41.0 Northeast edge of the *sand hills.*

47.3 The town of *Moscow.*

51.6 The town of *Cave.* Despite this town's name, caves are rare in this area. Few rocks crop out at the surface, and the unconsolidated materials that cover the ground are generally too loose to support any void spaces such as caves. The closest legitimate caves are in the Red Hills to the east.

53.5 The elevation here is 3,000 feet.

54.0–56.0 The *Cutter oil and gas field,* discovered in 1961.

56.1 *Stevens/Seward county line.* U.S. 56 cuts across the northwestern tip of Seward County for less than a mile.

56.9 *Seward/Haskell county line.*

57.5 *The Cimarron River* has its source in the lava-capped mesas of northeastern New Mexico; it flows eastward through the Oklahoma Panhandle, across the southeastern corner of Colorado, and into southwestern Kansas. Here it makes a big bend, similar to the Great Bend in the Arkansas River. In this area the Cimarron curves to the southeast and heads for Oklahoma. After returning to Kansas for a short distance in the Red Hills of Clark and Comanche counties, the Cimarron reenters Oklahoma, joining the Arkansas River west of Tulsa. Though this is the major river in southwestern Kansas, it is dry much of the time, carrying water mostly during spells of heavy precipitation. Although irrigation and lowered ground-water levels have probably contributed to the lessened flow, the Cimarron has historically been an undependable source of water. The elevation at the stream bed here is 2,800 feet.

As is apparent throughout southwestern Kansas, the Cimarron's stream bed is sandy and therefore is a good source of sand and gravel. A large sand and gravel pit operates here, along the north side of the highway on the east side of the river. This sand and gravel is Quaternary in age and is widespread in southwestern Kansas; it is often covered by a veneer of loess or windblown sand. Beneath the sands and gravels of Quaternary age are similar deposits of the Ogallala Formation, which was deposited in the Tertiary Period. The combined thickness of these deposits is more than 300 feet in this area. They are is partially saturated with ground water, which feeds the numerous irrigation wells in southwestern Kansas.

58.6 *Natural gas compressor station.* These stations, common throughout southwestern Kansas, are used to develop pressure in the pipe lines that carry natural gas out of the area.

61.8 The west edge of *Satanta,* which is named for a chief of the Kiowa Indians, who roamed southwestern Kansas until the mid 1800s. This is also the junction of northbound K-190. The town of Ryus is eight miles to the northwest.

62.9 East edge of *Satanta* and the junction with southbound K-190.

65.5 Wells south of the road are part of the *Victory oil and gas field,* discovered in 1960. In addition to this field, other oil fields in Haskell County have notable names, such as the Pollyana and the Pleasant Prairie.

69.3 *Roadside park* and junction with U.S. 160 and U.S. 83, which are described in chapters 1 and 4 of this book. Along U.S. 83, Liberal is 33 miles to the south, and Garden City is 34 miles north.

70.0–71.2 *Sublette,* the seat of Haskell County. Many of the town's streets are named after explorers, such as Zebulon Pike and Kit Carson. The town itself is named after the mountain man William Sublette.

74.0 This is the east edge of the *Hugoton gas field,* which underlies much of southwestern Kansas. Westbound U.S. 56 travels over the Hugoton from here to the Oklahoma border.

77.8 The town of *Tice.*

82.1 *Haskell/Gray county line.*

83.4 The town of *Copeland.*

87.2 Junction with K-144, which runs 17 miles west to intersect with U.S. 160 and U.S. 83.

87.8 *Crooked Creek,* which, in this area, is generally dry and gets its name from the angular course it takes downstream on its way through Ford and Meade counties. Crooked Creek and the Cimmaron River are the only named "streams" that U.S. 56 crosses in the 120 miles between Elkhart and Dodge City.

89.0 A *feed lot* on the south side of the road. Today, southwestern Kansas is one of the major beef-producing areas of the world, but in the 1800s it was home to vast herds of buffalo. The animals grazed on the short drought-tolerant buffalo grass that covered the plains before they were cultivated.

94.4 *Montezuma.*

100.2 Junction with *K–23*. Meade is 24 miles to the south on K–23, and Cimarron, the seat of Gray County, is 12 miles to the north.

101.5 The town of *Haggard.*

104.5 The elevation here is 2,750 feet.

106.5 The town of *Ensign.*

107.6 *Gray/Ford county line.* Ford County is named for James Ford, a colonel in the Colorado cavalry during the Civil War. Gray County is named after Alfred Gray, secretary of the Kansas State Board of Agriculture from 1871 to 1880. Much of today's Gray County was originally part of Foote County. In the 1870s, southwestern Kansas included several counties with names that have since been changed, including Kansas County (now Morton), Arapahoe (now Haskell), Sequoyah and Buffalo (now part of Finney), and Foote. For a time in the 1880s, these counties were absorbed into Finney County; later, most of them became separate entities. The two that didn't—Sequoyah and Buffalo—were combined to give Finney a shape that is unique among Kansas counties.

114.6 *Indian treaty boundary.* This boundary marks the western edge of lands that were granted to the Osage Indians in the 1800s. These lands extended eastward to the Cherokee Neutral Lands of present-day Cherokee and Crawford counties in southeastern Kansas. West of this boundary were the hunting grounds of the Comanches.

119.0 This point marks the approximate boundary along U.S. 56 between the High Plains, to the southwest, and the Arkansas River Lowlands to the northeast. The lowlands consist of the alluvium, terrace deposits, and windblown sand that the Arkansas River deposited.

119.6 *Abandoned irrigation canal* north of the highway. Lack of water has long been a problem in cultivating southwestern Kansas. The area's major stream, the Arkansas River, regularly carried water, but it was too far from most fields to do any good. In the 1880s, however, local developers decided to build an extensive canal system, which would take water out of the Arkansas and distribute it to farmers in Gray, Ford, and Edwards counties.

The source of capital for much of the canal construction was Asa T. Soule, a patent-medicine millionaire from Rochester, New York. Construction began in 1884 on a canal that ran from Ingalls to near the town of

Construction on the Eureka irrigation canal in the late 1800s (courtesy of the Kansas State Historical Society)

Offerle. Called the Eureka Canal, it meandered for 96 miles through the uplands north of the Arkansas River; the canal had 50 more miles of lateral extensions. During the height of the canal's use in the 1880s, it watered 500,000 acres of ground, but problems with low flow in the Arkansas and with the water's seepage into the porous ground led to the canal's demise. However, the path of this canal, along with several others, is still apparent here in southwestern Kansas, a reminder of an attempt to deal with water shortages before there was extensive development of ground-water-based irrigation.

119.7   South edge of *Dodge City,* which is famous as a cow town. But before the cattle drives, Dodge was known for its buffalo hunts. Fort Dodge was established east of here in 1864. The railroad arrived in 1872, and Dodge City was soon established. From 1872 to 1875 it was the headquarters for buffalo hunting in southwestern Kansas. Millions of bison lived on the plains west of here, and for three years, hunters made a living by shooting the animals for their hides. By 1875 the herds had been all but destroyed. A few years later, pioneers went back onto the plains to collect the bleached buffalo bones, which were used as fertilizer.

121.7 *Arkansas River.* The first European men to see the Arkansas River were members of Coronado's expedition, in search of the fabled land of Quivira. They reached the river in late June 1541, around St. Peter and St. Paul's Day; so one member of the party named it the River of St. Peter and St. Paul. The name Arkansas, as applied to the river, first appeared on a map in 1757.

122.0 Junction with westbound *U.S. 50.* Nine miles west of town, on U.S. 50, is a location where wagon ruts from the Santa Fe Trail are easily visible along a hillside. The east-west street here is Front Street, which runs west a few blocks to Boot Hill and Wyatt Earp Boulevard. Today, Boot Hill is a museum and memorial cemetery, built on the site of the original Boot Hill. Front Street is a reconstruction of the historic buildings that were once located here.

Dodge City is probably best known for the gunfighters who patrolled its streets. Many of the gunmen who were famous in the American West came through Dodge at one time or another. Bat Masterson was elected sheriff here in 1877, although he later moved to Tombstone, Arizona, where he worked with another former Dodge City law officer, Wyatt Earp. Other famous Dodge City gunfighters included Doc Holliday, Ben Tilghman, and "Prairie Dog" Dave Morrow.

Dodge City became just as well known among television viewers as the setting for the series "Gunsmoke," which starred James Arness and Amanda Blake. One of the series regulars, Milburn Stone (Doc Adams), was a native Kansan, born in Burrton, about 125 miles east of here.

123.4 *Roadside park,* overlooking a nearby feed lot. Dodge City's reputation as the Cowboy Capital of the World is based on the cattle drives that reached here in the 1870s and 1880s. In 1876 the Kansas Legislature moved the cattle quarantine line west of Wichita, and cowboys who were bringing cattle up the Chisholm Trail drifted west to Dodge City, to Hays, and up to Ogallala, Nebraska. The cattle trade in Dodge reached its peak in 1884, when 300,000 head were shipped out of town on the Santa Fe Railroad, and thousands of others came through the area on their way north.

In 1885 the legislature passed even-more-restrictive quarantine laws, which effectively ended the entrance of cattle into Kansas from Indian Territory or from Texas. That move ended the era of great cattle drives. But Dodge continues to be in the center of the cattle trade, particularly with the development of large feed lots, such as this one southeast of the highway.

124.3 *Indian Treaty boundary.* This boundary marks the north edge of a strip of land that was ceded to the Osage Indians when Kansas was Indian reservations.

125.6   Junction with Alternate *U.S. 50.* This highway loops nine miles north around Dodge City, rejoining U.S. 50 west of town. Dodge is one of the windiest cities in the United States, with an average wind speed of 14 miles per hour.

126.4   This county road runs three miles north to *Ford County State Park.*

126.7   *Elm Creek,* which drains into Ford County State Lake.

128.2   The town of *Wright.*

129.3   This point marks the eastern boundary of the old *Fort Dodge military reservation.* The fort was established about five miles southwest of here in 1864, to protect travelers from the Indians who were a constant threat on the trails through southwestern Kansas. The fort was named after Col. Henry I. Dodge, a territorial governor of Wisconsin and the source of the name of Dodgeville, a town west of Madison, Wisconsin. This fort was put in charge of Dodge's newphew, Grenville M. Dodge. Before the fort was abandoned in 1882, it was home to such famous figures as George Armstrong Custer and Philip Sheridan. Today, Fort Dodge is the Kansas Soldiers Home, and several stone buildings that date back to the 1860s are still in use.

This is also approximately the point at which Francisco de Coronado is believed to have crossed the Arkansas River during his exploration of Kansas in 1541.

131.0   The elevation here is 2,500 feet.

137.0   The town of *Spearville.* The Crustbuster factory, which produces farm implements, is on the north edge of town. The Eureka Irrigation Canal, which was to have stretched all the way to Kinsley, runs parallel to the highway from about milepost 132 to 137. Though the remnants of the canal are apparent in a few places—especially north and west of Dodge City—here they have largely been obscured by cultivation.

142.7   The town of *Bellefont.*

143.3   *Little Coon Creek.*

143.4   The county road here goes 7 miles south to *Windhorst.*

145.5   The oil well to the north is in the *Bellefont oil field,* discovered in 1979. Oil production and drilling reached a peak in Kansas during the 1950s, but suffered a general decline through the 1960s and 1970s. In the late 1970s and early 1980s, with increased oil prices, a new oil boom hit Kansas, and drilling reached record levels. Even production, which had declined for two decades, began slowly to rise. That spate of drilling produced a plethora of new fields, such as the Bellefont.

147.0   White Woman Creek enters Little Coon Creek about two miles north of here.

147.6   Ford/Edwards county line.

148.3   The town of *Offerle*.

150.0   The elevation is 2,250 feet.

151.7   The town of *Ardell*.

156.6   Junction with eastbound *U.S. 50*, which goes 10 miles to Lewis and 87 miles to Hutchinson.

156.9   Roadside park at the west edge of *Kinsley*. A sign here marks the approximate halfway point between San Francisco and New York, both of which are about 1,571 miles away. The Edwards County Historical Society museum and a replica of a sod house are also located at this intersection. Kinsley is the seat of Edwards County. The county courthouse, completed in 1929, is built of blond bricks and is located several blocks northwest of the central business district.

The Arkansas River is just east of Kinsley, and until the 1880s it was considered to be navigable this far upstream. Navigation is hard to imagine today, with the river's low flow and shallow, sandy bottom.

157.7   Junction with southbound *U.S. 183*. This road goes 25 miles to Greensburg and 48 miles to Coldwater. Greensburg is the site of the world's largest hand-dug well, constructed in the 1880s to provide water for the town and the railroad. The well, 32 feet in diameter and 109 feet deep, is cased with native stone. Close to the well is a museum that houses a large iron and stony meteorite, which was discovered east of Greensburg, near Haviland. This meteorite was found in a shallow depression, known as the Haviland Crater, which, until 1933, was considered a buffalo wallow. The discovery of meteorites surrounding this depression led to its excavation and the uncovering of the large meteorite and other fragments of meteorites. The discovery marked the first time that a meteorite had been recovered from a crater.

158.0   The north edge of *Kinsley*. Kinsley, founded in 1872, was originally named Petersburg, after T. J. Peter, a director of the Santa Fe Railroad. In 1873 it was renamed for the Boston philanthropist E. W. Kinsley. In 1878, thieves attempted to rob the safe at the Santa Fe train station, but eventually they were captured by Dodge City sheriff Bat Masterson.

160.0   *Coon Creek*. In 1848 the Battle of Coon Creek was fought near this point, when 140 soldiers, some of them bound for the Mexican-American War, were attacked by Comanche and Apache Indians. The soldiers managed to repulse the attack while suffering few casualties.

161.3   *Coon Creek* is immediately northwest of the highway.

164.6   *Edwards/Pawnee county line.* Edwards County was named after William Edwards, who served as Kansas secretary of state during the 1890s. Pawnee County is named after the Indian tribe.

165.7   Junction with northbound *U.S. 183,* which goes 31 miles to Rush Center and 36 miles to La Crosse, home to the Post Rock Museum—which documents the use of limestone throughout the Post Rock Country of north-central Kansas—and the Barbed Wire Museum.

166.0   Gravel pit east of the highway.

166.8   The *Arkansas River* is immediately southeast of the highway. Sand hills are on the other side of the river.

170.4   *Coon Creek.*

171.5   The town of *Garfield.*

171.8   *Roadside park* at the east edge of Garfield. In the park is a chapel with a bell and a cornerstone, which were donated by President James Garfield, after whom the town was named.

173.0   An obscured outcrop of the *Ogallala Formation* is visible in the road cut at this intersection. Along westbound U.S. 56, this is the last hard-rock outcrop. The Ogallala was deposited during the past few million years. It is composed of sand and gravels that were eroded off the Rocky Mountains, washed onto the Kansas plains, and cemented together in some locations. Here the Ogallala lies directly above the Dakota Formation, a Cretaceous sandstone that forms the hills northwest of the highway.

This is also a turnoff to Fort Larned National Historic Site, about 7 miles to the north. The fort, which was established on the banks of the Pawnee River in 1860, was used to defend travelers along the Santa Fe Trail from Indian attacks. During the 1860s, Gen. Winfield S. Hancock and Lt. Col. George Armstrong Custer operated from the fort in their campaigns against the Plains Indians. During the early 1870s, the fort protected workers as they built the Santa Fe Railroad through the area. The fort was closed in 1878. Today the fort is open as a museum, and most of the buildings, which are constructed of native Dakota Formation sandstone, are open to the public.

Two miles east of Fort Larned, on U.S. 156, is the Santa Fe Trail Center, a museum with exhibits about the Santa Fe Trail and about life on the Great Plains.

180.0   *Larned State Hospital* for the criminally insane is located two miles northwest, at the base of Jenkins Hill, which was originally called Lookout Hill. It was the original location of Fort Larned, before the fort was moved three miles to the northwest.

181.5   *Pawnee River* and the south edge of *Larned*. The Pawnee River joins the Arkansas River 0.5 miles southeast of here. Larned was named after the nearby fort, which was named for Col. Benjamin F. Larned, the paymaster general.

182.0   South of this intersection 0.7 miles is *K-19*, which runs south 8.5 miles to Zook and 15.5 miles to Belpre.

182.8   Junction with *K-156*. To the west, this highway runs to the Santa Fe Trail Center, Larned State Hospital, and Fort Larned. The town of Jetmore is 46 miles west of here.

183.2   This is the eastern edge of the Dakota Formation escarpment, which is composed of Cretaceous sandstone. Outcrops of the Dakota make up the hills in the town of Larned. On the west side of town, in an old quarry site, is a house that is built atop a steep bluff of Dakota Formation sandstone. The elevation at the base of the hill is 2,000 feet.

183.5   The east edge of *Larned,* the seat of Pawnee County. The daily newspaper here carries the distinguished name of the *Larned Tiller and Toiler.*

184.0–185.0   Wells here are in the *Larned oil field*, which was discovered in 1949. It has produced more than 7 million barrels of oil, making it the second-largest field in Pawnee County.

185.3   In the pasture immediately north of the highway is a prairie-dog town.

189.3   *Ash Creek.*

190.8   *Pawnee/Barton county line.*

191.2   The town of *Pawnee Rock*. This road leads 0.5 miles north to Pawnee Rock State Park.

192.0   *Pawnee Rock* is visible 0.7 miles to the west. This massive outcrop of Dakota Formation sandstone was a landmark along the Santa Fe Trail. Early explorers, such as Kit Carson, commented on Pawnee Rock on their way west. The outcrop was originally much larger, but settlers and railroad workers quarried stone from the site. Today, it is a state park, topped by a shelter house and a monument to the pioneers. This hill affords a sweeping view of the Arkansas River valley.

193.0   To the northwest is the *Bergtal South oil field*. These

Dakota Formation sandstone forms Pawnee Rock in Barton County.

wells are drilled into rocks in the Arbuckle Group of formations, about 3,500 feet underground.

194.0   Wells to the south are in the *Unruh South field.* Barton County is the second leading oil-producing county in Kansas; 4.7 million barrels were pumped there in 1984. This area is atop the Central Kansas Uplift, a subsurface geologic feature which is responsible for nearly half of all the oil production in Kansas. In 1984, Kansas produced 75.7 million barrels of oil; 34 million barrels came from fields in the Central Kansas Uplift.

195.0   Wells to the north are in the *Unruh oil and gas field,* which was discovered in 1945. It and several others in the area are noteworthy because they produce natural gas that contains helium. For many years the U.S. Bureau of Mines operated a plant at Otis, 15 miles to the northwest, which extracted the helium for use by federal agencies and industry.

197.0   The town of *Dundee.*

199.8   The *Great Bend Municipal Airport,* a former Army Air Force Base, is one mile north of the highway.

203.0   The west edge of *Great Bend,* named after the bow in the Arkansas River as it heads southeast toward Wichita and Oklahoma. This sweeping bend was noted by many early explorers, including Coronado. In 1806, Zebulon Pike's exploration party camped here and then split up, one group going downstream in canoes, while Pike's group headed west along the river and was later captured by Spanish authorities. Though this long bend is probably the most obvious feature in the Arkansas River, geologists remain uncertain about its cause.

203.8   Junction with westbound *K-96,* which goes 30 miles to Rush Center.

205.6   Junction with *U.S. 281* and Great Bend's Main Street. Hoisington is 11 miles and Russell is 41 miles to the north; St. John is 25 miles and Pratt is 50 miles south.

Great Bend is the seat of Barton County, the only Kansas county named after a woman, Clara Barton. The courthouse, in the city's central business district, was completed in 1918. It was built of Bedford limestone.

207.2   East edge of *Great Bend.* For a brief time during the 1870s, Great Bend was a railhead, and it developed some of the characteristics of a Kansas cow town. Today it is home to Barton County Community College, the Central Kansas Medical Center, and the Brit Spaugh Park and Zoo.

207.5–208.0   The oil wells to the north of this stretch of highway are in the *Red Brick field.*

208.5   *Walnut Creek.* This stream drains south into the Arkansas River. North of the highway is Fort Zarah Park, where Fort Zarah was established in 1864 by Gen. Samuel R. Curtis. Curtis named the fort after his son—Zarah Curtis—who was killed in the Baxter Springs massacre in southeastern Kansas in 1863. Zarah was one of a string of forts that were built along the Santa Fe Trail to protect travelers from Indians. The fort, which was constructed from native sandstone, was abandoned in 1869.

In 1864, along Walnut Creek, a government wagon train, carrying flour and wagon parts to New Mexico, was attacked by Indians. Ten teamsters were killed; two others, though wounded and scalped, survived. When Walnut Creek flooded in April 1973, it exposed the skeletons of ten bodies, which archaeologists determined were the remains of the victims of the Walnut Creek massacre.

209.0   West edge of the *Ft. Zarah oil field.*

209.4   Intersection with *K-156,* which runs northeast 23 miles to Holyrood, 35 miles to Ellsworth, and 46 miles to I-70. About six miles north of here is Cheyenne Bottoms Wildlife Refuge, a series of man-made ponds that have been built in a natural depression. This area has long been a

Cheyenne Bottoms, wildlife wetlands near Great Bend, north of U.S. 56 in Barton County

marshy swampland that attracted birds and other wildlife; levees were later built along the edges of the depression, so that it retained water most of the year. Cheyenne Bottoms, which now comprises 19,000 acres, is on the flyway for many North American bird species; it is an excellent spot for bird watchers. Most of the water in Cheyenne Bottoms comes from the Arkansas and Walnut rivers, as well as from two smaller tributaries; but because of lessened stream flows in recent years, the ponds of Cheyenne Bottoms have been more prone to drying up during the summer.

211.4   The escarpment north of the highway is in the Dakota Formation.

211.6   The town of *Dartmouth*.

212.0   The east edge of the *Ft. Zarah oil field*, which was discovered in 1951. It produces oil from formations that are 3,000 to 3,400 feet deep.

214.0   The wells south of the highway are part of the *Hammer oil field*, which has produced about 4.9 million barrels of oil since it was discovered in 1940.

215.0   The *Arkansas River* is 0.2 miles south of the highway.

215.2–216.2   The town of *Ellinwood*.

217.0   The west edge of the *Chase–Silica oil field*, discovered in 1930, which is named after two Rice County towns. It is the second-largest field in Kansas. Only the El Dorado field in Butler County has produced more oil. The field extends from here, in Barton County, into Rice and Stafford counties, covering nearly 40,000 acres. During its lifetime the field has produced more than 260 million barrels of oil, which amounts to more than three times the state's total production in 1984. Wells in the field produce oil from eight different units, beginning with the Winfield Formation, at about 1,400 feet, and extending down to the Arbuckle Group of formations, at 3,300 feet.

220.0   A mile to the south is a man-induced sinkhole, one of three located south of U.S. 56 in the Chase–Silica oil field. Known as the *Panning Sink,* this sinkhole developed on 24 April 1959, around an oil well that had been converted to dispose of the salt water that is pumped out of the ground along with oil. Because of environmental concerns, it is illegal to dispose of the salt water on the ground's surface or in underground aquifers that contain fresh water. Thus, salt water must be pumped into geologic horizons that already contain brine.

In much of Kansas, saline water is pumped into the Arbuckle Group of formations, a porous dolomite that was deposited during the Cambrian and Ordovician periods of geologic history. The Arbuckle lies below the highly soluble Hutchinson salt bed. Occasionally a disposal well fails, allowing water to flow outside of the well casing and to come into contact with the salt, where active dissolution creates a cavity. When the rock over the cavity collapses, it creates a sinkhole at the surface. Such was the case with the Panning Sink, which formed in just a few hours, reaching 300 feet in diameter and 100 feet in depth. Today the sink, which is filled with water, serves as a wildlife refuge.

221.0   The mound one mile south of the highway is capped by the Dakota Formation; it is 1,800 feet in elevation.

221.4   *Barton/Rice county line.*

221.7   West edge of a tract of *sand hills.*

222.6   The town of *Silica* is 0.3 miles north of here. Silica is another name for silicon dioxide, which is the principal component of quartz and sand. This town's name probably had something to do with its situation in the sand hills.

224.4   This county road runs five miles south to the town of *Raymond.*

226.0   The elevation here is 1,750 feet.

226.3   *Spring Creek.*

228.4   This is the east edge of the sand hills. The town of Chase is 0.2 miles north of here; a series of hills called Plum Buttes are three miles to the south.

229.0 Here the highway passes over a buried river valley, known as the *Chase Channel*. This valley was formed by a southeasterly flowing stream that emptied into the McPherson Channel (which U.S. 56 crosses about 35 miles east of here) during the Ice Age. Today the valley is filled with up to 200 feet of sand and gravel that contains ground water. The town of Chase, northwest of here, uses the Chase Channel for its water supply.

230.0 The east edge of the *Chase–Silica oil field*. In 1936, Rice County was the leading oil-producing county in the state. It is still traditionally among the top ten oil-producing counties in Kansas, having pumped 2.1 million barrels of oil in 1984. The Chase–Silica is responsible for much of that production. It is the leading producing field in the history of Rice County, and it was the county's top producing field in 1984, when it pumped more than 700,000 barrels of oil.

230.9 *Spring Creek,* named for the dependable springs that occur along its course in this area.

231.2 *Roadside park.*

231.4 *Cow Creek.* Cow Creek is a major tributary of the Arkansas River. The two streams come together near Hutchinson. The Santa Fe Trail crossed Cow Creek about a mile southeast of here; a ranch at that spot was called Beach Valley. William Mathewson, who worked at the ranch from 1859 to 1862, earned the nickname "Buffalo Bill" by killing buffalo for starving settlers in this area during the drought in 1860. The nickname was also given to the more familiar "Buffalo Bill" Cody.

232.5 This cross is a monument to *Father Juan de Padilla,* who accompanied Coronado on his exploration of Kansas in 1541. The Franciscan monk, who returned to New Mexico with Coronado, came back to Quivira in 1542 and was killed by Indians. Padilla is considered the first Christian martyr on land that now belongs to the United States, and there are monuments to him throughout Kansas. This location is reasonable, because it is near the site of a large village in which Indians were living during Coronado's trek through Kansas. The Rice County Historical Society's museum, located in Lyons, contains a number of artifacts that are related to Coronado's trip, area Indians, and other local history.

234.4 *Little Cow Creek.* This point marks the approximate boundary along U.S. 56 between the Arkansas River Lowlands, to the west, and the southernmost extent of the Smoky Hills to the east. Alluvium, terrace deposits, and windblown sand are the most common surface materials to the west; to the east, Cretaceous sandstones and shales have been eroded to form a rolling, upland topography.

234.8 Wells here are part of the *Lyons West oil field.*

235.4 West edge of *Lyons.*

236.4 The *Rice County Courthouse,* which occupies the square block south of the highway, was completed in 1911. It is built of red brick, with limestone trim. Rice County is named for Gen. Samuel Allen Rice, an Iowa native, who died of wounds received at the Battle of Jenkins Ferry in 1864.

236.5 Junction with *K-14,* which runs north and south, and *K-96,* which runs to the south. Ellsworth is 28 miles north of here, and Sterling is 9 miles to the south.

236.6 *Salt Creek.* Salt is mined in many locations in central Kansas, including several in Lyons. The first mine, dug in 1887, struck salt at about 800 feet, and today's mine takes salt from about 1,000 feet. Lyons salt is used mostly for melting ice on winter roads and for industrial uses. In nearby Hutchinson, salt is mined by solution, pumping down fresh water that dissolves the salt away and brings it the surface in the form of brine, which is then evaporated. All of these mines take salt from the Hutchinson Salt Member of the Wellington Formation, a bed of salt that underlies much of central Kansas and is as much as 400 feet thick in places.

During the late 1960s, one of the abandoned Lyons salt mines was considered for another use: the disposal of radioactive waste. The Atomic Energy Commission (AEC) studied the site for possible emplacement of high-level nuclear waste. However, research indicated that the Lyons site had several problems, particularly from old oil wells that had been drilled in the area. Because there were no records of many of those wells, their locations could not always be determined, and scientists were concerned that they might allow ground water into the mine. Concerns about the mine, as well as political opposition, caused the AEC to abandon plans to dispose of waste in Lyons.

237.6 East edge of *Lyons,* the seat of Rice County and the county's largest town. During the 1870s, Lyons was known as Atlanta; it vied with nearby Peace (now Sterling) to become the county seat. Settlers decided to establish the seat at the center of the county, on land owned by Truman J. Lyons. Atlanta was moved to that location and was renamed.

237.8 *Owl Creek.* A number of species of owls are common in Kansas, including the Great Horned Owl, the Barn Owl, the Screech Owl, and the Burrowing Owl, which is sometimes known as the Prairie Dog Owl, because it nests in prairie-dog or badger burrows. Owls are one part of a group of birds of prey known as raptors. Red-tailed hawks are probably the most common raptors in Kansas, particularly in the eastern end of the state. Red-tailed hawks, or "chicken hawks" as they are known locally, are rust-red in color and have a lighter belly. Their diet primarily consists of rodents.

238.5   The wells here are part of the *Lyons oil field,* and to the south is the Lyons Gas Storage Area, where natural gas is stored in underground formations.

241.9   *Jarvis Creek.* South of here, along Jarvis Creek, is a location known to treasure hunters as the spot where, in 1843, the Spanish trader Don Antonio Jose Chavez was murdered. Chavez was carrying gold from Santa Fe to Independence, Missouri, in order to buy goods to trade with back in New Mexico. Local settlers believed that Chavez buried his load of gold along the banks of this creek before he was killed. Though Chavez's murderer was caught and tried, the treasure was never found, and natives have searched for it ever since.

242.5   The town of *Mitchell* is 0.5 miles to the north.

243.0   About five miles north is the site where a village of Wichita Indians lived from about 1500 to 1700, including the time of Coronado's visit. The spot has been intensively studied by the Smithsonian Institution, and archaeologists believe that some of the council circles in the village may have been located so as to line up with the sun during the summer solstice. The site has produced thousands of Indian artifacts, including arrowheads, pottery, stone grinders, beads, and human and animal skeletons. Nearby is an outcrop of Dakota sandstone that has Indian petroglyphs carved on it.

243.7   A *gas storage area,* similar to the one at milepost 260.

245.0   This road runs 1.5 miles north to a small lake constructed on a tributary of the Little Arkansas River. On the south side of the lake, the Marquette Sandstone Member of the Kiowa Formation forms a steep bluff known as *Spriggs Rocks,* which includes several petroglyphs that were carved by Indians. At the base of the rock cliffs is a spring that helps to feed the lake.

247.0   Junction with K–46, which runs 1.5 miles north to the town of *Little River,* which was named for the Little Arkansas River. A salt mine was dug on the north edge of the town in 1895 and operated for a time.

248.4   *Little Arkansas River.* This river flows to the south, joining the Arkansas—locally known as the Big Arkansas—at Wichita.

250.0   About three miles south of here is a gravel pit where

The type section of the Stone Corral dolomite, south of U.S. 56 in Rice County

limestone and dolomite are quarried, mostly from the Stone Corral Formation. Dolomite closely resembles limestone, except that it is composed of calcium magnesium carbonate; limestone is calcium carbonate, or calcite. To distinguish between the two, geologists place a drop of dilute hydrochloric acid on the rock. The acid fizzes on limestone, but not on dolomite. The type section of the Stone Corral Formation dolomite is near the quarry.

The Stone Corral Formation is named after a stone corral that was built near the spot where the Santa Fe Trail forded the Little Arkansas River. William Mathewson lived here in 1857 and 1858. In 1859, William Wheeler built the corral, which was 400 feet long and 200 feet wide. The walls were eight feet high and 30 inches thick, with slits from which to fire at Indians. The corral was later torn down, and the stone was used in construction elsewhere.

    250.3   *Salt Creek.*

    250.9   North Fork of the *Little Arkansas River.*

    252.0   *Rice/McPherson county line.* McPherson County is named for James B. McPherson, a Civil War general who was killed in the battle of Atlanta. McPherson is one of twenty-two Kansas counties that were named after Civil War generals. To the south, a county road follows the Rice/

McPherson County line, eventually joining Plum Street in Hutchinson. In its path through northern Reno County, Plum Street passes through an area of sand dunes, deposited between the Little Arkansas and the Arkansas rivers, that support extensive growths of sandhill, or Chickasaw, plums, a fruit that is picked to make jelly. North of the intersection of U.S. 56 and Plum Street is the *Welch–Bornholdt oil field*, discovered in 1924, which extends to the west into Rice County. It consisted of more than 200 wells in 1984 and has produced nearly 40 million barrels of oil since it was discovered.

252.6   The town of *Windom*, north of here, is named after an Ohio lawyer. Windom was originally called Laura, but the name was changed because it resembled Larned, about 70 miles southwest along U.S. 56.

253.5   *Lone Tree Creek.*

256.0   About here, U.S. 56 crosses the point where rocks of Cretaceous age come into contact with underlying rocks of early Permian age. The younger Cretaceous rocks extend west of here, while the older Permian rocks extend east. This contact between rocks of vastly different geologic ages is called an unconformity, which represents an interruption in the geologic record as it has been preserved in the rocks. After the deposition of Permian rocks, which are shales in this area, the land was exposed above sea level, and a period of erosion occurred. Later, this area dropped below sea level, and the deposition of shales resumed during the Cretaceous. This unconformity is not conspicuous at the surface; even where it is exposed in outcrops, it is not striking, because rocks above and below the contact are both shales. Yet this unconformity represents a missing chapter in geologic history, which includes the Jurassic and Triassic periods and spans more than 100 million years.

257.0   *Butane plant.*

257.3   *Underground gas-storage area.* Throughout this area, small pipes mark the location of wells that store natural gas in underground cavities called "jugs." These oblong-shaped voids are created by using fresh water to dissolve away cavities in an underground salt layer. At this location the top of the salt layer is about 380 feet below the ground, and the jugs are created between 380 feet and 600 feet below the surface; they have a capacity of between 25,000 and 40,000 barrels and are used to store liquified petroleum gas, a form of natural gas that has been refined out of oil and is kept in a liquid state under great pressure. Because Kansas produces natural gas and is centrally located, it is the crossing point for many pipe lines, several of which intersect at Conway.

259.5   *Conway.* Natural-gas storage is not without problems.

During the late 1970s, one of the storage cavities began to leak natural gas into the subsurface, and evidence indicated that it had gathered in basements and cisterns of homes in Conway. Residents and houses were moved out of Conway, and the town is all but deserted today.

260.0 *Storage wells* on the north and south sides of the road. Like those at milepost 257.3, these wells store gas in the Hutchinson Salt Member of the Wellington Formation. The Hutchinson salt, which underlies 27,000 square miles of Kansas, is is up to 400 feet thick. Here it is about 380 feet below the surface and about 300 feet thick. This salt was deposited about 250 million years ago, during the Permian Period, after the deposition of the Permian rocks that form the Flint Hills and before the deposition of rocks that make up the Red Hills. Even though salt is easily dissolved by water, the Hutchinson salt bed remains largely intact because it is protected by a thick layer of shales, which are generally impermeable to water, thus preventing precipitation from moving through the ground and washing away the salt.

However, the salt has been dissolved in several areas. Around here, cavities have been purposefully created to store gas in the resulting void space. In other areas, the salt was penetrated by wells used in searching for oil in formations below the salt. In some locations, poorly constructed wells have allowed water to drain underground and to dissolve void spaces in the salt, which can lead to slow or sudden collapse at the surface. Such sinkholes are found throughout central Kansas. In addition, salt has been dissolved away naturally where the formation approaches the surface. At this point, called a solution front, saltwater may leak into streams and create pollution problems.

260.8 Here the highway passes from the *Smoky Hills physiographic region,* to the west, into the *McPherson Lowlands,* to the east. The McPherson Lowlands are underlain by deposits of silt, sand, and gravel, which are known as the Equus Beds. This formation is of Pliocene and Pleistocene age, less than five million years old. *Equus* is the Latin word for horse, and these beds are named after the horse fossils that they contain. During the Pleistocene, in addition to horses, other large mammals roamed this area and left their remains in the Equus Beds. These animals included mastodons, elephants, bison, camels, and ground sloths. The elevation here is 1,500 feet.

261.0–263.0 *Big Basin,* which covers about four square miles, is an area of internal drainage caused by subsidence of the earth's surface. Big Basin is part of a linear trend of sinkholes and depressions, many of which contain lakes, that extends from a few miles north of here, southward to near Colwich, northwest of Wichita. Lake Inman, eight miles to the south, is part of this trend. At 0.5 miles in diameter, it is the largest natural lake in Kansas. This line of lakes and sinkholes marks the east edge of the Hutchinson salt, a location where the salt bed is being actively dissolved away by ground water, forming new areas of subsidence and possibly new lakes, while older ones disappear as they are filled with sediments.

Natural lakes are generally short-lived features, geologically speaking, since the ongoing process of weathering and erosion works to fill them with silt or to attack their outlets and drain them. Thus, the presence of natural lakes indicates that the geologic process that created them either is still operating or has operated in the recent geologic past. For instance, glaciers formed the Great Lakes, as well as many of the lakes in Minnesota, Wisconsin, and southern Canada, during the Pleistocene Epoch of the last million years.

In Kansas, the most common natural lakes are oxbows, which form along major streams when they shift their course, abandoning and sealing off meander loops and leaving them full of standing water. Lake Inman and other lakes in this area were formed by a different process—the solution of the Hutchinson salt and the resulting subsidence of overlying rocks.

261.4 Because the land here is tabletop flat, water would collect and turn Big Basin into a marshy swampland if it were not for the *drainage canal* that the highway crosses here. This canal begins north of U.S. 56 and runs south, where it joins Blaze Fork Creek and eventually runs into Lake Inman in southern McPherson County. While this ditch takes care of most of the runoff, it does not reach some parts of the Big Basin, where the ground is often marshy.

263.0 Southeast edge of *Big Basin*. This Big Basin should not be confused with a much smaller subsidence feature by the same name in western Clark County.

264.0 Here the road crosses over the *McPherson Channel*, the former course of the Smoky Hill River to the south toward the Arkansas River. Today the Smoky Hill makes a U-turn after it leaves Kanopolis Lake in Ellsworth County; it flows through northwestern McPherson County and then heads north toward Salina. During the Pleistocene Epoch, the Smoky Hill flowed south through a broad, deep valley, joining the Arkansas River near Wichita. This valley, now occupied by the McPherson Lowlands, was formed by the solution of the Hutchinson salt bed and by the collapse of overlying rocks. This basin was later filled in, as the ancestral Smoky Hill River flowed through the McPherson Channel, carrying silt and sand that had been eroded from rocks in central and western Kansas, as well as debris from melting Pleistocene glaciers in the Colorado Rockies. This accumulation of silt, sand, and gravel, called the Equus Beds, is more than 250 feet thick; in many places it is saturated with ground water. The Equus Beds are a major source of water for communities in this area, including McPherson, Newton, and Wichita.

265.2 *Bull Creek* and the west city limits of *McPherson.* Immediately south of the highway, Bull Creek is crossed by a covered bridge.

266.2 *McPherson County Courthouse,* which was recently remodeled. It was built in 1894 of Cottonwood limestone. In the park to its west is a statue of Gen. James B. McPherson, purportedly the only life-size bronze equestrian statue in Kansas.

267.1 *Dry Turkey Creek.*

267.8 East edge of *McPherson.* For a time during the 1960s and 1970s, McPherson billed itself as the Light Capital because its downtown was so brightly illuminated. Today the town of 11,000 is home to McPherson College, a private school located north of here, and Central College of McPherson.

268.8 *I–135 overpass.* To the north, I–135 goes to Salina and eventually becomes U.S. 81, which runs to the Nebraska border. To the south, I–135 joins I–35 in Wichita. Chapter 5 describes the geology along I–135 and U.S. 81.

269.0 *The Johnson oil field,* which is not visible from the road, lies south of the highway; to the northwest is the *Chindberg oil field.* These are part of a 40-mile-long trend of oil and gas fields that stretches from northeastern McPherson County southwest into Harvey and Reno counties. These fields lie along a subsurface feature called the Voshell Anticline. An anticline is an arched fold of rocks; in some cases, oil is trapped in the peak of the anticline's arch, which makes anticlines attractive for oil producers. The Voshell Anticline is parallel to and smaller than the Nemaha Anticline, about 50 miles to the east. Unlike the Nemaha Anticline, which has faults running along its eastern flank, the Voshell Anticline is faulted on the west.

    This anticline also corresponds to areas of unusually high gravity and magnetic measurements, which have been used to define the Midcontinent Geophysical Anomaly (MGA). The MGA probably represents an ancient rift system that stretched from Lake Superior, through Iowa and

Nebraska, into this part of Kansas. The rift was a cracklike valley that developed about a billion years ago, partially splitting the continent apart. When the rift stopped growing, it was filled by igneous rock, intruded into the rift, and by volcanic rock that poured out onto the ancient land surface. These igneous rocks have a higher density than do the surrounding rocks, thus producing a higher gravitational pull in this zone. Likewise, the higher concentration of magnetic minerals, such as magnetite, that occurs in the igneous rocks produces a higher-than-normal magnetic field along the ancient rift. Exact measurements of the earth's gravity and magnetic fields are commonly used by geophysicists in determining the earth's composition and structure; they are also used during exploration for oil and other minerals.

270.9 *Turkey Creek.* Wild turkeys are native to Kansas, but their populations were greatly reduced with the arrival of Europeans. Turkeys have since been reintroduced into the state; there is now a hunting season in the fall in southern and northwestern Kansas, and there will soon be a statewide hunting season. Two types of turkeys are found in Kansas. In the western part of the state, the Rio Grande turkey has been reintroduced. Indigenous to the southwestern United States, this bird is more suited to the semiarid conditions of western Kansas than is the eastern turkey, which is common in the woodlands of the eastern United States. The eastern turkey, which provided the main course at the first Thanksgiving, is now being introduced to the woodlands of eastern Kansas. These two birds also produce a hybrid where their ranges overlap.

273.4 To the north is *Galva,* which is named after a town in Illinois.

273.5–278.5 To the south are wells in the *Ritz–Canton oil field,* which was discovered in 1929. It had produced more than 70 million barrels of oil by the end of 1984. These wells penetrate several zones of oil-producing rocks that are between 2,300 and 3,400 feet deep.

Oil here in McPherson County comes from an ancient underground arch of rocks that separates the Sedgwick Basin from the Salina Basin. A geologic basin is an underground bowl-shaped depression in the earth's crust which has been filled with sedimentary rocks. The Sedgwick Basin is responsible for oil production in counties south of here, including Sedgwick, Sumner, and Barber counties, as well as Marion County to the east. The Salina Basin, which underlies all of north-central Kansas, has produced little oil.

275.7 *Running Turkey Creek.*

279.4   K–86 runs north to *Canton,* which is famous for its "hot" and "cold" water towers. From Canton a county road runs about 6 miles north to the Maxwell Game Preserve, home to one of the state's largest herds of buffalo and elk. The buffalo make possible a buffalo-chip throwing contest, held every summer in Canton. Elk were once common throughout Kansas—herds in the mid 1800s numbered in the thousands—but they were eventually extirpated in the state. In addition to this confined herd in McPherson County, an unconfined herd has been reintroduced in the Cimarron National Grassland in Morton County. Elk are more correctly known as wapiti, a Shawnee Indian word meaning "pale rump."

The game preserve itself is located in pastures that are strewn with boulders of Cretaceous sandstone from the Dakota Formation. Most of the surface rocks in McPherson County are of recent geologic origin, deposited during the past million years of geologic history. The Dakota Formation is considerably older, and much of it crops out in a line from roughly central Rice County to northeastern Washington County. However, a small patch of Dakota juts out to the east of that line, underlying the hilly pastures immediately north of Canton and extending into western Marion County, where it has been highly eroded into unusual-shaped sculptures. Eastern outcrops of the Dakota indicate that these older rocks once covered far more of Kansas than they do today, but have since been eroded away. Still, it is not clear if the formation once covered all of eastern Kansas—and has since been eroded away—or if it was never deposited over the eastern end of the state to begin with. Time and erosion have removed much of the evidence, so the argument is difficult to settle. However, geologists have found some evidence of the Dakota Formation as far east as Johnson County, indicating that it may once have covered nearly all of the state.

280.0   Wells in this area are part of the *Bitikofer North oil field,* discovered in 1946, which produces oil from Mississippian formations that are about 2,900 feet deep.

280.7   *Santa Fe Trail crossing.* U.S. 56 roughly parallels the course of the Santa Fe Trail across Kansas, and this is one of several locations where the old and new roads intersect. The old trail is difficult to discern here, however, because cultivation and time have obscured the evidence. Where the trail can be seen, its ruts form large, trench-like formations, rather than the flat road that one usually associates with a trail. As thousands of wagons moved over the Santa Fe, they wore deep ruts into the ground, particularly when the trail was muddy. The resulting terracelike mounds may be several feet deep. In some locations, the trail can by pinpointed by looking for abrupt changes in vegetation where wagons wheels disturbed native prairie. One of the flowers that thrived in that disturbed soil was the sunflower. In fact, the seeds of the sunflower, which was designated the Kansas state flower in 1903, may have been brought into Kansas from the southwest on the wheels of freight wagons as they moved along the trail during its early days.

282.4   *McPherson/Marion county line.*

283.0   At about this point, U.S. 56 crosses the boundary separating the *McPherson Lowlands,* to the west, and the *Flint Hills Upland* to the east. The geology changes markedly at this boundary. To the west are the geologically young silts, sands, and gravels deposited by wind and streams during the last million years or so. This is largely a flat landscape in which older bedrock is buried beneath younger sedimentary debris. To the east is a much different landscape, one formed by the erosion of ancient rocks that were deposited in the seas of the Permian Period, more than 250 million years ago. These rocks have been uplifted to their present position above sea level and have been tilted slightly toward the west. Thus, to the east, U.S. 56 descends through progressively older rocks, mostly limestones and shales. Some of these limestones contain layers of chert, or "flint," which gives the Flint Hills their name. The elevation here is 1,560 feet.

284.7   *K-15* goes south from here to Newton and passes the small town of Goessel, site of the Mennonite Heritage Complex, buildings that display exhibits related to Mennonite settlement in Kansas. This part of central Kansas is home to many Mennonites, the result of German immigration here during the 1870s. Some of these immigrants were families that moved from Germany to Russia during the 1760s, after they were promised religious freedom and permanent military deferment. In 1794, Czar Alexander II decreed that all residents of Russia were subject to military conscription, and many of the Mennonites moved to the United States. Today the Mennonite church is still known for its strong belief in religious freedom and for its opposition to war.

285.3   *South Cottonwood River.*

286.0   To the north is the *Lehigh oil field,* discovered in 1946.

286.3   K-168 to *Lehigh.*

289.5   *K-15* runs north from here to Abilene and Clay Center. It also passes through the small town of Elmo, about 23 miles to the north. An area south of Elmo is known for Permian fossil insects, including the remains of dragonflies, lice, and stoneflies. The wings of these insects were often the best-preserved parts and are used in identification.

292.4–293.5   *Hillsboro*. Both Hillsboro, here in central Kansas, and Hill City, further west, are named after men named Hill, not for any rise in the topography. Hillsboro, a Mennonite community, is headquarters of the Mennonite Brethren of North America and is home to Tabor College, a private Mennonite school. An adobe house, built by Mennonite settlers in the 1870s, is preserved in the city park.

Mennonites played an important role in the Kansas economy when they imported Turkey Red winter wheat to Kansas in 1874. That importation helped change the state's wheat production from soft to hard grain and helped to revolutionize the state's agriculture. Today, wheat is the dominant crop produced in Kansas. Kansas farmers plant more than thirty varieties of winter wheat in the fall, usually in September or early October; then they rely on winter snows to insulate the newly sprouted wheat from the cold. In some locations, farmers use the young wheat for pasturing cattle during the winter; the wheat recovers once the cattle have been removed.

The Kansas wheat crop usually matures, depending on the weather, in June or July. Wheat harvest is the busiest time of year in the agricultural community. Harvest usually begins first along the southern edge of the state, where warmer weather allows the crop to mature earlier, and then proceeds north across Kansas. In general, the wheat crop ripens about one day earlier for every 10 to 15 miles traveled south in Kansas. That is, the wheat 50 miles south of here will probably be ready to cut about five days before it is ready here.

Once the wheat is ripe, farmers try to harvest the crop as quickly as possible. A hailstorm can destroy a year's work in a matter of seconds, and even an ordinary rainstorm can make fields too muddy for cutting the wheat and can delay harvest for weeks. Wheat that stands too long in the field may bleach and lose protein in hot, sunny weather. Many Kansas farmers cut their own wheat, but others hire custom cutters, who may own several combines and trucks and who make a living by following the harvest from Texas north to Montana.

Farmers also keep a close eye on the moisture, protein content, and weight of their wheat. The higher the moisture content, the less valuable the wheat, because it must be dry enough to store. Wheat that is cut early is often higher in moisture than wheat that is taken later in harvest. Conversely, high protein content makes the wheat more valuable. Kansas wheat generally averages between 11 and 13 percent protein. In 1980 the average protein content for wheat in Marion County was 13 percent, slightly above the statewide average of 12.3 percent. Finally, heavier wheat is considered more valuable. During the 1980 harvest, Kansas wheat averaged 61.2 pounds per bushel, somewhat below the ten-year average of 61.4 pounds per bushel.

    294.1   The elevation here is 1,400 feet.

    297.7   The town of *Canada* is 0.5 miles north of here.

    298.7   County road to *Marion Reservoir.*

    299.3   *South Cottonwood River.*

    299.6   *Hollenberg Limestone Member of the Wellington Formation* on the north side of the road.

    302.0   *South Cottonwood River.* The Cottonwood is dammed in central Marion County to form Marion Reservoir. It then runs through Chase County and joins the Neosho River east of Emporia.

    302.2–303.4   *Marion,* the county seat and largest city in Marion County. The county courthouse, located south of the central business district, was built of native Cottonwood limestone in 1907. On the east edge of the city park is a historic spring, which was discovered in 1860 and still runs, except during times of very dry weather.

    304.4   The hills to the north are capped by the *Nolans Limestone Formation,* a Permian rock that is common around Marion and in the Flint Hills. The highway encounters Permian formations around Marion, but to the north and west of the town it passes through much younger rock. In places, ground water has eroded through the Nolans and has created small sinkholes. One such sinkhole to the north, near the town of Lincolnville, drains about 20 acres of ground.

    305.6   Junction with *U.S. 77,* which runs south to El Dorado and Winfield and north to Junction City and Lincoln, Nebraska. Nolans Limestone is exposed in road cuts here. About 8 miles to the south is the town of Florence. Since 1920, Florence has depended on nearby Crystal Springs for its water supply. The city's pump can produce 370 gallons per minute from the spring, with the excess running into a nearby creek. The Florence water tower advertises that its product is 99.96 percent pure; recent tests show that this claim isn't far off. This spring alone could probably provide enough water for all 13,504 residents of Marion County, if it could somehow be distributed throughout the county. This intersection also marks the south edge of the *Lost Springs oil field,* whose wells are visible along the highway north of here.

306.6   *K-150.*

307.1   *Herington Limestone* on the west side of the road. In places the Herington has a pitted appearance; it may contain geodes, concretions, and cauliflowerlike masses of chert and quartz. In places the Herington is dolomitic—that is, it is composed of calcium magnesium carbonate, as opposed to ordinary limestone, which is composed of calcium carbonate only.

309.4   A field several miles west of here has a *flowing well* that produces water from the Barneston Limestone Formation at the rate of 900 gallons per minute, or more than a million gallons per day. The water is high in sodium, chloride, and sulfates. It has a noticeable odor, probably from hydrogen sulfide. Sulfur precipitate forms a white crust around the edges of the pool.

Marion County has many springs, including nearby Chingawassa Springs, which was earlier known as Carter's Spring. This consists of a series of twenty to thirty springs that emerge from a bluff into a nearby creek. Chingawassa Springs was a popular picnic spot during the 1870s, and in 1889 a railroad was built from Marion to the springs. A hotel and spa were established; they operated until 1893, when both went bankrupt. The concrete abutments for the train trestle are still visible near the springs.

According to recent geological studies of Marion County, Chingawassa Springs produces mostly fresh water from the Winfield Limestone, although at least three springs produce mineralized water that contains hydrogen sulfide. Recent measurements show that the springs produce as much as 600 gallons per minute. Along with several artesian wells down stream, they add more than 2,000 gallons per minute to the flow of Clear Creek.

311.4   *Nolans Limestone.*

311.7   The town of *Antelope* is about a mile west of here. Before settlers appeared, pronghorn antelope were common over the western two-thirds of Kansas. This town, which is about at the eastern edge of their range in Kansas, may represent the area where they were first spotted by travelers on the Santa Fe Trail. Though antelope have excellent vision and can run fast—some have been clocked at close to 60 miles per hour—pronghorns made easy prey for hunters and were therefore all but eliminated in Kansas. During the 1970s, a survey estimated that only fifty to sixty pronghorns survived in the state. They have since been restocked, although they seem to thrive best in northwestern Kansas, particularly in Wallace, Sherman, and Thomas counties. There is now a short hunting season for pronghorns, although licenses are issued only to Kansas

A flowing well west of U.S. 56 in Marion County produces more than 900 gallons of water per minute.

residents. Pronghorn are characterized by a bright white rump, and they disperse quickly when frightened. Usually, they go under, rather than over, obstacles; and they can duck under a barbed-wire fence while maintaining an astonishing speed.

312.1 *East Branch of Clear Creek.* Clear Creek flows south of here into the Cottonwood River. The highway and a nearby railroad follow the creek valley to the north.

312.5 Oil wells in this vicinity are part of the *Lost Springs field,* which was discovered in 1927. The field stretches southwest almost to the town of Canada, and as far north as the outskirts of Herington, covering more than 12,000 acres. It has produced more than 23 million barrels of oil from four different formations, which range in depth from 1,190 to 2,674 feet. Oil made Lost Springs a boom town; during the late 1920s, its population jumped from 250 to nearly 1,000. As production began to taper off, so did the population of Lost Springs. Only 96 residents remain.

312.8 and 313.0   On the east side of the road, *Nolans Limestone,* which ranges in thickness from 22 to 40 feet.

313.6   *Nolans Limestone* and quarry on the east side of the road.

313.9   *Clear Creek.*

314.0   Four miles west of here is the small town of *Pilsen,* which was probably named for Pilsen, Czechoslavakia, a city famous for the spring water that it uses in brewing a lager beer called pilsner.

315.6   *Lincolnville.* Lincoln is one of the most popular place-names in the state; more Kansas townships are named after the Civil War president than after any other person.

Not far from Lincolnville there are several good sites for collecting the crystal form of limonite. Limonite is a compound of iron, oxygen, and water. It usually has a dark, noncrystalline form. However, in some locations, limonite chemically replaces pyrite and keeps pyrite's crystal form. Geologists say that such limonite is a pseudomorph after pyrite.

317.6   To the east is *Centre High School,* one of the schools created in Kansas as smaller towns consolidated their school systems. During the 1960s and 1970s, as the number of students in some Kansas schools declined, towns often joined together to form consolidated schools in one location. In some cases, they built new schools in a location central to two towns. Centre High School, for example, is about equidistant from Lincolnville and Lost Springs.

320.8   The town of *Lost Springs* is about a mile to the west, at the junction of the Rock Island and Santa Fe railroads. It is named after a spring about 4.5 miles west of here, just off the county road that runs through the town of Lost Springs. Called Lost Springs because it occasionally dried up and was thus "lost," the spring is usually visible along the banks of Cress Creek. Lost Springs was another stop on the Santa Fe Trail; it is about 15 miles, or a day's travel, from Diamond Springs, in Morris County.

321.9   *Santa Fe Trail crossing.* Although the remains of the old trail aren't visible, this is the approximate crossing point of the Santa Fe

Trail as it wound from Diamond Springs west to Lost Springs. The old trails often followed crests of ridges where the ground was even and there were fewer streams to cross. The elevation along this ridge is about 1,500 feet.

322.5 *Arkansas/Missouri river drainage divide.* Precipitation to the south of this divide drains into the Arkansas River system; to the north it drains into the Missouri River system.

323.9 *Marion/Dickinson county line.* U.S. 56 cuts across the southwestern corner of Dickinson County, covering less than five miles of the county before exiting into Marion County, to the south, and Morris County to the east.

324.8 The oil wells to the west mark the northern extent of the *Lost Springs field.*

326.0 *Herington Limestone Member* of the Nolans Limestone Formation. Named after a nearby town, this limestone ranges in thickness from 6 to 10 feet in northern Kansas and up to 30 feet in southern Kansas.

326.3 and 326.7 *Herington limestone,* which often has a pitted and pocked appearance that is especially common in this area. Those pockmarks are probably caused by geodes composed of calcite crystals that have weathered away following exposure to the elements.

327.4 *Lime Creek,* which runs through Herington, probably received its name from the limestone outcrops common along its banks.

327.8 *Paddock shale,* overlain by *Herington limestone,* both of which are members of the Nolans Limestone.

328.3 *Junction with U.S. 77.* The town of *Herington,* immediately to the west, is named after Monroe Herington, a prominent landowner in Dickinson County who persuaded the railroad to pass through his land. For many years, Herington was an important rail center, with two lines of the Rock Island and one line of the Missouri Pacific converging here. During the late 1970s and early 1980s, however, the railroads abandoned many of their routes in Kansas, and Herington's economy suffered because of it. Herington's city park contains a monument honoring Father Juan de Padilla, a Franciscan friar who accompanied Coronado in his exploration of southwestern United States in 1540 and 1541.

328.4 *Dickinson/Marion county line.* Both Morris and Dickinson counties are named after United States senators. Daniel S. Dickinson was a senator from New York during the 1840s, and Thomas Morris was a senator from Ohio during the 1830s. Morris was staunchly opposed to slavery, and the county's name reflects something of the sentiment of the Kansas Legislature.

331.6   *Nolans Limestone.*

333.5   These oil wells are part of the *Grandview* and the *Grandview South oil fields,* two recently discovered fields that produce oil from formations that are just over 2,000 feet deep.

334.4   *Threemile Creek.*

335.0   *Delavan.* About three miles north of here is the *Herington Army Air Field,* one of several air fields built in Kansas during World War II for the training and formation of Army Air Force flying units. The Herington Air Field had three concrete runways, 150 feet wide and totaling nearly 4 miles in length. At its peak, it was home to more than 2,200 officers and enlisted men. It was used primarily as a staging base where B–17, B–24, and B–29 bomber crews were processed, assigned to aircraft, and given final briefings and training before being assigned overseas.

335.8   *Mile-and-a-Half Creek.* Many of the creeks in Morris County are named after distances. This one and Threemile Creek, to the west, are named according to their distance from Diamond Springs along the Santa Fe Trail, which parallels U.S. 56 about 2 miles south of this area.

338.0   Down this valley 2.5 miles is Diamond Springs, a well-known stop on the Santa Fe Trail. The spring was originally named the Diamond of the Plains, probably because early reports said that the water in the spring bubbled and sparkled like diamonds. The town of Diamond Springs, today only a railway siding, was about 5 miles south of the spring.

338.5   *K–149* goes north to the town of White City. The area north of White City was the site of an unusual discovery of hydrogen gas in the early 1980s. Wells that were drilled for oil produced gas with a high percentage of hydrogen, more than 20 percent in some cases. Traces of hydrogen are common in Kansas wells, but usually not in the amounts found here. Hydrogen can be used as a fuel, so the wells may have a commercial application.

The real mystery is the source of the hydrogen, although scientists who are studying the wells have several theories. One theory is that the hydrogen may be created when blocks of granite are crushed against each other during faulting. Active faulting in this area is related to the Humboldt Fault Zone, which runs across eastern Kansas and produces small earthquakes from time to time. Scientists aren't sure, however, if faulting could produce the volumes of hydrogen that are found in Morris and Geary counties. A second theory is that the gas was created during a chemical process called serpentinization. When the mineral serpentine is created, hydrogen gas may be given off. Because serpentine is found in kimberlite, igneous rock formations north of here in Riley County, this process may have produced the gas. A third possibility is called outgassing, or the movement of gas to the surface from deep under the mantle. Because the kimberlites popped up to the surface about 100 million years ago, this part of eastern Kansas may represent a zone of crustal weakness and may present an avenue for hydrogen to work its way to the surface. Hydrogen is a very diffuse gas; it is rarely trapped securely in geological formations in the way that oil, for example, might be. Thus the hydrogen must be renewed almost continually, so that the escaping hydrogen is replaced by newly produced gas.

340.2 The beginnings of *Haun Creek,* named after a local minister who was one of the first settlers of the area. It runs northeast into the Neosho River, which cuts southeast across Morris County.

342.5 *Wilsey* is about a mile south of here.

343.6 *Florence limestone,* which is one of the chert-bearing limestones that provide the flint in the Flint Hills. It is characterized by brachiopod, pelecypod, bryozoan, and fusulinid fossils.

344.0 and 345.1 *Florence limestone,* which ranges from 12 to 45 feet thick. It is named for a nearby town. The town, in turn, was named after the daughter of Samuel J. Crawford, who was governor of Kansas from 1861 to 1863.

346.5 The wells lining the road are part of the *Bosch oil field.* Morris County is not a major oil producer; about 8.7 million barrels had been pumped there by the end of 1984.

348.2 *Florence limestone.*

348.7 *Fort Riley limestone,* which has long been used for building in Kansas. Quarried around Junction City, it was used in the construction of the Kansas Capitol's east wing and in many of the buildings at Fort Riley. The same rock is mined in Cowley County, where it is sometimes called the Silverdale limestone. In places the Fort Riley forms a layer of rock so hard and strong that it cannot be cut directly from the ledge, but must be drilled and blasted away from the outcrop. The Morris County Courthouse in Council Grove and the Cowley County Courthouse in Winfield are built of Fort Riley limestone.

Also at this point, a county road leads a mile north to *Council Grove City Reservoir,* which covers about 500 acres. This lake was built on Canning Creek, and the overflow runs into a 3,000-acre reservoir called Council Grove Lake, built by the U.S. Army Corps of Engineers, at the confluence of the Neosho River, Canning Creek, and Munkers Creek.

349.6 *Florence limestone,* a bluish-gray flint, is visible in this cherty outcrop.

350.1 and 350.4 *Threemile limestone.* This rock layer ranges in color from light gray to nearly white. In places it contains abundant chert, but some of the lower and middle parts contain no chert at all. The formation ranges from 6 to 33 feet thick; it was used in the construction of foundations and buildings in Council Grove.

351.3 West edge of *Council Grove.* In 1825 the government negotiated a treaty with the Osage Indians, guaranteeing safe passage across their lands on the way to Santa Fe. Those negotiations took place under a massive oak tree—the trunk remains just off Main Street—which gave Council Grove its name. The grove served as a jumping-off point on the Santa Fe Trail, which ran along Main Street, and as the headquarters for the Kansa Indian Reservation from 1846 to 1872. The Kansa originally roamed throughout northeastern Kansas, but in 1846 they ceded away their lands along the Kansas, Solomon, and Saline rivers for a 250,000-acre reservation near Council Grove. The move to the Neosho River valley, with its proximity to settlers and traders moving through on the Santa Fe Trail, all but spelled the end for the Kansa. Before the move to Morris County, their population had held steady at about 1,500. By the time they left Morris County in 1872 for an even smaller reservation in Indian Territory, their numbers were down to 600. During one disastrous winter, a smallpox epidemic claimed 400 members of the tribe.

Established on the banks of the Neosho River, which cuts a north-south path through the town, Council Grove preserves its history in various sites, including an old jail, the Last Chance store, and the old Kaw Indian Mission and museum. Charles Curtis, who was vice-president of the United States from 1929 to 1933, was a member of the Kansa Indian tribe; he attended the mission as a child. On Main Street of Council Grove is the Hays House. Seth Hays built a log cabin at this site in 1847 and began to trade with the Kansa Indians. Today the Hays House is a restaurant—advertised as the oldest continuously operating restaurant west of the Mississippi—but the building has been put to a variety of uses over the years, including a court, a church, a printing shop, and a hotel, whose guests included Gen. George Custer and Jesse James.

353.5 Eastern edge of *Council Grove.* In 1846, Susan Shelby Magoffin, traveling with her husband who was a Santa Fe trader, made the following journal entry at Council Grove: ''We are now at the great rendezvous of all traders. Council Grove may be considered the dividing ridge between the civilized and barbarous, for now we may look out for hostile Indians. Each company coming out generally stop here a day or so to repair their wagons, rest the stock, get timbers for the remainder of the journey.''

The road cut here is composed of Funston Limestone, overlain by Speiser Shale, overlain by Threemile limestone.

354.0 Top of the *Threemile limestone,* which is one of several limestones and shales that are lumped together in the Chase Group of formations, named for Chase County, to the south of here. Chert-bearing limestones in the Chase Group crop out to form the Flint Hills. In southwestern Kansas, where they lie deep underground, these same rocks are the source of natural gas in the Hugoton Field.

354.5 *Big John Creek.* "Big John" Walker, a Sac Indian from Wisconsin, guided one of the early surveying missions along the Santa Fe Trail. Walker discovered several springs in this area. To the east is Little John Creek, which empties into Big John Creek and should more properly be called Little Big John. After they join together, the streams empty into the Neosho River.

354.9–357.9 Outcrops of *Threemile limestone.*

358.1 *Speiser Shale,* overlain by *Threemile limestone.* Speiser Shale comes in a variety of colors, although red is probably the most common. This layer varies from 18 to 35 feet in thickness.

359.5 *Rock Creek.*

359.8 *Morris/Lyon county line.* U.S. 56 cuts across the northern edge of Lyon County, paralleling the Wabaunsee County line to the north.

360.5 and 361.2 *Funston Limestone,* a light-gray to bluish-gray limestone that occasionally contains layers of shale and chert. The Funston is another of the limestone formations used for construction. Near the town of Onaga, north of here, the Funston is about 5 feet thick, somewhat thicker than most occurrences in the state. There it is quarried and is locally called the Onaga limestone. In addition to being used for residential construction, this limestone was used in the Eisenhower Museum in Abilene and in the Topeka Public Library.

362.0 To the north is the *Kizler North oil field,* which was discovered in 1972.

362.5 *Bluff Creek.* The massive blocks of limestone that are exposed in the creek bed are part of the Cottonwood limestone.

362.8 Located in the hills to the south are wells in the *Kizler oil field,* discovered in 1969.

363.0 *Funston Limestone.*

363.3 *Speiser Shale,* overlain by *Threemile limestone.* A rock layer that is found again and again across a wide area is said to be "laterally persistent." That is, the rock layer "persists," rather than thinning out and disappearing. Threemile limestone is persistent throughout this area, as is evidenced by the number of times the highway passes through its outcrop. The Florence and Fort Riley formations are also extremely laterally persistent.

363.6–364.1 *Threemile limestone,* which is particularly cherty in this area.

364.6 *Kinney limestone.*

364.7 *Schroyer limestone,* another of the chert-bearing formations that crop out in the Flint Hills.

364.9 The county road here runs two miles south to *Bushong.* According to some sources, several towns in this area are named for major-league baseball players from the late 1800s. Bushong may have been named after Albert ("Doc") Bushong, who played for Cleveland, St. Louis, Brooklyn, and several other teams in a career that lasted from 1875 to 1890. East of here, just across the Morris County line, is Comisky Cemetery, all that remains of a town that may have been named after Charles Comiskey, who played for St. Louis and Cincinnati, managed the White Sox, and supplied the name for Comiskey Park in Chicago.

365.0 *Havensville shale.*

365.5 *Speiser Shale,* overlain by *Threemile limestone.* Speiser Shale is noticeably red at this road cut, and the Threemile limestone is cherty.

366.1 *Crouse Limestone,* which includes two layers of limestone separated by a few feet of fossiliferous shale. At this road cut, a thin, platy shale covers the massive limestone below. Both layers are probably part of the Crouse.

366.8 *Allen Creek.*

367.5 *Cottonwood limestone,* which, except where it thins out along the southern part of its outcrop, maintains a remarkably constant thickness of about 6 feet. At this location it is full of fossils, containing bits and pieces of many marine invertebrates that flourished on the floor of Permian seas.

368.3 *Neva limestone.* This rock layer contains both limestone and shale. The limestone contains a variety of fossils, including fusulinids, brachiopods, algae, and echinoids, which are related to starfish and are

ESCARPMENT
GENTLE SLOPE

SHALE
LIMESTONE
SHALE
LIMESTONE

Cuesta topography, common in eastern Kansas, is formed on alternating hard and soft rocks.

represented in modern oceans by sea urchins and sand dollars. In the Permian seas where the Neva limestone was formed, echinoids were generally globular in shape, with a flexible outer skeleton composed of hundreds of small bony plates that protected the internal organs.

370.0   *Allen.*

370.9   *Northern Heights High School* north of the road.

371.5   *Hill Creek.*

373.9   Intersection with *K–99,* which goes south to the town of Admire, named after one of the town's founders, Capt. Jacob V. Admire of Indiana.

374.0   *One Hundred and Fortytwo Mile Creek.*

377.3   *Interchange with the Kansas Turnpike,* which goes north to Topeka and south to Wichita. Northeast of the road there is a small waterfall, which has formed along the spillway of a pond. The pond's overflow falls over beds of the Stormont limestone and the Wamego shale.

At about this point, westbound U.S. 56 enters the Flint Hills, which get their name from the erosion-resistant, chert-bearing limestones that were deposited in the Permian seas about 250 million years ago. Today these limestones crop out in a line from northern Kansas into Oklahoma.

Eastbound U.S. 56 enters the Osage Cuestas here. *Cuesta* is a Spanish word meaning hill or slope; it is used by geologists to describe a hill or ridge, such as those in eastern Kansas, where gently dipping, erosion-resistant rock layers form the landscape. Cuestas have a gentle slope in the direction in which the rocks dip, but they have a steep, clifflike face where the resistant rocks crop out. Eastern Kansas is made up of a series of these cuestas, arranged one above the other like broad, slightly dipping stair steps. The resistant rock units that form the cliff face are usually limestones, and the intervening gentle slopes are usually shale outcrops.

Eastern Kansas cuestas that are formed by resistant limestones can be traced from northern Kansas southward into Oklahoma. The steep eastern faces of these cuestas are sometimes called escarpments, or scarps; they are named by geologists for the rock layer that forms them. For instance, the Oread escarpment is formed by the Oread Limestone. The Flint Hills are actually a large escarpment formed by several resistant limestones—some bearing chert, or "flint"—and are sometimes referred to by geologists as the Flint Hills escarpment.

379.2 *Elm Creek.*

379.7 *K-78* goes to the town of Miller.

381.8 The limestone that forms the creek bed here is highly jointed, crisscrossed by straight cracks. These joints represent fractures in the limestone, In places such as waterfalls, the limestone erodes naturally along these joints. As it breaks off, it forms a clean, straight cut that looks almost man-made. This layer of limestone is part of the Tarkio Limestone Member of the Zeandale Limestone Formation.

382.0 *Lyon/Osage county line.*

383.0 *Salt Creek.* This stream gets its name from salty springs or seeps that occur along its course. There are no salt beds in this part of Kansas, but there is much salt water in Pennsylvanian formations and in older rocks that underlie this area. This brine is actually ancient sea water, which saturated the marine sediments as they were deposited. As these sediments were buried, compressed, and formed into sedimentary rocks, some of the sea water was squeezed out. But much of this water, which geologists call connate water, remained in the formations, sealed in by impervious rocks above and below. Today these salty ground waters are found in the subsurface throughout the state; it is almost impossible to drill a deep well in Kansas without encountering salt water. In eastern Kansas, south of the glaciated region, good-quality ground water is often restricted to alluvium and occasionally to sandstone bodies. Locally in southeastern Kansas, good water is obtained from a deep aquifer, but the remaining rocks yield only salt water, the salt content of which often increases with depth.

384.7 *Swede Creek.*

386.7 *Kibbee Creek.* An old coal-mine dump is visible on the east bank of the creek on the south side of the road. The number of creeks in this area is the result of climate and topography. Precipitation in eastern Kansas is much more abundant—almost twice as much on an average annual basis—than in western Kansas. Much of that precipitation flows down the eastern face of the Flint Hills, which rise up to the west of here. The elevation here, for example, is just over 1,100 feet, compared to 1,400 to 1,500 feet throughout the Flint Hills.

387.5 South of the road is an old *coal-mine dump*, produced during the days of coal mining in Osage County. These mine dumps are located near old mine shafts and mark the locations of abandoned tipples. A tipple is an apparatus near the mouth of a coal mine, where coal cars or buckets were tipped and emptied of their load. Coal was then screened into different sizes, and unmarketable coal waste and fractured rock were tossed onto waste piles, or gob piles, which dot the landscape in this area.

These gob piles contain iron pyrite, which is sometimes called fool's gold because of its yellow metallic luster. Pyrite is iron sulfide, one of the sources of sulfur that occurs in most Kansas coals. When pyrite is exposed to oxygen and water, as it is in these mine dumps, it undergoes a chemical reaction that produces sulfuric acid, iron oxides, and hydroxides. Sulfuric acid is the culprit in acid drainage from mines, one of the main environmental threats to streams and lakes where coal is mined. The iron oxides and hydroxides that are formed in this chemical reaction are similar to common rust, and they color these waste piles red. This red cast is visible not only in the mine dump but also on the shoulders of U.S. 56 west of here, indicating that the dumps were probably the source of some road-bed material.

388.1 Intersection with *K-31*, which goes south to *Osage City*. About two miles south of Osage City there is an old limestone quarry, which produced footprints of a Pennsylvanian-age amphibian called *Lumnopus vagus*. The footprints were discovered by Benjamin Franklin Mudge, the first director of the Kansas Geological Survey, when Mudge was crossing Kansas Avenue in Topeka in July 1873 and noticed a set of "clearly defined footprints." He tracked down the source of the flagstone used in the sidewalk; then he took the next train to Osage City, where he found more tracks. Mudge sent the fossil footprints to Prof. O. C. Marsh at Yale University, who identified them twenty years later. In 1956, Walter Schoewe, another staff member at the Kansas Geological Survey, traced Mudge's trail to the Osage County quarry, where he determined that the tracks were in the Utopia Limestone Member of the Howard Limestone Formation. Schoewe found additional footprints in the same formation, several miles from the old quarry.

388.7   About 0.5 miles east of the road there is another old *coal mine dump.* The coal here, from the Nodaway coal bed, is a bituminous coal that is generally bright, shiny black, and moderately hard. The Nodaway coal, which underlies much of eastern Kansas, is up to 36 inches thick in Osage County, though it averages 16 to 20 inches in thickness and occasionally contains pyrite or calcite.

Over the years the Nodaway has been mined in nine eastern Kansas counties, producing nearly 12 million tons of coal; 97 percent of that production came from Osage County. It is not mined today because of its sulfur content, as much as 6 to 8 percent.

The heyday of coal mining in Osage County was from 1885 to 1892. In 1889, Osage County had 118 coal mines, which employed more than 2,200 people and produced almost 400,000 tons of coal. More than 25,000 people lived here in 1890, compared to just over 15,000 in 1980. For many years, this coal-mining area was the main fuel supply for the Santa Fe Railroad, whose main line passes through here. Between the nearby towns of Scranton and Burlingame, a distance of five miles, there were at least fifty mines, and one report estimated that there were probably another fifty mines within a one-half-mile radius of Osage City.

Those early coal mines were of three types. In surface mines simply, the overlying shales and limestones were simply removed, and then the coal was stripped away, the way that strip mines are operated today in southeastern Kansas. Shaft mines went down until coal veins were hit, and then the miners tunneled out horizontally, leaving behind pillars to support the ground above. Finally, at drift mines, the miners followed an outcrop of coal horizontally—or drifted—into the ground where the coal was overlain by other rocks.

Will coal mining ever return to Osage County? According to recent studies, the thickness and heating quality of Nodaway coal make that possible, in spite of the coal's high sulfur content. Certainly substantial coal resources remain; their use will depend on the price and availability of energy during the coming years.

388.8   *Gob pile* on the east side of the road.

388.9   *Smith Creek.*

389.5   One mile east of here, a few remaining houses mark the site of *Peterton,* a town that sprang up with the coal-mining industry and once had a population of 600.

391.4   The hill about a mile west of the road is known as *Elkhorn*

Knob, whose elevation at its highest point is over 1,200 feet, compared to 1,040 feet here along the highway. This hill and others west of the highway are capped by the Bern Limestone; they mark the eastern edge of the Bern escarpment. The most prominent member of the Bern is the Burlingame limestone, which was first described in this area.

391.8 *Dragoon Creek.*

393.7 To the east is an old *gob pile.*

394.0–395.0 *Burlingame,* another of the towns along the Santa Fe Trail, which ran down Burlingame's main street. During the Civil War, William Quantrill, the proslavery sympathizer who was famous for his bloody raid on Lawrence, planned a similar attack on Burlingame while most of the men were away fighting for the Union. The women of the town constructed a fort of rocks and held off Quantrill and his raiders for six weeks, until Union soldiers came to their aid.

395.3 *Switzler Creek.*

396.3–399.6 *Old mine dumps* on either side of the road.

400.0 The town of *Popcorn* once lay 2.5 miles to the south.

401.0 South edge of *Scranton,* which is named for a Pennsylvania coal town. The main part of Scranton is north of here, but a few streets and houses are scattered on this edge of town, the remains of the boom years of the 1880s and 1900s, when the population was around 1,700.

Scranton's connection with Pennsylvania goes beyond its name. The rocks in this part of Kansas were deposited about 300 million years ago, about the same time as rocks that are commonly found at the surface in Pennsylvania, where they were first described. Pennsylvania is the source of the geological name for the period when these rocks were deposited, a time of swamps and shallow seas in eastern Kansas. The swamps produced the coal that is found here and in Pennsylvania and that is the source of another name for this period, the Coal Age.

Geologists outside of North America do not recognize the Pennsylvanian Period or the preceding period, which is called the Mississippian, after rocks that occur in the Mississippi Valley. Instead, rocks of these two periods are lumped together and are called Carboniferous, the Latin term for coal-bearing.

402.2 *Strip pit* north of the road.

403.1 *Hundred and Ten Mile Creek.*

405.0 *Intersection with U.S. 75.* Carbondale, about 2 miles north of here, is another former coal-mining town. It is surrounded by abandoned strip mines, many of which are now filled with water. Carbondale was once known for its mineral springs, such as Merrill Mineral Spring, which brought salty and sulfate-bearing water close to the surface, where it was pumped into a large bath, which had a nearby pavilion. A hotel and sanitarium for the treatment of nervous disorders once operated near the spring. North of Carbondale 1.5 miles is an area that is still shown on maps as Mineral Springs. Here springs and shallow wells produce mineralized water, which was once bottled and shipped out for medicinal purposes. This is also the intersection of U.S. 75, which goes north to Topeka and south to Lyndon, the Osage County seat. The area where the two roads intersect is called the Four Corners.

405.3 *Strip pits,* overgrown with trees, line both sides of the road on top of this hill. The coal that was mined in this area was from the Nodaway coal bed in the Howard Limestone. The swamps that produced these coal beds stretched across the entire north-south distance of Kansas and on into Oklahoma, Missouri, Nebraska, and Iowa. Geologists believe that these swamps bordered the coast line of the Pennsylvanian sea for several hundred miles. In many cases the swamps produced coals that are overlain by limestones, indicating that the coal-forming swamps were close to sea level and were occasionally flooded by limestone-depositing sea waters. Geologists aren't sure if the swamps that formed the coal were in fresh or brackish water, but it seems clear that the limestones that covered the coal were deposited by seas, because of the marine fossils that are found in the limestone.

405.8 *Osage County State Park and County Lake* are directly south of here.

406.8 *Santa Fe Trail High School* south of the road.

408.7 *Plummer Creek.*

411.0 *Topeka Limestone.*

412.0 *Overbrook.* This town sits on a ridge, elevation about 1,200 feet, that serves as the drainage divide between the Wakarusa and the Marais des Cygnes rivers. However, Overbrook may be named after a town in Pennsylvania, as are nearby Scranton and Carbondale.

413.5 *Rock Creek.*

413.6 *Ervine Creek limestone.* Although this member of the Deer Creek Limestone Formation is thin in this area, it thickens to more than 30 feet in southern Kansas, where it is quarried.

414.3 *Ozawkie Limestone Member* of the Deer Creek Limestone

Formation. Although fossils are abundant in a few outcrops of this limestone, they are generally sparse in most exposures.

414.9 *Osage/Douglas county line.* Douglas County is named for Stephen A. Douglas, the abolitionist senator who is probably best remembered for his debates with Abraham Lincoln. Douglas County and the city of Lawrence, in particular, were strongholds of abolitionist sentiment during the 1850s and 1860s. Lawrence, which was established by the New England Emigrant Aid Society, bore the brunt of attacks by proslavery forces from eastern Kansas and Missouri. Probably the most famous of those attacks was by William Quantrill, who rode through Lawrence in 1863, burning much of the town and killing more than one hundred citizens.

415.2 With an elevation of 1,200 feet, the hill south of the road is the *highest point in Douglas County.* About 5 miles to the north are two hills known as Twin Mounds. Their elevation, 1,108 feet, is somewhat lower than the elevation here, but they still stand about 100 feet above the surrounding countryside.

420.3 *Globe.* About all that remains of Globe is a general store. It was originally called Washington, the first location in the state to be named after George Washington. Later the name was changed to Marion and finally to Globe. The county road here runs about three miles north to Lone Star Lake, built on Washington Creek, and on to Clinton Lake, a much larger federally constructed reservoir on the Wakarusa River. Also, at about this point the highway crosses Eightmile Creek.

421.8 The pond south of the road is actually an old *quarry,* probably in the Kereford limestone, that has filled with water.

422.8 At this point U.S. 56 passes over the *Worden Fault,* named for the small town 1.2 miles east of here. This fault, which is not visible from the highway, has been mapped southward to the Franklin County line and north of U.S. 56 a short distance, where it curves to the east and roughly parallels the highway all the way to Baldwin City. The fault apparently moved when the Oread Limestone, the surface rock here, was being deposited about 250 million years ago. This faulting may account for the absence of some members of the Oread in this vicinity, notably the Toronto limestone.

The Worden Fault is not considered to be tectonic in origin. That is, it was not caused by crustal movements associated with uplift and mountain building. Rather, it is apparently the result of sediment adjustment shortly after deposition. This adjustment may be caused by a thick body of sandstone, the Ireland sandstone, which fills a deep buried valley in this area. After sediments are deposited and buried by younger sediments, the sea water is squeezed out of them, and they became compacted. Some rocks, such as sandstones, compact much less than others, especially shales. Where two such rocks are near each other, as they are along the margin of the Ireland sandstone, compaction occurs at different rates, and the sandstone compacts much less than does the adjoining shale. This differential compaction can lead to deformation and even to faulting, as in this case, in the overlying rocks.

423.8   *Worden.*

424.9   *Leavenworth limestone* on both sides of the road. This rock layer is found throughout much of eastern Kansas and, except south of here in Franklin County, has a uniform thickness between 1 and 2 feet.

425.4   *Leavenworth limestone* in the creek bed south of the highway.

426.0   *Plattsmouth limestone* on the north side of the road.

426.8   *West Fork of Tauy Creek.*

427.3   Intersection with *U.S. 59*. This road goes north to Lawrence and south to Ottawa. In the park here, there is an outcrop of the Ireland sandstone, which has been highly eroded by a branch of Tauy Creek. In places the sandstone has large holes and steep bluffs that have been exposed by the creek. The Ireland, a massive, thick sandstone, was deposited in an old stream channel that cut across southern Douglas County and northern Franklin County during the Pennsylvanian Period. In places, that river channel was several miles wide and more than 100 feet deep, so that the Ireland may be as much as 160 feet thick in places.

The sandstone was probably deposited by a river that drained into a Pennsylvanian sea west of here. The direction that water flowed in a stream can often be discerned by a pattern of angled or wavy lines, called cross-bedding, that are still evident in the sandstone. By determining the direction of the lines, geologists can tell which way the water was flowing in the stream that deposited the sand. In a location downstream from here, called Hole in the Rock, a particularly thick exposure of the Ireland is so highly cross-bedded that the sandstone has a ropy, wavy appearance.

Where the Ireland sandstone is present fairly close to the surface, it contains fresh water and is the source of water for many farms in this area, as well as for Baldwin City, four miles to the east, and several other towns along its trend. The name of the Ireland may be related to W. E. Ireland, a member of the Kansas Legislature from Yates Center, an area

The Ireland sandstone at Hole in the Rock in Douglas County. The sandstone is so highly cross-bedded, with angled lines, that it takes on a wavy appearance.

where Ireland sandstone is also exposed. During one particularly troublesome legislative session in the 1930s, the director of the Kansas Geological Survey, R. C. Moore, reportedly named the Ireland after Ireland Creek, which was on the legislator's farm, in an attempt to improve the chances that the Survey's appropriation would be passed by the Kansas Legislature. Stratigraphic units are generally named after locations, such as rivers and towns, and not after people.

    430.1   *Middle Fork Tauy Creek.*

    431.0   This hill, with an elevation of 1,100 feet, is capped by an outlier of Oread Limestone.

The same formations that produce oil in eastern Kansas are far deeper to the west. The small pump jack is in Douglas County in eastern Kansas, and the larger pump jack is in Comanche County in southwestern Kansas. The deeper the oil, the larger the equipment necessary to extract it.

431.8   *Baldwin City*, which was originally named Palmyra, grew up on the Santa Fe Trail, along with other nearby settlements, such as Prairie City to the west and Black Jack to the east. Baldwin City is home to Baker University, a Methodist school, which is the oldest four-year college in Kansas. Baker also houses the Quayle Bible Collection. Two miles northeast of here, by county roads, is the Douglas County State Lake.

434.1   The oil wells in this area are part of the *Baldwin field*, discovered in 1921. They produce oil from a sandstone formation called the Knobtown, named for a small town southeast of Kansas City, Missouri. This sandstone is in the Pleasanton Group of rocks, which crop out near the town of Pleasanton, Kansas, about 50 miles southeast of here. Oil is also produced from the Squirrel sandstone, in the Cherokee Group of rocks, which crop out in southeastern Kansas. These oil-producing formations are about 500 feet deep here, compared to oil-producing formations in south-western Kansas, which may be more than 6,000 feet deep. That is why these pump jacks are much smaller than those found farther west in Kansas; in general, the deeper the oil deposit, the larger the pump.

435.1   *Captain Creek*.

435.2   *Ireland sandstone* is exposed on the north side of the road. South of the road is Black Jack Park, and immediately south of the park is the Ivan Boyd Memorial Prairie Preserve. This area of virgin tall-grass prairie contains wagon ruts from the Santa Fe Trail. The Battle of Black Jack took place nearby on 2 June 1856. In May 1856, proslavery forces attacked Lawrence—the Free-State stronghold in territorial Kansas—destroying printing presses and burning buildings. John Brown and his abolitionist followers avenged the raid by killing five proslavery men south of here, on Pottawatomie Creek. In retaliation, a proslavery band raided Palmyra—now Baldwin—taking three prisoners and later making camp here in a grove of black jack oaks. On the morning of June 2, John Brown and his followers attacked the camp, wounding several proslavery sympathizers, who finally surrendered. This battle received much publicity; it was just one of many that occurred in the days of "bleeding Kansas."

438.1   *K–33* leads four miles south to Wellsville and I–35.

438.5   *Tonganoxie sandstone*. This reddish-brown sandstone is similar in appearance to the Ireland sandstone, which is exposed at milepost 427.3. Like the Ireland, the Tonganoxie was deposited in a broad alluvial valley during the Pennsylvanian Period. The Tonganoxie valley, which extended southwest into south-central Kansas, is up to 140 feet deep. The Tonganoxie is also an important aquifer where it lies close to the surface, providing water supplies for farms and towns along its trend. In areas where these sandstones crop out at the surface, water moves back into the formation and recharges the aquifer. Also, streams that cut their valleys into these sandstones may contribute water directly into the water table, helping to maintain the ground-water supply in the aquifer.

439.0 *Martin Creek.*

439.1 *Douglas/Johnson county line.* Johnson County is named for Rev. Thomas Johnson, a Methodist missionary who established the Shawnee Methodist Mission in 1830 in what is now Fairway, Kansas.

440.1 and 440.9 *Martin Creek.*

441.7 *Edgerton.* This town was named for a chief engineer on the Santa Fe Railway. About 8 miles north of here is the Sunflower Army Ammunition Plant, which produces an artillery propellant called nitro-guanidine. The plant, first used in World War II to make weapons propellant, closed at the end of the war, and then reopened during the Korean War, the Vietnam War, and again in 1984. It produces 8 million pounds of nitroguanidine per year, and is the only such plant in the country.

441.9 *Martin Creek.*

443.4 *Bull Creek,* which flows southward into Miami County. About 10 miles downstream it has been dammed by the Army Corps of Engineers to form Hillsdale Lake.

444.5 *Gardner oil and gas field.* Production in this field comes from a shallow sandstone in the Marmaton Group of rocks, which crop out in a band southwestward from Fort Scott to near Coffeyville in southeastern Kansas.

446.8–448.0 *Gardner.* Near this point, the Santa Fe and Oregon trails diverged after following the same route westward from Independence and Westport (Kansas City). This fork in the road was marked by a board with the words "Road to Oregon" pointing the way. While the Santa Fe Trail began as a trading route that connected the east with the important Old Mexico market of Santa Fe, the Oregon Trail was an emigrant trail from the beginning, with settlers following it to lands in the Pacific Northwest. In later years, emigrants also followed the Santa Fe Trail to settle in the Southwest. During the 1840s, stagecoach service began along the Santa Fe Trail; daily departures began in the 1860s. The trip from Independence to Santa Fe took two weeks, and each coach carried eleven passengers. The fare, including meals, was $250.

448.7 To the north is the *Johnson County Industrial Airport,* formerly the Olathe Naval Air Station. This airport opened in 1942 as a training station for navy pilots. It was closed by the navy in 1969.

449.0 *Little Bull Creek.* This creek joins Bull Creek in what is now Hillsdale Lake.

450.6 The town of *Clare.*

453.5 *Cedar Creek.* A mile upstream, on a tributary of Cedar Creek, is Olathe Lake; downstream one mile is New Olathe Lake. Both provide part of the water for Olathe, with the balance being taken from wells in the Kansas River alluvium near the mouth of Cedar Creek, about 10 miles northwest of here.

453.6 To the south are limestone quarries in the *Stanton Formation.*

454.0 West edge of *Olathe.* Olathe, the seat of Johnson County, was organized in 1856. William Quantrill, notorious for his raids against opponents of slavery, claimed a homestead south of here in 1857, but he couldn't hold the claim because he was not of legal age. While his most famous attack was against Lawrence in 1863, Quantrill also raided Olathe on 6 September 1862, killing two citizens and destroying property, including the town's two newspaper offices. A month later he and his raiders destroyed the town of Shawnee.

455.7 Intersection with *K–7*, which runs north through Olathe, Bonner Springs, and Leavenworth, and continues to the Nebraska border. A mile south of here, K–7 joins U.S. 169, leading southwest to Paola, Osawatomie, Coffeyville, and Oklahoma.

456.2 *South Bend Limestone*, the uppermost member of the Stanton Limestone, is exposed in a small road cut.

456.8 *U.S. 56/I–35 intersection.* To the northeast, U.S. 56 joins I–35 for the next 11.8 miles, which are described in chapter 7, between miles 217 and 228.8. To the southwest, U.S. 56 leaves I–35.

469.2 *U.S. 56/I–35 intersection.* To the west, U.S. 56 joins I–35, while eastbound U.S. 56 leaves I–35 here. Northbound U.S. 169 also leaves I–35 at this interchange.

469.8 One-quarter mile to the north is a granite marker denoting the location of the *Shawnee Friends Mission.* In 1825, when the United States government began to relocate eastern Indian tribes to the western wilderness, the Shawnee Indians were moved to a 2,500-square-mile tract of land here in northeastern Kansas. Religious groups established missions in the area in order to educate the Indians. One of these, the Shawnee Friends Mission, was opened by Quakers in 1837 and operated continuously until 1869.

470.1 This point marks the city limits between *Merriam,* to the west, and *Overland Park* to the east, two Johnson County suburbs that are part of the metropolitan Kansas City area. Johnson County and Wyandotte County, north of here, have a combined population of more then 400,000, and most of those people live in fourteen towns and villages. Overland Park, with a population of 81,784, is the largest suburb in Johnson County; the smallest is Mission Woods, with a population of 213. The 400,000 people in these two counties account for nearly half of the population in the metropolitan Kansas City area, and about one-fifth of the population of Kansas.

470.6 *Metcalf Avenue.* U.S. 169 exits to the north and leads to I-635 and Kansas City, Kansas. A two-acre cemetery in downtown Kansas City, Kansas, is devoted to members of the Wyandot Indian tribe. During the 1930s, when an attempt was made to move the cemetery, Lydia B. Conley fought the removal. In arguing the case, she became the first woman to appear before the United States Supreme Court.

471.7 The city limits between *Overland Park* and the suburbs of *Countryside,* to the west, and *Mission,* to the east. The headquarters of the National Collegiate Athletics Association is on the south side of the highway. The first newspaper that was published in Kansas was printed in what is now Mission.

472.5 *Rock Creek* and the city limits dividing *Mission,* to the west, and *Fairway* to the east.

473.0 The elevation here is 900 feet.

473.4 *Mission Road.* One-quarter mile to the north is Shawnee Mission State Park, in which buildings of the old Shawnee Methodist Mission have been restored and opened to the public. The mission was founded in 1830 by Rev. Thomas Johnson; it was moved to its present site in 1839. A school where Indian children learned English, manual arts, and agriculture was part of the mission; it had an enrollment of 200.

The mission had other uses. In 1854 it housed the executive offices of Andrew Reeder, the first territorial governor of Kansas. In 1855 the first territorial legislature met here. Also known as the "Bogus Legislature," this body was elected by proslavery Missourians, and it passed laws promoting slavery. During the Civil War the mission provided barracks for Union troops. Rev. Thomas Johnson's son, Alexander, who was born here in 1832, was the first white person to be born in Kansas.

473.6 A small cemetery on the south side of the highway includes the burial plot of Thomas Johnson.

474.0   The city limits dividing *Fairway,* to the west, and *Westwood* to the east. South of the highway is Mission Hills.

474.2   The city limits dividing *Westwood,* to the west, and the small city of *Mission Woods* to the east. The even-smaller city of Westwood Hills is one-quarter mile north. Although it covers slightly less than 40 acres, Westwood Hills has a population of 422, giving it a population density of 6,400 per square mile.

474.5   *Kansas/Missouri state line.* U.S. 56 parallels Brush Creek immediately to the south and crosses the Kansas–Missouri border. To the east is the Country Club Plaza shopping district, which is considered the country's first suburban shopping center. This area was devastated in September 1977, when a slow-moving storm dumped more than a foot of rain, transforming normally placid Brush Creek into a torrent that caused more than $60 million in damage and killed twenty-five people. Brush Creek's source is about 5 miles southwest in Overland Park, at an elevation of more than 1,050 feet. Its elevation at the state line is 840 feet, a drop of more than 200 feet in 5 miles, giving an average stream gradient of more than 40 feet per mile, which is comparable to the gradients of some mountain streams.

REFERENCES

We relied on many publications while writing this book. The following list does not include every that source we used; it includes only those that we consulted most often. Most noticeably, this list includes only a few publications of the Kansas Geological Survey, even though we relied heavily on the survey's books and maps, particularly the survey reports on the geology in specific Kansas counties. For a list of all publications that are available from the survey, one should contact the Kansas Geological Survey, on the west campus of the University of Kansas.

We consulted many periodical publications, particularly those of the Kansas State Historical Society and the Kansas Academy of Science. Although we have not listed every one of those articles, we would like to acknowledge our use of the *Kansas Historical Quarterly* (now *Kansas History*) and the *Transactions of the Kansas Academy of Science*. This list also does not include publications of the Kansas Geological Society in Wichita. For many years the society has sponsored field trips into various parts of Kansas, and we occasionally consulted the field guides that the society has produced.

Finally, we have not listed the many maps that we used. Most especially, we relied on highly detailed 7.5-minute topographic maps produced by the U.S. Geological Survey. Those maps provided us with much information about the natural features in Kansas; they were the guides that we used in determining place names. Those topographic maps are also available from the Kansas Geological Survey.

Bates, Robert L., and Julia A. Jackson, eds. *Glossary of Geology*. Falls Church, Va.: American Geological Institute, 1980.

Baughman, Robert W. *Kansas in Maps*. Topeka: Kansas State Historical Society, 1961.

Bee, James W., and others. *Mammals in Kansas*. Lawrence: University of Kansas Museum of Natural History, 1981.

Buchanan, Rex C., ed. *Kansas Geology*. Lawrence: University Press of Kansas, 1984.

Clark, John G. *Towns and Minerals in Southeastern Kansas*. Special Distribution Publication no. 52. Lawrence: Kansas Geological Survey, 1970.

Collins, Joseph T. *Amphibians and Reptiles in Kansas*. 2d ed. Lawrence: University of Kansas Museum of Natural History, 1982.

Collins, Joseph T., ed. *Natural Kansas*. Lawrence: University Press of Kansas, 1985.

Cross, Frank B., and Joseph T. Collins. *Fishes in Kansas*. Lawrence: University of Kansas Museum of Natural History, 1975.

Darton, N. H., and others. *Guidebook of the Western United States: The Santa Fe Route*. Bulletin no. 613, pt. C. Washington, D.C.: U.S. Geological Survey, 1916.

Dort, Wakefield, Jr., and J. Knox Jones, Jr., eds. *Pleistocene and Recent Environments of the Central Great Plains*. Special Publication no. 3. Lawrence: University of Kansas Department of Geology, 1970.

Haywood, Robert C. *Trails South: The Wagon-Road Economy in the Dodge City-Panhandle Region*. Norman: University of Oklahoma Press, 1986.

Kansas State Board of Agriculture. *The Sixty-Fourth Annual Report and Farm Facts*. Topeka: Kansas State Board of Agriculture, 1980.

Kansas Water Resources Board. *Kansas Water Atlas*. Topeka: Kansas Water Resources Board, 1967.

Kuchler, A. W. *Potential Natural Vegetation of the Conterminous United States*. Special Publication no. 36. New York: American Geographical Society, 1964.

Merriam, Daniel F. *The Geologic History of Kansas*. Bulletin no. 162. Lawrence: Kansas Geological Survey, 1963.

Miner, Craig. *West of Wichita: Settling the High Plains of Kansas, 1865–1890*. Lawrence: University Press of Kansas, 1986.

Miner, H. Craig, and William E. Unrau. *The End of Indian Kansas: A Study of Cultural Revolution, 1854–1871*. Lawrence: Regents Press of Kansas, 1978.

Muilenberg, Grace, and Ada Swineford. *Land of the Post Rock*. Lawrence: University Press of Kansas, 1975.

O'Neill, Brian. *Kansas Rock Art*. Topeka: Kansas State Historical Society, 1981.

Paul, Shirley E., and Douglas L. Beene. *1984 Oil and Gas Production in Kansas*. Energy Resources Series no. 25. Lawrence: Kansas Geological Survey, 1985.

Reichler, Joseph L., ed. *The Baseball Encyclopedia*. New York: Macmillan Publishing Co., 1985.

Richmond, Robert W. *Kansas: A Land of Contrasts*. St. Louis: Forum Press, 1974.

Rydjord, John. *Kansas Place-Names*. Norman: University of Oklahoma Press, 1972.

Schruben, Francis W. *Wea Creek to El Dorado: Oil in Kansas, 1860–1920*. Columbia: University of Missouri Press, 1972.

Steeples, Don W., and Rex C. Buchanan. *Kansas Geomaps*. Educational Series no. 4. Lawrence: Kansas Geological Survey, 1983.

Stephens, H. A. *Woody Plants of the North Central Plains*. Lawrence: University Press of Kansas, 1973.

Unrau, William E. *The Kansa Indians: A History of the Wind People, 1673–1873*. Norman: University of Oklahoma Press, 1971.

Webb, William P. *The Great Plains*. Boston: Ginn & Co., 1931.

West, Ron R., ed. *Stratigraphy and Depositional Environments of the Crouse Limestone (Permian) in North Central Kansas*. Sixth Annual Meeting of the South-Central Section of the Geological Society of America. Manhattan: Kansas State University Department of Geology, 1972.

Wortman, Julie A., and David P. Johnson. *Legacies: Kansas' Older County Courthouses*. Topeka: Kansas State Historical Society, 1981.

WPA Guide to 1930s Kansas, The. Reprint, Lawrence: University Press of Kansas, 1984.

Zeller, Doris, ed. *The Stratigraphic Succession in Kansas*. Bulletin no. 189. Lawrence: Kansas Geological Survey, 1968.

GLOSSARY

Acre-foot—The amount of solid or liquid necessary to cover an acre of ground 1 foot deep. This term is generally used in reference to water, in which 1 acre-foot equals 325,851 gallons.

Aggregate—Generally used to denote a mixture of substances. In construction, an aggregate is a hard, inert material, such as sand or gravel, that is mixed with cementing material to make concrete.

Alluvium—The unconsolidated sediment—usually sand, gravel, and clay—that has been deposited by running water, such as rivers and streams. The alluvium along many Kansas rivers is a ready source of water.

Ammonite—A fossil occasionally found in Kansas rocks. This is an extinct mollusk that is related to today's nautilus.

Anhydrite—A mineral composed of calcium sulfate, usually deposited by the evaporation of seas. It is common in Permian deposits in south-central Kansas, especially in the subsurface, where it occurs in association with gypsum, another calcium-sulfate material.

Anthracite—A hard, glassy coal that contains 92 to 98 percent carbon. Virtually no anthracite is mined in Kansas.

Anticline—A fold in a layer of rocks that generally arches upward, with older rocks at its center. In Kansas, anticlines often contain oil, which may be trapped at the apex of the anticline's fold.

Aquifer—A formation that is capable of holding and yielding significant amounts of ground water. In most aquifers, water is held in the pore spaces between particles of rock.

Arroyo—A deep gulley that has been cut by an intermittent stream. Common in southwestern Kansas.

Artesian well—A well in which the water level rises above the level of the aquifer, whether or not it flows at the surface. A flowing well is one in which the water does exit at the surface.

Barite—The mineral barium sulfate. Though common in many Kansas rock formations, barite seldom occurs in great quantity. In Saline County, it forms the cementing material between grains of sand, thus creating unusual concretions called barite roses.

Basin—At the surface, a basin is generally a topographically depressed area, such as the Big Basin in Clark County. In the subsurface, a basin is a bowl-shaped depression in the earth's crust, which has been filled with sediments, such as the Salina Basin of central Kansas. A drainage basin is a hydrological term for the area that is drained by one stream, such as the Kansas River Basin.

Bedrock—The solid rock that underlies any unconsolidated sediment or soil. Limestone, for example, is the common bedrock in eastern Kansas.

Bituminous—The soft, most common grade of coal. All coal that is mined in southeastern Kansas is bituminous.

Bogus Legislature—The territorial legislature of Kansas that met in 1855. Elected primarily by Missourians who crossed over into Kansas to vote, the Bogus Legislature had pronounced Southern sympathies and produced a number of Kansas place-names that were changed by later, more abolitionist legislatures.

Brachiopods—Shelled, clamlike animals, fossils of which are commonly found in Pennsylvanian rocks in Kansas.

Bryozoans—Colonial animals that, like corals, built protective skeletons of calcium carbonate. Unlike corals, they did not generally build large reefs. They produce a netlike fossil that is common in Kansas rocks.

Calcite—The mineral calcium carbonate, which is the principal component of limestone. One of the most common minerals in Kansas.

Caliche (pronounced ka-leech'-che)—A rock or soil horizon, just below the surface, which is cemented by calcium carbonate and is common in western Kansas.

Cambrian Period—The period of geologic history that occurred 570 to 500 million years ago. In Kansas, rocks that were deposited during the Cambrian are found only in the subsurface.

Carbonate rock—Rocks composed of carbonate minerals, such as calcite. Common Kansas carbonate rocks are limestone and dolomite.

Cenozoic Era—One of the four major divisions of geologic time. The Cenozoic was the most recent; it includes the Tertiary and Quaternary periods. In Kansas, the Cenozoic is generally represented by unconsolidated sands and gravels, volcanic ash, loess, and other sediments.

Concretion—An accumulation of mineral material, within a rock mass, that is harder or denser than the surrounding rock. As the softer rock weathers away, the concretion remains; most are spherical in shape. Examples are the large, round sandstone features at Rock City and at Mushroom Rocks State Park in central Kansas.

Cone-in-cone—Although it may look like a fossil, this is a sedimentary structure that resembles a stack of inverted cones, taking on a scaly appearance. Easily mistaken for petrified wood, it is composed of impure calcium carbonate. In central Kansas, it occasionally weathers out of the Kiowa Shale.

Confluence—The point at which two streams flow together.

Contact—In geological terms, this is the surface that separates two different types or ages of rocks. For example, the location where limestone and an overlying sandstone meet is called a contact.

Corals—Bottom-dwelling marine invertebrates that may grow as individuals or in a colony, sometimes creating reefs. They create external skeletons that are common fossils in Kansas.

Cretaceous Period—The period of geologic history from 138 to 63 million years ago, when much of Kansas was covered by a shallow sea that deposited shale, sandstone, limestone, and chalk, including many of the spectacular vertebrate fossils of western Kansas.

Crinoids—Marine invertebrates with a cup-shaped head attached to a jointed stalk. Also known as sea lilies, their fossils are common in eastern Kansas, particularly the disc-shaped pieces of their stem.

Cross-bedding—Inclined layers of sediment that have been deposited at an angle when compared to rocks above or below. Cross-bedding occurs in sedimentary rock that has been deposited by wind or moving water; it is apparent in sandstone concretions in central Kansas and in several of the sandstones and limestones in eastern Kansas.

Cuesta (pronounced kwe'-sta)—A ridge with a steep face at one end and a gently sloping face at the other. Such topography characterizes the Osage Cuestas area of eastern Kansas.

Devonian Period—In geologic history, this is the period from 410 to 360 million years ago. In Kansas, rocks of Devonian age are found only in the subsurface.

Dip—In geology, dip is the angle of a rock layer as inclined to the horizontal. Generally it is the angle at which a layer enters the ground. In eastern Kansas, most formations dip, or get deeper, as you move from east to west. Thus, formations at the surface in eastern Kansas are deeply buried in the west.

Displacement—The relative movement of rocks on either side of a fault. For example, a rock layer that is 6 feet lower on one side of a fault, as compared to the other side, is said to have a displacement of 6 feet.

Dolomite—A mineral composed of calcium magnesium carbonate. This is also the name for rock that is composed largely of the mineral dolomite. It is common in the subsurface of Kansas, and it crops out in central Kansas.

Drainage divide—The boundary between two drainage basins.

Epoch—A division of geologic time. Epochs make up geologic periods. For example, the Pleistocene Epoch is part of the Quaternary Period.

Era—The largest division of geologic time. Eras are made up of geologic periods. The recognized geologic eras are the Precambrian, the Paleozoic, the Mesozoic, and the Cenozoic.

Erratic—A large fragment of rock that has been transported into an area from far away. Erratics differ markedly from the bedrock in the area in which they are now found. For example, quartzite boulders in northeastern Kansas are erratics that were transported into the state by glaciers.

Escarpment—A steep slope or cliff.

Evaporites—Sediments that are deposited as a result of evaporation. Gypsum, salt, and anhydrite are three common evaporites, which were deposited in Kansas when a shallow sea evaporated during the Permian Period.

Fault—A fracture in the earth's crust where rock layers on two sides have moved relative to each other. Movement along faults often produces earthquakes.

Flagstone—A rock that splits readily into thin, hard slabs. Limestone and sandstone are common flagstones. In parts of Kansas, flagstone is used for making sidewalks and for building houses.

Formation—A layer of rock that is, or once was, horizontally continuous. In statigraphic terms, a formation is the basic unit that is used to describe rocks; formations can be either lumped together into groups or subdivided into members.

Fusulinid—A group of extinct one-celled ocean dwellers. They were shaped like grains of wheat; their fossils are common in the Pennsylvanian and Permian rocks of eastern Kansas.

Galena—Lead sulfide, a mineral that is the principal ore of lead. Its crystal form is a cube. It can be found in southeastern Kansas, where it was mined from about 1870 to 1970.

Geode—A round, hollow body that measures from one to twelve inches in diameter. The inside is filled with crystals, usually quartz, that point inward. They are most commonly found in limestone, less often in shale.

Gob pile—The waste material that is separated from coal after mining and is piled near the mine shaft or entrance. Gob piles are common in Osage, Crawford, and Cherokee counties.

Graben—A linear crustal block that is bounded on its sides by faults and has been downdropped relative to the rocks on either side.

Ground water—Underground water that is generally found in the pore space of rocks. These water-saturated formations are called aquifers.

Group—A large stratigraphic division. Groups are composed of formations.

Gypsum—Hydrated calcium sulfate. Gypsum is a soft mineral that is common in sedimentary rock. It is one of the first minerals formed when sea water evaporates. Layers of gypsum are found in rocks of Permian age in south-central and north-central Kansas.

High Plains Depressions—Broad, shallow sinkholes that are scattered across the High Plains of western Kansas. Geologists are uncertain about what causes them.

Hogback—A ridge that ends in two steep, equal slopes. A hogback is formed by the outcrop of a steeply dipping erosion-resistant rock unit.

Holocene Epoch—The most recent period in geologic time, the Holocene Epoch includes the time from the present back to the end of the Pleistocene Epoch, about 10,000 years ago. Geologists sometimes refer to the Holocene as the Recent period of time.

Igneous rock—One of the three general classes of rocks. It is formed by the solidification of molten rock or magma. Although rare at the surface in Kansas, it is far more common in the subsurface.

Joints—In the geological sense, these are natural fractures in rock. They are especially common in limestones.

Jurassic Period—The period of geologic history from about 205 to 138 million years ago. Rocks that were deposited during the Jurasic are found in the subsurface in western Kansas. They crop out in only a few locations in southwestern Kansas.

Kimberlite—A type of igneous rock found in volcanic pipes that have forced their way to the surface from miles underground. Kimberlite often contains garnets; it is also the only known source of diamonds. A cluster of kimberlites is found in Riley County.

Lignite—A brownish-black soft coal that has about 70 percent carbon content. Lignite was an important fuel in Kansas during the late 1800s and early 1900s; it was mined extensively in central and northern Kansas.

Limonite—A compound of iron that is formed by the weathering of iron and iron minerals. It is common in the Cretaceous sandstones of central Kansas. In some locations it takes on the crystal form of pyrite. Ordinary rust is the most familiar form of limonite.

Loess (pronounced luss)—A silty, dusty sediment that has been deposited by the wind. It is often rich in clay minerals. It is found throughout Kansas, particularly in steep bluffs in northeastern and northwestern Kansas.

Massive—A term used to describe rocks that have a homogeneous, even structure, not banded or layered. It is also used to describe rocks that occur in thick layers, such as some limestones of eastern Kansas.

Meander—A broad, semicircular curve in a stream bed. Meanders are common in the Kansas, Saline, Smoky Hill, and other rivers of the state.

Member—The smallest stratigraphic subdivision. It is used to label rock layers. Members make up formations, and formations make up groups of rocks.

Mesozoic Era—One of the four large divisions of geologic time. It includes the Triassic, Jurassic, and Cretaceous periods and includes the time from about 240 to 63 million years ago.

Metamorphic rock—One of the three general classes of rocks. This describes rock that has been significantly changed by heat, pressure, or chemical processes. Metamorphic rocks are rare at the surface in Kansas, except in a small area of Woodson County. Others have been transported into the state or occur in the subsurface.

Meteorite—A mass of mineral or rock matter that falls from outer space to the earth's surface. Meteors are similar, but they burn up before they reach the earth's surface.

Microearthquake—An earthquake that is too small for most people to feel. It generally registers less than 3.0 on the Richter Scale.

Mississippian Period—The period of geologic history from about 360 million to 330 million years ago. In Kansas, rocks that were deposited during the Mississippian are found at the surface only in the extreme southeastern corner of the state, although they underlie nearly all of Kansas.

Mortarbeds—The popular name for cemented rock material that is part of the Ogallala Formation and occurs at the surface throughout western Kansas.

Mosasaur—An extinct marine reptile that grew to a length of 15 to 20 feet. Mosasaur fossils are found in Cretaceous formations of western Kansas.

Moss opal—A type of opal that includes branching lines of manganese oxide. It is found in the Ogallala Formation of western Kansas.

Oolites—Spheres of carbonate material, usually about the size of a grain of sand. Oolites are formed by chemical precipitation in a warm ocean, and they can make up some types of limestone that are common in eastern Kansas.

Ordovician Period—The geologic period from 500 to 435 million years ago. Rocks that were deposited during this time are found only in the subsurface in Kansas. In Kansas, geologists often refer to rocks from the Ordovician together with rocks from the Cambrian as Cambro-Ordovician deposits.

Outcrop—The surface expression of a geologic formation.

Outlier—Portions of a rock formation that are detached or away from the main body. Castle Rock in Gove County, for example, is an outlier that is away from a nearby bed of Niobrara chalk.

Overburden—The rock and soil that lie atop a rock layer and must be removed in order to mine the underlying rock.

Oxbow lake—A crescent-shaped lake that is created when a stream abandons a river bend and changes its course. The Kansas River has created an oxbow lake near Lawrence.

Paleozoic Era—One of the four large divisions of geologic time. The Paleozoic includes seven geologic periods, from about 570 million to 240 million years ago.

Pelecypods—Salt-water and fresh-water clams, the fossils of which are common in Kansas rocks.

Pennsylvanian Period—The period of geologic history from about 330 to 290 million years ago. Pennsylvanian-age rocks are common in eastern Kansas.

Period—A division of geologic time. Periods are made up of epochs; periods constitute eras. For example, the Holocene and Pleistocene epochs make up the Quaternary Period, which, together with the Tertiary Period, makes up the Cenozoic Era.

Permian Period—The period of geologic history from about 290 to 240 million years ago. Permian-age rocks are found at the surface in Kansas roughly in the same area as the Flint Hills and the Red Hills.

Pleistocene Epoch—The geologic term for the time from about 2 million to 10,000 years ago. Much of the loess and dune sand, which is found along several Kansas rivers, was deposited during the Pleistocene. It also includes the time when glaciers moved into northeastern Kansas.

Precambrian Era—The longest and most ancient of the four major divisions of geologic time. The Precambrian spans all of geologic time from the formation of the earth to the beginning of the Cambrian Period, or about 4.6 billion years ago to 570 million years ago. In Kansas, Precambrian rocks are found only in the subsurface; they are generally igneous and metamorphic, although cobbles and boulders of Precambrian rocks have been carried into Kansas by streams from the west and by glaciers from the north.

Pterosaurs—Extinct flying reptiles. Their fossils have been found in Cretaceous rocks of western Kansas.

Pyrite—Iron sulfide, also called fool's gold. It is commonly found in yellow cubes in parts of Kansas.

Quartzite—A rock that has been formed by the metamorphism of sandstone. Red quartzite boulders, brought into Kansas via glaciers, are common in the northeastern part of Kansas.

Railroad ballast—The coarse gravel that is used to form the roadbed that railroad tracks are laid upon.

Recharge—The replenishment of ground water in an aquifer. Recharge in western Kansas is low, only a few inches per year at most; whereas in other parts of Kansas it is much higher.

Red beds—The term applied to red sedimentary rocks, usually sandstones or shales. The red sediments of the Red Hills of south-central Kansas are often called red beds.

Rift valley—A valley that has formed along a rift, which is a deep fracture or break in the earth's crust along which a continent breaks apart and separates. Geologists believe that a Precambrian rift may have formed in north-central Kansas but that it stopped and was later covered by sediments.

Road cut—An excavation of a hilltop or hillside through which a road is built. The material that is removed is often used to fill adjacent valleys and low spots. Recent road cuts provide fresh exposures of rock formations, which are valued by geologists.

Section—In geological terms, this is a vertical exposure of rock strata. The term is often applied informally to a drawing of the different layers in such a cross section.

Sedimentary rock—One of three general classes of rock. It is formed by the deposition of sediments by ice, water, wind, and chemical precipitation or by the accumulation of the secretions or remains of organisms. Characteristically deposited in layers or beds, sedimentary rocks are the most common rocks on the surface of Kansas. They cover about 75 percent of the earth's land surface.

Selenite—A coarsely crystalline transparent variety of gypsum. It is common in the Kiowa, Carlile, and Pierre shales of Cretaceous age in western Kansas.

Shoestring sandstones—Sandstone bodies that often contain oil in eastern Kansas. These sandstones, which were probably deposited in river channels, are so named because they resemble wavy lines when they are plotted on maps.

Siding—A short section of railroad track that is adjacent to and parallel to the main railroad track. A train may pull off the main track and onto a siding in order to load or unload frieght or passengers or to let another train pass. Many Kansas ghost towns are today only represented by railroad sidings.

Silurian Period—The period of geologic time from about 435 to 410 million years ago. In Kansas, rocks from the Silurian are usually limestones that are found only in the subsurface.

Sinkhole—A funnel-shaped depression in the land's surface, which has been caused by the slow or sudden collapse of the room of a subterranean cavern. Sinkholes are common in areas of central and southern Kansas that are underlain by salt or gypsum and in southeastern Kansas where the ground has been undermined for lead, zinc, or coal. Big Basin in Clark County is a sinkhole that is more than a mile in diameter.

Sphalerite—Composed of zinc sulfide, sphalerite is the most important ore of zinc. Sphalerite was the primary zinc ore in the mines of southeastern Kansas. The best specimens of this mineral come from abandoned mines in Cherokee County.

Stream piracy—The capture of the upper end of a stream by the growth of another stream. This occurs as a river expands its drainage basin by eroding away soft materials and by capturing additional drainage.

Subsidence—The sinking of part of the earth's crust. Subsidence creates sinkholes.

Syncline—Generally, a downward fold in a layer of rocks in which younger rocks are found at the center of the fold. A syncline is the opposite of an anticline.

Tailwater pit—A pit that irrigators use to collect the runoff from irrigation. The water in those pits is generally pumped out and used again for irrigation.

Terraces—In geological terms, these are flat, broad benches of land that are bounded on either side by steep rises and drops. Rivers often leave behind terraces, which are evidence of former river channels that existed at higher levels.

Tipple—A device at the mouth of a coal mine where mine cars are tipped and unloaded. This also is where coal may be cleaned, sorted, and loaded onto railroad cars or trucks.

Topography—The shape of the land's surface. Kansas topography, for example, consists of rolling hills, in the eastern and central parts of the state, and plains in the west.

Triassic Period—The period of geologic history from 240 to 205 million years ago. Rocks from the Triassic are generally found in Kansas in the subsurface.

Trilobites—A group of marine invertebrates that had a jointed outer skeleton. Their fossils are sometimes found in Pennsylvanian rocks in eastern Kansas. They became extinct at the end of the Permian Period.

Type section—A location where a sequence of rocks was originally described. It is the standard to which other sections of the same rocks in other locations are compared. The geographic name of the type section's locality is used in naming the rock unit that is described there. For instance, the type section for the Oread Limestone is on Mount Oread in Lawrence.

Unconformity—An erosion surface that separates younger rocks from older rocks. An unconformity, for example, may occur in an area in which sediments were deposited by a sea, followed by erosion when the land's surface was lifted up, followed in turn by additional sediment deposition at a later time. An unconformity thus represents a gap in the geologic record.

Underclay—A layer of clay found beneath a coal seam. It is sometimes found in southeastern Kansas.

Volcanic ash—The fine-grained material, usually glass, that is thrown out during a volcanic eruption. By definition, ash is composed of particles that are smaller than four millimeters in size. Ash deposits are common in western and central Kansas.

INDEX

Abilene, 17, 97-98, 192, 195, 261, 309
Admire, Capt. Jacob V., 321
Admire, 278, 321
Admire Group, 277
Agra, 144
Allen, 320
Allen County, 127, 233, 258
Allen Creek, 277, 320
Alligator snapping turtle, 60
Almena, 140
Altamont Limestone, 234
Amazonia Limestone Member, 166
Amoret Limestone Member, 234
Anadarko Basin, 24, 28, 55, 176
Anderson County, 71, 242
Andover, 267
Antelope, 312, 313
Antelope Creek: Barber Co., 53; Clark Co.,
 40; Ottawa Co., 204; Sumner Co., 58
Apache Indians, 34
Arapaho Indians, 44, 52
Arbuckle Group, 61, 70-71, 295, 298
Ardell, 292
Argentine, 128
Argentine Limestone Member, 126-27
Argonia Creek, 56
Arikaree River, 131, 133
Arkalon, 31
Arkansas City, 62, 261-62, 265-66, 269
Arkansas River, 11, 13, 25-26, 39, 57, 59-62,
 60 (fig.), 107, 149, 152, 171-76, 173 (fig.),
 192-94, 196, 198, 200, 221, 228-29, 248,
 257, 261, 263-67, 274, 277-78, 282, 286,
 289-94, 296-99, 301, 305, 315
Arkansas River Lowlands, 61, 194-95, 266,
 288, 299
Arma, 228
Armadillos, 74
Arness, James, 290
Artesian Valley, 36
Ash Creek: Pawnee Co., 294; Washington Co.,
 155
Ashland, 39, 42-43
Ashland-Englewood Basin, 39-40, 43
Ashland oil field, 42
Asphalt rock, 235
Assaria, 201
Atchison, 164
Athol, 144-45
Attica, 54
Atwood, 137, 189
Atwood Lake, 137
Aubrey, 241
Auburn Shale, 118
Augusta, 62
Axtell, 160

Bachelor Creek: Elk Co., 70; Harper Co., 54
Bader Limestone, 112-14, 275
Badger, 219
Badger Creek: Butler Co., 267; Lyon Co., 247
Baileyville, 160
Baker Canyon, 46
Baker University, 331
Bald Mound, 71
Baldwin City, 279, 327-28, 331
Baldwin Creek, 122
Baldwin oil field, 331
Bandera Quarry Sandstone Member, 234
Bandera Shale, 70, 233-34, 236
Barbed Wire Museum, 293
Barber County, 6-7, 41, 45-47, 49, 53, 96
Barite, 202, 203 (photo)
Barneston Limestone, 63-64, 108-10, 110
 (fig.), 159, 271-72, 312
Bartlett Arboretum, 264
Barton, Clara, 296
Barton County, 98, 192, 269, 294, 296, 298
Barton County Community College, 296
Battle Canyon, 179
Battle Creek, 204
Bavaria, 202
Baxter Springs, 211-12, 215, 218, 228, 263,
 296
Bazaar, 274
Bear Creek, 20, 22-23, 27
Bear Creek fault, 20-22, 21 (fig.)
Beattie, 159
Beattie Limestone, 112, 115-16
Beaver Creek: Decatur Co., 190; Rawlins Co.,
 136-37, 140; Republic Co., 150; Scott Co.,
 178; Sumner Co., 57-58
Beckys Knob, 41
Beil Limestone Member, 249
Bellaire, 146
Bellefont, 291
Bellefont oil field, 291-92
Belle Plain, 59, 264
Belleville, 191, 209
Belleville channel, 209-10
Belleville Formation, 209-10
Beloit, 147, 206-7
Belpre, 294
Belvidere, 46
Bender, Kate, 75
Bender Mounds, 75
Bennington, 204
Bergtal South oil field, 294
Bern Limestone, 118, 248, 280, 325
Bethany College, 199
Bethany Falls Limestone Member, 234, 238
Bethel College, 195
Beto Junction, 249

Longton oil field, 69
Longton Ridge, 69
Lookout Mountain, 164
Lost Creek, 59
Lost Springs, 314-15
Lost Springs oil field, 311, 313-15
Louisburg, 241, 253
Louisburg oil field, 239
Louisiana Purchase, 37
Love-Three oil and gas field, 57
Lovewell, 150
Lovewell Reservoir, 149-50
Lucas, 101
Lyndon, 326
Lyon, Gen. Nathaniel, 275-76
Lyon County, 247-48, 252, 273, 275, 279, 319
Lyons, 300
Lyons oil field, 301
Lyons West oil field, 300

McConnell Air Force Base, 266
McCoy, Joseph, 107
McCune, 77
McDonald, 136
McDowell Creek, 111
McFarland, 116
McGuire-Goemann oil field, 53
McLains, 195
McPherson, Gen. James B., 196, 302, 306
McPherson, 191, 196-97, 209, 306
McPherson Channel, 299, 305-6
McPherson College, 306
McPherson County, 196-200, 269, 302-3, 305-8
McPherson Formation, 209
McPherson Lowlands, 195-96, 198, 304-5
Magoffin, Susan Shelby, 318-19
Manhattan, 112, 114-16, 118, 156, 166
Mankato, 147
Manter, 23
Marais des Cygnes River, 235-38, 248, 252-53, 256, 279
Marion, 311
Marion County, 64, 158, 307-9, 311-12, 315
Marion Reservoir, 311
Marmaton Group, 332
Marmaton River, 230-31
Marquette, Father Jacques, 111
Marquette, 198
Marquette Sandstone Member, 301
Marsh, Prof. O. C., 89-90, 183, 323
Marshall, Col. Frank, 158
Marshall County, 156-60, 272, 275
Martin Creek, 332
Marymount College, 202
Marysville, 121, 157-59, 168
Masterson, Bat, 290
Matfield Green, 271-72
Matfield Shale, 64, 109-110, 112, 272
Matheny, James, 137
Mathewson, William ("Buffalo Bill"), 52, 299, 302

Mattoon Creek, 127
Maxwell Game Preserve, 308
Maxwell Spring, 160
Mayfield, 57
Meade, Gen. George G., 33
Meade, 27, 33, 35-36, 288
Meade County, 31, 35, 38, 287
Meade salt sink, 35
Medicine Lodge, 24, 28, 47, 52-53
Medicine Lodge Gypsum Member, 48
Medicine Lodge River, 45-46, 51
Meek, F. B., 155
Melrose, 215
Melvern, 250
Melvern Reservoir, 248, 250, 279
Melvin Creek, 155
Menninger Foundation, 118
Mennonite Heritage Complex, 309
Mentor, 202
Merriam, Daniel, 2
Merriam, 258, 334
Merriam Limestone Member, 258
Merrill Mineral Spring, 326
Meteorites, 143-44, 292
Miami County, 127, 234, 238-41, 253-54, 332
Miami Indians, 238
Miami Trough, 214-15
Midcontinent Geophysical Anomaly, 154-55, 306
Middle Beaver Creek, 145
Middle Branch Hackberry Creek, 187
Middle Buffalo Creek, 147
Middle Cedar Creek, 144
Middle Creek, 239
Middle Creek Limestone Member, 234, 238-39
Middle Emma Creek, 195-96
Middle Fork Beaver Creek, 66
Middle Fork Chisholm Creek, 193
Middle Fork Sappa Creek, 67
Middle Fork Tauy Creek, 328
Middle Limestone Creek, 147
Middle Oak Creek, 146
Middle Spring, 283
Midway, 228
Milan, 57, 262
Mile-and-a-Half Creek, 316
Milford Reservoir, 109, 133
Mill Creek: Wabaunsee Co., 116-17; Washington Co., 155-56; Wyandotte Co., 126
Miltonvale, 206
Mine Creek, 235
Mine Creek Battlefield, 235
Mineral coal bed, 80-82, 228
Mineral Springs, 326
Minersville, 208
Mingo, 87
Minneapolis (Kans.), 13, 102, 205
Minneola, 39
Mission, 258, 334
Mission Creek: Doniphan Co., 164; Shawnee Co., 117-18
Mission Hills, 335

Tumtum & Nutmeg
The Rose Cottage Adventures

Stories by Emily Bearn

With pictures by Nick Price

LITTLE, BROWN AND COMPANY
NEW YORK   BOSTON

Text copyright © 2009, 2010 by Emily Bearn
Illustrations copyright © 2009, 2010 by Nick Price

Little, Brown and Company

Hachette Book Group
237 Park Avenue, New York, NY 10017
Visit our website at www.lb-kids.com

Little, Brown and Company is a division of Hachette Book Group, Inc.
The Little, Brown name and logo are trademarks of Hachette Book Group, Inc.

The publisher is not responsible for websites (or their content) that are not owned by the publisher.

First U.S. Paperback Edition: February 2013
First published in the U.S. in hardcover as *Tumtum & Nutmeg: The Rose Cottage Tales*
in October 2010 by Little, Brown and Company
A Christmas Adventure first published in 2009; *A Seaside Surprise* first published in 2010; and *A Circus Adventure* first published in 2010 in Great Britain by Egmont UK Limited

ISBN 978-0-316-08599-1 (hc) / ISBN 978-0-316-08598-4 (pb)

10 9 8 7 6 5 4 3 2 1

RRD-C

Printed in the United States of America

CONTENTS

A Christmas Adventure

Chapter One

Christmas was coming, and Nutmouse Hall looked very splendid.

Tumtum and Nutmeg had put up a tree in the library, with a big pile of presents underneath. And the drawing room and the ballroom and the dining room had all been adorned with tinsel.

General and Mrs. Marchmouse were coming for lunch on Christmas Day, and Nutmeg had been baking and bottling and pickling since early October.

When Tumtum peeked in the refrigerator, his tummy started to rumble. There was a glazed ham and a pork pie and a plum pudding...and there were mince pies and jellied fruits and sugared ants, and a big white cake with chocolate reindeer on top.

He could hardly wait for Christmas Day to come.

"Only two more nights to go," he said hungrily.

"Yes, and we've still a lot to get through," Nutmeg fussed.

They had just finished supper, and Nutmeg was sitting at the kitchen table, making a long list of things to do.

"Now let's see . . ." she said, nibbling the end of her pencil. "We have the Christmas crackers to stuff, and the pears to poach, and the silver to shine, and we must polish the ballroom floor, and hang the mistletoe. . . . Oh, and gracious me if I'm not forgetting to glaze the marzipan fruits!"

"I'm sure we'll manage," Tumtum said. "Tomorrow's Christmas Eve, so we've still a whole day to get everything ready. And you've been baking for weeks. Just think what a feast it's going to be!"

Tumtum was very excited. But Nutmeg looked worried.

"I hope Arthur and Lucy have a nice Christmas, too," she said. "When I looked under their tree this morning, there were no presents at all. And their refrigerator was almost bare. Would you believe, I don't think Mr. Mildew's even made a plum pudding!"

2

"Well, I'm sure Father Christmas will bring them something nice," Tumtum said.

"I do hope so," Nutmeg replied. "But I bet their Christmas stockings are full of holes." She looked anxiously at her watch. "Let's go up to the attic and I'll see if they need darning," she said. "It's nearly ten o'clock—the children are sure to be asleep by now."

Tumtum agreed, so he took his flashlight, then they let themselves out of Nutmouse Hall and tiptoed through their front gates into the Mildews' kitchen.

Tumtum went first, cautiously twitching his nose. But there was nothing to fear. The lights were out, and everything was very quiet. He turned and beckoned Nutmeg to follow.

They ran through to the hall, then clambered upstairs to the landing.

They could hear footsteps in the study as Mr. Mildew paced around, trying to think of another silly thing to invent.

They crept past his door and started heaving themselves up the steep flight of wooden steps to the attic.

Puff! Pant! Wheeze! It was hard work for two little mice.

Finally, they reached the top and crept out onto Arthur and Lucy's bedroom floor.

They could hear the children snoring softly, but everything else was still. The curtains had been left open, and the floor was pale with moonlight.

"Look, there are the stockings!" Nutmeg said, pointing across the toy train track.

The children had dug them out of the toy chest, ready to hang up on Christmas Eve, and now they were lying in a bundle on the floor.

The mice ran over to them and stretched them out flat on the carpet. They were very nice stockings. Each was made of red felt, with a white rim, and a Christmas tree embroidered on the front.

"This one's got a hole in the toe, just as I suspected," Nutmeg tutted. She ran over to the dollhouse and fetched her sewing basket from the cupboard under the stairs.

Then she sat at the foot of the stocking and started to darn.

Swoosh! Swish! Swipe!

4

Nutmeg always darned at lightning speed.

While she was working, Tumtum shone his flashlight around the room to see if there were any toys that needed to be repaired. Then, suddenly, he noticed a big white envelope propped up on the dresser.

He stepped back and craned his neck up to see who it was addressed to. "Look, dear, they've written a letter to Father Christmas!" he exclaimed. "I wonder what they've asked him for."

Nutmeg looked worried. "Do you think it will reach him in time?" she said. "Tomorrow's Christmas Eve. If they don't go out straight after breakfast, they'll miss the last post."

"I think we had better see what it says," Tumtum said. "Then perhaps we could send Father Christmas a separate letter by Royal Mouse Post, telling him what the children want."

"Good idea," Nutmeg agreed—for the Royal Mouse Post was very efficient. Even if the Nutmouses posted their letter tonight, Father Christmas would be sure to receive it in time.

So Nutmeg abandoned her sewing, then she and Tumtum

hurriedly climbed to the top of the dresser, scrambling up by a pair of tights hanging from the top drawer.

They carefully pried open the envelope and tugged out the two letters from inside. Then they stood at the bottom of the page, reading them by the light of Tumtum's flashlight. The first one said:

> *Dear Father Christmas,*
>
> *If you have one on your sleigh, please may I have a toy car with lights and a horn and an engine that goes* Vroom! —*like I wanted last year. I'm sorry you couldn't come then, but I hope you can come this Christmas instead.*
>
> *Thank you very much.*
>
> *Love,*
>
> *Arthur Mildew*

"Well, I'm sure Father Christmas will be able to give him a car!" Nutmeg said. "But I wonder why he didn't give him one last time."

But when they read Lucy's letter, everything became clear. It said:

Dear Father Christmas,

I'm very sorry that you couldn't come last year because the chimney was bricked up. We asked Pa to unblock it, so you could get down this time, but he said he couldn't, because if he did the wind would come in and we'd all be very cold. We hope you'll find another way in. And if you do, I'd like a box of magic tricks, please — like I asked for last year.

Love,

Lucy Mildew

"Oh, Tumtum! What are we to do?" Nutmeg cried. "If Mr. Mildew's blocked the chimney, then there's no hope of Father Christmas coming! And the children won't get any presents at all!"

Tumtum looked upset, too. He wouldn't enjoy opening his own presents knowing that the children had none.

Christmas would be spoiled for everyone. Unless . . .

"*We* shall have to find them a toy car and a box of magic tricks," Nutmeg said.

"But how, dear?" Tumtum asked helplessly. "It's Christmas Eve tomorrow — and the nearest toy shop is five miles away! And besides, I don't suppose they would serve mice!"

"I wasn't suggesting we should go to the toy shop!" Nutmeg cried. She'd had another idea. But it was such a bold one it made her paws feel clammy.

"We must go and see Baron Toymouse," she said.

Tumtum turned very pale. "*Baron Toymouse!*" he stammered. "Gracious! Do you think that's wise?"

Chapter Two

Tumtum and Nutmeg had never met Baron Toymouse. But they had heard lots of terrible things about him. For Baron Toymouse was infamous for being horrible.

His temper was said to be so bad it could make cupboards shake and doors rattle. And when he got really angry, you could see smoke coming from his nostrils! By all accounts he looked very frightening.

One mouse who had spotted him said that he had eyes as black as night. And another mouse said he had silver claws . . . and another mouse said that he had teeth as sharp as daggers . . . and one mouse said he was taller than a full-grown rat!

So no wonder Baron Toymouse's name filled every mouse with dread. There must have been a reason why the Baron was so horrible. But no one could understand what it was—for if you met him, you would think he was a very lucky mouse indeed.

He lived in the nursery of a big, rambling house just up the lane from Rose Cottage. The children of the house had grown up long ago, and no human had set foot in the nursery for many years. So Baron Toymouse had the toys all to himself.

And what wonderful toys they were! Think of any toy you like, and the Baron was sure to have it. And since he was just a mouse, the toys didn't seem like toys. To him, they were life-size.

All day long, he roared about the nursery in his tanks and trains and racing cars and whooshed through the air in sleek model planes. At night, the Baron slept in a big toy castle, with knights in armor stationed around the walls. And when he was hungry, he helped himself to delicious things from his toy candy shop.

Well, a mouse doesn't get much luckier than that!

"The Baron's certain to have a toy car and a box of magic tricks to spare," Nutmeg said. "And if we go and see him, and tell him that Arthur and Lucy won't have any presents on Christmas Day, then I'm sure he'll want to help."

"And I'm sure he won't!" Tumtum said. "You know what everyone says — Baron Toymouse is the meanest mouse alive. Remember what happened last year, when the poor postmouse went to the nursery to deliver him a Christmas card? The Baron frightened him half to death with a toy snake — then he charged at him in a toy tractor!"

Nutmeg sighed. It was true, the Baron was famously ill-tempered. And he was said to guard his toys very fiercely.

But if they wanted to find presents for Arthur and Lucy, then Baron Toymouse was their only hope.

"I'm sure we can win him over, dear," she said brightly. "It is Christmas, after all. Perhaps he'll be in a more festive mood."

Tumtum doubted it. But he could see that Nutmeg's mind was made up.

"All right," he said reluctantly. "We'll go and see him first thing in the morning."

"Oh, thank you, dear!" Nutmeg said.

They put the letters back in the envelope and brushed away any telltale hairs. Then, as soon as Nutmeg had finished her

darning, they hurried back to Nutmouse Hall and tumbled into their four-poster bed. Tomorrow was a big day, and they needed rest. But they both slept fitfully that night. Tumtum dreamed that Baron Toymouse had locked him up in a toy pigsty, and Nutmeg dreamed she was being chased around the nursery by tin soldiers.

She sat up with a fright. And when she looked out of the window, she saw that it was already light.

She turned around and gave Tumtum a shake. "Wake up, dear!" she said nervously. "It's time to visit the Baron!"

14

Chapter Three

The Nutmouses had a big breakfast to keep their spirits up. Nutmeg made porridge and pancakes and a fruit salad with cherries on top. And they washed it down with hot chocolate.

"Well, we'd better be setting off," Tumtum said, wiping the crumbs from his whiskers. He always felt braver when his tummy was full.

"We might be gone all morning, so I'll make a picnic lunch," Nutmeg said.

She packed a basket with cold meats and a chocolate cake, and two slices of cockroach pie left over from dinner.

"And we'd better take a Christmas present for Baron Toymouse," she said. "I know! Let's give him that box of strawberry creams that I made for Christmas Day. He's sure to like those."

Tumtum's face fell. He had been looking forward to the strawberry creams all week. But Nutmeg was right, they must take the Baron a present. And Nutmeg's creams were so delicious — just think how pleased he would be!

"All right," Tumtum said. And he went off to find some tissue paper and a piece of ribbon to wrap them up with.

Finally, everything was ready. Nutmeg tucked the picnic basket under her arm, and they went to fetch their coats from the cloakroom. Then off they went.

The Baron lived in a big house just down the lane from Rose Cottage, but for Tumtum and Nutmeg it was a long journey, and there was no time to dillydally.

They tiptoed swiftly across the Mildews' kitchen floor and wriggled under the back door into the garden.

It was bitterly cold, and the ground was covered with frost.

"Brrrr!" Nutmeg said, rubbing her paws together.

"Come on," Tumtum said. "We'd better get a move on. I wouldn't be surprised if it starts snowing before the day's out."

They pulled their scarves over their ears and hurried down

the garden path and out the front gate. Then they made their way down the narrow lane that weaved through the village. They could see a cat prowling, so they kept to the bank, walking under the cover of the hedgerow.

Eventually, the hedge stopped and they came to a pair of huge black gates with a sign on them that read GRIMSBY HALL.

Beyond the gates was a long drive, and at the end of it was a tall, dark house full of turrets. In one of those turrets was the nursery where Baron Toymouse lived.

Nutmeg trembled. Grimsby Hall looked like a horrible place. Tumtum squeezed her paw, then they hurried on up the drive, huddling together to keep warm.

Eventually, they came to the front of the house. It looked cold and deserted. There was a creeper snaking over the front door, and the windows were boarded with gray shutters.

"The humans must have gone away for the winter," Tumtum said. "No wonder Baron Toymouse thinks he owns the place!"

"Let's try around the back. Maybe there's a way in there," Nutmeg said.

They ran along the front of the house, past a bay tree and a stone lion. Then they went around a corner, and through a vegetable patch, and around another corner — until eventually, on the far side of the house, they came to a pair of old glass doors leading onto an overgrown terrace. The frame of the door was rotting, and one of the glass panes had come away, leaving a big hole at the bottom.

"Come on, dear," Tumtum said — and they crept inside.

They came out in a long, paneled drawing room, full of tapestries and marble busts. It smelled cold and damp, and the furniture was covered in dust sheets.

They knew that the nursery must be at the top of the house, for that is where nurseries always are. So they started looking for the stairs.

"Let's try through there," Tumtum said, pointing to a door on the far side of the room.

They ran over and peeked through it. And what a sight!

In front of them was an enormous hall, as big as a basketball court, and covered in shining marble.

And leading from it was the biggest staircase the Nutmouses had ever seen. Up and up it went, and on and on,

twisting and turning high into the distance. It had no carpet, and the wood was dark and shiny.

"I suppose we'll have to climb up it," Tumtum groaned.

And so they did.

Puff! Pant! Wheeze! Up and up they went, step by slippery step. And no sooner did they reach the top of one flight of stairs than another flight began.

Eventually, when they had climbed five flights and crossed four landings, they arrived in a long corridor full of doors.

They explored it one way, then the other. And at last they came to a pale blue door that had been left open a crack.

And at the bottom of the door, in writing so small only a mouse could have read it, there was a sign that read:

BARON TOYMOUSE'S TOY KINGDOM. TRESPASSERS WILL BE EATEN BY A MECHANICAL CAT.

"Oh, dear," Nutmeg said. For it didn't sound at all welcoming.

Then they heard a strange noise coming from inside:

Clank! Clonk! Clack!

Clank! Clonk! Clack!

They crept around the door and peeked inside. And what an extraordinary sight it was!

They were standing in a vast room, full of the most magnificent toys they had ever seen. It was like a magic kingdom. In front of them was a huge toy farm, with tractors and barns and pigsties, and fields full of cows and sheep.

And beyond the farm was a sprawling model village, with shops and houses and a toy school, and a white palace with beads twinkling from the turrets. And on the far side of the nursery there were a toy fairground and an electric train and an army base crowded with tanks. There was even a toy airport, with a whole fleet of fighter planes.

And in the middle of the room there was a cold black castle, with a gold flag flying from the tower.

Everywhere the room was teeming with dolls and knights

and tin soldiers. They all looked very busy and purposeful—almost real.

But there was no sign of Baron Toymouse.

"Baron Toymouse!" Tumtum called nervously.

No answer.

"Baron Toymouse! I say, are you there?"

Still no answer.

But then they heard the strange *Clank! Clonk! Clack!* again, and this time it was much louder.

The Nutmouses looked around anxiously, wondering where the noise was coming from.

Then all at once a huge voice boomed around the room:

"WHO . . . GOES . . . THERE?"

Tumtum and Nutmeg gulped.

They could not see who was speaking. But they knew it must be Baron Toymouse. They had heard that he always talked through a loudspeaker.

"If you please, Baron, it is Mr. and Mrs. Nutmouse here, from Nutmouse Hall," Tumtum said politely. "If you would be so kind, we have come to ask you a favor."

"*A FAVOR!*" the Baron snorted, as though he didn't like favors one bit. Then: "STAND OVER THERE, BY THE JACK-IN-THE-BOX!" he ordered. "PUT YOUR PICNIC BASKET ON THE FLOOR AND PUT YOUR PAWS ABOVE YOUR EARS!"

The Nutmouses did as they were told.

The next moment they heard a horn blast—then, to their horror, they saw a big backhoe crashing across the floor.

It was bright yellow, and it was coming straight for them. And perched high up in the driver's seat was the most extraordinary-looking mouse they had ever seen.

He was as big as a rat, and he had pink fur, and snarling fangs, and eyes like black pins. He was wearing a long purple cape and a gold crown—and his claws were a dazzling silver.

"Run!" Tumtum cried.

24

Chapter Four

Tumtum and Nutmeg flew across the room and hid in the toy police station. The big backhoe crashed after them and came to a stop just in front of the door.

Baron Toymouse stood up behind the wheel and addressed them through his loudspeaker.

"COME AND SALUTE ME!" he shouted.

The Nutmouses crept out, looking very shaken.

They could see now that the Baron's fur wasn't *really* pink—he had dyed it with a felt-tip pen. And his claws weren't *really* silver—they were just painted with glitter.

The Nutmouses thought he looked very silly. But they didn't want him to run them over in his big backhoe, so they did just as he said.

Tumtum gave a salute, and Nutmeg bobbed a curtsey.

Then Tumtum tried to reason with him.

"Now, listen here, Baron Toymouse," he said. "We don't mean you any harm. We've come here because there are two little children living in our cottage, and all they want for Christmas is a toy car and a box of magic tricks.

"But their father's too poor to spend any money on presents, and Father Christmas can't visit them, because their chimney's blocked. So we thought *you* might help."

The Baron glared.

Then he tipped back his head and gave a cold, mocking laugh:

"*ME!* HELP? HA! HA! HA! HA! HA! WHY SHOULD I HELP? WHY SHOULD I CARE? HA! HA! HA! HA! HA! HA!"

The Baron's laugh was so loud it made the whole nursery shake.

The Nutmouses pressed their paws to their ears. They were astonished that any mouse could be so awful.

"Let's give him his present," Nutmeg whispered. "That might put him in a better mood."

She ran back to the picnic basket and pulled out the box of strawberry creams.

"WHAT ARE YOU DOING? STOP WHERE YOU ARE!" the Baron boomed.

"We've got a present for you," Nutmeg said, holding it up so he could see.

The Baron's black eyes glinted greedily. He loved presents — as long as they were for him.

"GIVE IT TO ME!" he demanded.

Nutmeg walked back to the big backhoe and handed the box up to him. The Baron snatched it and ripped off the paper.

"CANDY!" he cried gleefully, tearing open the lid.

He opened his mouth and crammed in three at once. But when he tasted the soft strawberry centers, he spat them out with fury.

"STRAWBERRY CREAMS! YUCK! I *HATE* STRAW-BERRY CREAMS!" he cried.

The Nutmouses looked at him in amazement. What a truly horrible mouse he was!

But the Baron hadn't finished yet. He opened his cape and took out a slingshot. Then he loaded it with a cream. And . . .

THWACK!

It hit Tumtum smack on the nose. The candy shell burst open, and a big glob of cream slopped down his coat.

"How dare you!" Tumtum shouted.

But then another piece of candy whizzed through the air and smacked Nutmeg on the forehead.

"Ouch!" she cried.

"I'LL TEACH YOU TO COME TRESPASSING IN MY KINGDOM!" the Baron roared, reloading his slingshot.

PING!

THWACK!

SPLAT!

The strawberry creams kept coming.

"Run!" Tumtum cried.

He took Nutmeg by the arm, and they fled across the floor with pieces of candy whizzing past their ears.

They tried to find the nursery door so they could escape

down the corridor. But the room was so big that they couldn't see where they had come in.

They ran on blindly, through the farm and the fairground, then into the toy village, and past the toy school. . . .

All the buildings were locked. There was nowhere to hide.

"YOU CAN'T ESCAPE!" Baron Toymouse boomed behind them.

"This way!" Tumtum shouted, pulling Nutmeg around the back of the church and over a chessboard.

Then at last they saw the door, just a few feet ahead of them.

They ducked their heads and sprinted toward it.

But then they heard a terrifying noise—and when they looked behind them, they each let out a shriek.

They were being chased by an army of mechanical cats!

And sitting astride the biggest cat of all was Baron Toymouse, twirling a lasso made of strawberry licorice.

"HA! HA!" the Baron boomed. "I SHALL TAKE YOU CAPTIVE AND MAKE YOU MY SLAVES!"

The Nutmouses fled for the door.

They could hear the cats clattering after them, faster and faster, and the Baron crying, "HA! HA! HA!" And just as they were about to escape, they heard the lasso swish through the air — and the next moment the loop came down over Tumtum's head, and clutched him around the tummy.

Chapter Five

Tumtum wriggled and squirmed—but the licorice was pulled tight, pinning his arms to his chest.

"Let him go!" Nutmeg pleaded, trying to tug him free.

The Baron gave a bitter laugh.

"HA! HA! I SHALL LOCK YOU UP IN MY CASTLE AND MAKE YOU SCRUB THE TIN SOLDIERS!" he sneered, clanking toward them on his tin cat.

"Run!" Tumtum shouted, pushing Nutmeg away. "For goodness' sake, run—or he'll catch you, too!"

Nutmeg shot out of the door just before the Baron could grab her. Then she flew along the corridor and started tumbling helter-skelter down the stairs.

Finally she reached the hall and collapsed on the floor. She sat there shaking, listening to her heart going *thump*.

Oh, what a fool I was to think Baron Toymouse would help us, she thought wretchedly.

Tumtum was right, we never should have come. It's all my fault! And if I don't rescue him, Tumtum will spend Christmas Day scrubbing the Baron's soldiers!

She sat there miserably, wondering what to do. Then suddenly a name came to her like a flash:

General Marchmouse!

For if there was any mouse who could rescue Tumtum, then surely it was he. The General was one of the great military heroes of the day. The *Mouse Times* had once described him as "undefeatable"—and that was everyone's view.

He'll know what to do, Nutmeg thought. And she decided to summon him at once.

The General and his wife lived in the gun room of the Manor House, which was at the far end of the lane, beyond Rose Cottage. It would take her some time to get there. And when she looked at her watch, she saw that it was nearly noon.

34

If she was to get Tumtum home in time for Christmas, then there was no time to lose.

She got to her feet and raced back across the drawing room, then shot outside through the hole in the glass door. Her legs were trembling and her breath was coming in painful gasps, but she dared not stop. On and on she ran—back down the drive and along the lane and around the duck pond—until eventually she stumbled under the fence surrounding the Manor House garden.

She went the way she and Tumtum always did when they came to visit the Marchmouses—across the croquet lawn, through the apple orchard, then up the clematis on the side of the house and in through the broken windowpane in the cloakroom.

She climbed down onto the cold tiled floor, then crept down the corridor to the gun room.

General and Mrs. Marchmouse lived in the gun cupboard, which was a handsome oak chest, nearly six feet tall. There were no guns in it now because Mr. Stirrup, who owned the Manor House, had given up shooting when his eyesight started to fail.

So the Marchmouses had the cupboard to themselves.

"Thank goodness, they're at home!" Nutmeg said, seeing light spilling through the keyhole.

She raced around to the back of the cupboard, where the General had carved out his front door, and gave an exhausted *Rap! Tap! Tap!*

A moment later, Mrs. Marchmouse appeared. She was holding a pair of scissors and a roll of ribbon. She had spent a quiet morning wrapping her presents for Christmas Day. But when she saw Nutmeg she could tell at once that something was very wrong.

"Whatever has happened?" she asked.

But Nutmeg was too breathless to speak. So Mrs. March-mouse sat her down in the kitchen with a ginger cookie and a strong cup of tea. Then she fetched the General from his study, where he had been polishing his medals.

And when Nutmeg felt a little calmer, she told them both the whole extraordinary story, from beginning to end.

"So if we don't rescue him, Tumtum will miss Christmas completely," she sobbed. "And he'll spend the rest of his days

locked up in the castle, scrubbing the Baron's toys! And he's never scrubbed anything before and ... oh dear ... I know he won't like it one bit!"

Mrs. Marchmouse was very shocked.

But the General's eyes were ablaze. *An adventure!* he thought, feeling his blood stir.

Adventures were his first love, and he had been sitting in his gun cupboard all winter long, wishing one would occur.

He had never met Baron Toymouse before — but he had heard all about his magnificent nursery, and he had always longed to go there.

And now here was his chance!

"Do you think you could help?" Nutmeg asked tearfully.

"Help? Of course I can!" the General cried. "I shall summon my brother officers, and we shall take the nursery by force! And, oh, how that mean old Baron will tremble when he sees us coming! We'll seize his big backhoe and tie up his mechanical cats. We'll storm the fire station and pelt the castle with toy grenades! Just you wait, Mrs. Nutmouse! He'll regret the day he ever threw strawberry creams at *you*!"

Chapter Six

To show that he meant business, the General slammed his fist on the table and made the teacups shake.

Nutmeg looked very reassured.

"I shall go and fetch my officers!" the General cried, pulling on his camouflage jacket.

He knew exactly where to find them, for they were all lunching with Captain Snip, who lived just down the corridor in the Manor House drawing room. The General had been invited, too. He had been about to leave when Nutmeg arrived.

He let himself out of the gun cupboard and set off for the drawing room at a run.

Captain Snip lived inside a big wicker sewing basket. It

had been given to Mrs. Stirrup years ago as a birthday present, but she didn't like sewing, so she had tucked the basket away behind the sofa, and now the Captain lived there undisturbed. He had a grand bed made from spools of thread, and his dining room table was a big blue button.

The General hurtled toward the basket and leaped inside.

The other officers were already assembled. There were eleven of them in total, all sitting on pincushions, sipping thimbles of cocoa.

They were dressed in their Royal Mouse Army uniforms, which were bright red and green. And since it was a celebratory lunch, they had all oiled their whiskers. Everyone looked very neat and handsome.

Captain Snip, who was tall and thin with speckled hair, was just passing the peanuts to Colonel Acorn when the General suddenly burst in.

"TO ARMS! TO ARMS!" he cried excitedly. "Mr. Nutmouse has been seized by Baron Toymouse and locked up in his toy castle! And if we don't rescue him, he shall spend Christmas Day scrubbing the Baron's tin soldiers!"

Every officer leaped to his feet. Mr. Nutmouse, seized by Baron Toymouse! It was a terrible thought.

"We shall storm the nursery!" shouted Colonel Acorn, gulping his cocoa.

"Hear! Hear!" everyone cried.

They all ran around the sewing basket, arming themselves the best they could with pins and needles, and brass buttons to use as shields.

They set off as soon as they were ready, marching—"Left! Right! Left!"—down the corridor.

Nutmeg and Mrs. Marchmouse heard them from the gun cupboard. "We're coming, too!" they cried. Then they pulled on their coats and hurried to catch up with the officers.

The General led the way. He marched everyone very fast, and it seemed like no time until they reached the gates of Grimsby Hall.

They hurried down the drive, then Nutmeg guided them to the back of the house and showed them the way in through the glass door.

The officers all groaned when they saw the enormous

staircase they had to climb, but after much puffing and panting they finally reached the top floor.

"That's it," Nutmeg whispered, pointing along the corridor to the nursery door.

The officers crept up to it and read the Baron's sign:

BARON TOYMOUSE'S TOY KINGDOM.
TRESPASSERS WILL BE EATEN BY
A MECHANICAL CAT.

"Pah! We'll show him what to do with his mechanical cats!" the General snorted. "Come on, fellows—let's go in and ambush him!"

"Hear! Hear!" everyone cried.

Nutmeg and Mrs. Marchmouse waited in the corridor while the officers tiptoed around the door.

And when they saw inside the nursery, they all gasped. They had never seen such beautiful toys—and everything was mouse-sized.

What a splendid life Baron Toymouse must lead!

The General pointed across the room to the black castle. "Look! That must be where Mr. Nutmouse is locked up," he whispered.

They all studied it through their field glasses. There were toy soldiers firing from the arrow slits, and the entrance was barred with a black gate.

"How will we get in?" the Brigadier said.

Everyone thought for a moment. Then Colonel Acorn pointed to the toy fire station. "I have an idea. Let's seize those fire engines and smash through the gate," he said.

"Good idea," the General agreed. And they all started tip-toeing toward the station.

But then all of a sudden, just as they were cutting through a dolls' tea party, they heard an engine roaring overhead. And when they looked up, they saw a crimson fighter plane swooping toward them. And leaning out of the cockpit was an enormous pink mouse with black eyes and silver claws and a horrible snarl on his face.

44

Chapter Seven

The Baron swooped his plane over the dolls' tea party, flying so low he made the teapot shake.

"BOMBS AWAY!" he cried, hurling down a sherbet candy.

"Duck!" the General shouted.

But the next moment the candy smashed against a milk jug — and a thick cloud of powder filled the air.

The officers yelped. There was sugar everywhere, blinding their eyes. And they could hear the roar of the Baron's plane as he looped back toward them.

"Run!" the General shouted.

They stumbled over the cups and saucers, coughing and spluttering. Finally, they saw a white palace looming before them through the haze of sweet powder. It had a flower fairy at the gate and there were pink beads twinkling on the turrets.

"Quick! Let's hide in there!" the General shouted.

The officers staggered toward it and tumbled through the arch. But when they entered the courtyard, they all jumped— for staring down at them was a huge winged dragon, as big as a cat.

"Run!" cried the Colonel, for it looked very fierce.

"Oh, don't be silly," the General said, thwacking the dragon with his pin. "Look! It's only made of tin!"

Everyone laughed, for of course it was very silly to be frightened by a dragon made of tin.

But suddenly they heard a strange noise.

Clank! Clonk! Clack!

"What's that?" everyone said.

They rushed to the window. And when they looked out they saw a chilling sight.

The Baron had wound up his army of mechanical cats—and sent them all clattering toward the palace.

CLANK! CLONK! CLACK!

46

CLANK! CLONK! CLACK!

CLANK! CLONK! CLACK!

The palace was surrounded on every side.

"Lock the gates!" the General shouted.

The Colonel and the Brigadier flew across the courtyard and drew the bolts. Then they heard the Baron booming through his loudspeaker, "GIVE YOURSELVES UP AND COME OUT WITH YOUR PAWS ABOVE YOUR EARS!"

When they looked back through the window, they could see him standing on the ramparts of his castle, beside a toy cannon.

He looked huge and pink, and his claws were sparkling.

"We're not frightened of your silly cats!" the General shouted back at him. "Now listen here, Baron Toymouse. You're going to set Mr. Nutmouse free and find us a toy car and a box of magic tricks to give to Arthur and Lucy. And if you don't, we'll make you sorry!"

But Baron Toymouse didn't care. He loaded a marble into his cannon, and fired it across the room.

KABOOM!

It smashed straight into the side of the palace and made all

48

the walls shake. The officers cowered under their brass buttons. There were bits of plaster tumbling from the roof.

Finally, the trembling stopped. The General jumped to his feet. "Come on, let's go and teach him a lesson!" he said furiously, brushing the dust off his jacket.

"Hang on," the Brigadier said. "We can't go out on foot— not with all those cats hopping around the palace."

The other officers agreed. It would be much too dangerous. Then Colonel Acorn had an idea.

"I know! Let's ride on the dragon!" he said. He ran across the courtyard and pointed to its tummy. "Look! It's got a switch here to turn it on. It must run on batteries. I'll bet it goes at quite a gallop."

Everyone gathered around to see. They hadn't noticed the switch before.

It was certainly a very fine-looking dragon—there would be room for all of them on its back. It even had a tin saddle.

"Hah! Those mechanical cats will clatter out of the way fast enough when they see *this* coming!" the General said excitedly. "Come on, Colonel. Let's see how she goes."

The Colonel and the Brigadier crawled under the dragon's tummy and wrenched the control to ON.

The dragon shook one leg, then another. It had not been exercised for a long time, and at first it felt rather stiff.

"Quick! Jump aboard!" the General cried, and they all scrambled into the saddle.

The dragon raised its head and gave a long growl. "Grrrrrrr!"

It didn't feel stiff anymore. It reared up on its hind legs and thrashed its tail. Then it tore out of the palace gates and smashed aside a mechanical cat.

"CHARGE!" the General cried, swishing his pin in the air.

Chapter Eight

The Baron heard a crash. And when he looked down from the ramparts, he saw the dragon thundering across the nursery floor.

"Help!" he cried, hurriedly loading another marble into his cannon. But there was no time to fire — for the next moment the dragon crashed through the castle gate.

The Baron crouched down and hid behind a toy troll.

"Come out, Baron! The game is up!" the General shouted. But the Baron kept very still. The officers stomped around the courtyard trying to find him.

"Where are you?" the General barked.

Then a frightened voice piped up from the corner:

"Over here! I'm over here!"

They all spun around. But it wasn't the Baron they saw. It was Tumtum! And what a sorry sight he was. He was tied to a post and had been left with a bowl of soapy water and a pile of tin soldiers to scrub. You could see he had been having a miserable time.

"Oh, I am so glad you came!" he cried as the Colonel cut him free. "Oh, dear, oh, dear! Do you know, when I'd finished scrubbing the soldiers, the Baron was going to make me clean his toy pigs!"

Everyone agreed that the Baron was very wicked.

"How did you get here?" Tumtum asked anxiously. "Did Nutmeg fetch you? Is she here?"

Before anyone had time to explain, the Brigadier pointed across the castle courtyard and let out a cry. And when the others looked around, they saw Baron Toymouse creeping out of the gate.

"Catch him!" the General shouted.

They raced after him. But by the time they got out of the castle, the Baron had disappeared.

"Where in the world can he be?" the General raged, peering around the nursery through his field glasses.

Then suddenly Nutmeg and Mrs. Marchmouse came running over. They had been watching from the nursery door and had seen what had happened.

"He went that way," Nutmeg cried, pointing to the toy village. "We saw him running past the church!"

"Come on! After him!" the General shouted.

Everyone tore across the room.

"There he goes!" the Brigadier barked, seeing the Baron fleeing down the village street.

"I'll catch him!" Colonel Acorn cried. He leaped into a police car and raced after him with the siren screeching.

The Baron lumbered past the school and the toy candy shop, then dived into the little wooden forest. He had given Colonel Acorn the slip—but the next moment, the General spotted him clambering across a game of solitaire.

"He's over there!" the General shouted.

Then—"This way!" the Captain cried, seeing him duck behind the toy pigsty.

The Baron ran and ran, clutching his crown to his head as he searched desperately for somewhere to hide.

Eventually, he came to the toy fairground and threw himself over the gate. He saw the puppet show tent up ahead and dived inside it, pulling the velvet curtains shut tight. Then he crouched there very still, listening to his heart pounding.

He heard the police car stop and the door slam. Then suddenly there were lots of voices, shouting all around him.

"He's here somewhere!"

"Look in the Ferris wheel!"

"Search the bumper cars!"

The Baron held his breath. "Surely they won't find me here!" He trembled. But the next moment the curtain was ripped open — and he saw everyone peering in at him. They all looked very angry.

The General stepped forward. He looked angriest of all. "Come out, Baron Toymouse. Your reign is over," he said coldly.

The Baron knew when he was defeated. He stepped out of the tent, drooping his ears. He had lost his loudspeaker when he was running through the forest. And even though he was much bigger than the other mice, he suddenly felt very small.

"How dare you pelt Mrs. Nutmouse with strawberry creams, and chain Mr. Nutmouse to a post, and try to ruin Christmas?" the General thundered. "You should be ashamed of yourself! Colonel, remove his crown."

Colonel Acorn stepped forward, reached up, and scooped the crown off Baron Toymouse's head.

The Baron had behaved very badly — and he fully deserved to have his crown taken away. Still, he looked so wretched, it was hard not to feel a little sorry for him.

"Now look here, Baron — you've still got a chance to make things better," Tumtum said. "All you need to do is find us a nice car for Arthur, with lights and a horn and a big shiny engine. And a box of magic tricks for Lucy. Then we won't be angry with you anymore."

"That's right," the General said. "Or at least — we won't be as angry as we are now."

The Baron considered this for a moment. He didn't want to help — but he could see he was heavily outnumbered.

"See that big brown sack over there?" he said sulkily, pointing across the room to the fireplace. "Father Christmas left it here last year—and if you look inside you might find just what you need."

"Father Christmas? Whatever was he doing here?" everyone asked. For there had been no children in the nursery for years.

"He came by accident," the Baron said. "There was a snowstorm on Christmas Eve, and it became so blustery he couldn't drive his sleigh. So he dived down the nearest chimney he could find—which happened to be this one."

Everyone drew in their breath. How exciting to think that Father Christmas had been here!

"Were you awake? Did you see him?" Nutmeg asked eagerly.

"Oh, yes, I saw him all right," the Baron said, showing off. "I was hiding in the castle, and I had a good long stare. And I can tell you he was much bigger than he looks in the picture books!"

"How long did he stay?" Tumtum asked.

"Only an hour or two," the Baron said. "As soon as the storm had passed he went on his way again because he had lots of presents to deliver. But he left that sack behind, with two presents still inside. I heard him mutter something about how he'd tried to deliver them to some children down the lane— but he couldn't get into their cottage because the chimney was blocked. That's why he dumped them here."

Nutmeg clapped her paws in delight. "They must be the presents Father Christmas got for Arthur and Lucy last year!" she cried. "Come on, everyone, let's see what they are!"

The Baron had feared the officers might tie him up, or do something horrible. But they were so excited they forgot all about him.

They raced off toward the fireplace, leaving him standing by the puppet show tent. Then they clambered inside the sack.

The presents were still wrapped, and each one had a label tied to it. One said ARTHUR and the other said LUCY.

The mice tore off the paper. Tumtum and Nutmeg used their teeth and claws, while the officers slashed at the wrapping with their pins. And when they saw what was inside they all let

out a cheer. For it was just what the children had asked for — a toy car and a box of magic tricks, the ones that Father Christmas had tried to deliver last year!

And how splendid they looked. The toy car was a brilliant red, with padded leather seats, and a wooden dashboard with all sorts of mysterious knobs and buttons. And it had a big silver horn, and headlights that gleamed bright yellow.

And the box of magic tricks was full of curious things. There were coins with two heads, and a revolting-looking plastic thumb, and pots of magic potions. And there was a big book of spells called *What to Do to Annoying Brothers*.

"Oh, Tumtum, how wonderful!" Nutmeg cried. "Arthur and Lucy will get their presents after all!"

Chapter Nine

The officers all thumped each other on the back, delighted that the adventure had turned out so well. But Nutmeg suddenly looked anxious.

"However will we get the presents back to Rose Cottage?" she asked Tumtum. "The car's much too big for us to carry."

"Oh, we won't carry it!" Tumtum replied. "We'll load up the magic tricks, then we can drive all the way home!"

Tumtum beamed. He loved driving toy cars, and it had been a long time since he had last done so.

The Colonel and the Brigadier helped to push the car out of the sack. Then they found a long piece of licorice in the toy candy shop and used it to tie the box of magic tricks to the roof rack.

"I'll take her for a practice run!" Tumtum said, jumping

into the driver's seat. He gave the horn a loud toot. Then—
vroom!—off he went, racing around the floor.

"Wait for me!" shouted the General—and he quickly
climbed into a fire engine and started chasing after him. Then
the other officers scrambled aboard the toy train, and the big
backhoe, and the tractors and tanks—and soon every one of
them was tearing around the nursery floor.

Vroom! Vroom!

Toot! Toot!

"Merry Christmas!" they cried.

What a lovely time they were having, to be sure!

Nutmeg and Mrs. Marchmouse stood by the fireplace,
laughing at them for being so silly. "Really, you wouldn't think
they were officers," Mrs. Marchmouse said. "They're just like a
class of schoolmice!"

Nutmeg nodded. Tumtum was usually such a calm mouse.
She had never seen him as excited as this.

"I wonder where the Baron's gone," she said suddenly.
Everyone was having such a nice time they had forgotten all
about him. She looked back toward the fairground, wondering

if he was still there. And to her astonishment, she saw him standing at the gate, sobbing.

"*Sob! Sob! Sob!*"

His huge shoulders were shaking, and he was crying so hard his tears had made a puddle on the floor.

Nutmeg hurried over to comfort him.

"Don't be so sad, Baron. It's thanks to you that we've found Arthur's and Lucy's presents. Just think how happy they'll be," she said.

But the Baron cried even more.

"Why don't you come and play with the others?" Nutmeg suggested, thinking it might cheer him up a little.

"But I can't play!" the Baron wept. "I don't know how to!"

Nutmeg was very surprised. "But you have more toys than any other mouse in the whole world. You must know how to play," she said.

"Well, I don't!" the Baron replied miserably. "I've never played. And I've never laughed—at least, not a real laugh, not like that." And he pointed toward all the officers, who were giggling and whooping as they played with his toys.

"But why ever not?" Nutmeg asked.

"Because I was never taught to!" the Baron sobbed.

"But you don't have to be taught how to laugh and play!" Nutmeg said. "It's just . . . well, it's just something mice do."

"Well, it's not something that I've ever done!" the Baron said crossly. "And you wouldn't have done it either—not if you'd been born in a wastepaper basket, like I was, and if you'd had to run away from home after you'd seen your mother and your father and your five brothers *and* your seven sisters all whisked away by a cat! You wouldn't have found much to laugh and play about then!"

Nutmeg was very shocked. No wonder the Baron was so ill-tempered.

"How awful for you," she said.

"Yes, it was!" The Baron sniffed. "But I'm a survivor! And I tell you this, Mrs. Nutmouse. Survivors don't laugh and play. Survivors *fight*!"

Nutmeg looked thoughtful.

"Would you *like* to learn how to play?" she asked eventually.

Another big tear rolled down the Baron's nose. "Of course I would!" he sobbed. "I'd give every toy in my kingdom if only someone would teach me. But it's too late. I've been a grouch all my life. Nobody wants to play with *me*!"

"Oh, but how wrong you are!" Nutmeg cried. "Every schoolmouse in the village has heard about Baron Toymouse and his wonderful collection of toys. They'd give their first whiskers to be invited to come and play with you!"

The Baron looked surprised.

"Would they really?" he asked.

"Of course they would!" Nutmeg said. "Just think what fun they'd have! You could teach them how to fly a plane, and drive a tank. And they'd teach you how to laugh and play soon enough! I have an idea. You could invite all the schoolmice here tomorrow, for tea on Christmas Day. What a fine time you'd all have!"

The Baron gave his nose a loud blow. He looked very thoughtful. "Do you think they'd come?" he asked anxiously.

"Come? Of course they would!" Nutmeg replied. "The officers can go around to all the mouse holes tonight, leaving an invitation in every schoolmouse's stocking. How excited they'll be when they find them in the morning!"

"And I could give them all toy swords to play with, and candies for lunch!" the Baron said eagerly.

In his mind's eye, he began to see the young mice swooping and whooping in his fighter planes, and crashing through the dolls' tea party in his toy tanks. And suddenly it all seemed so funny that Baron Toymouse started to laugh.

But this was a very different sort of laugh than the one he used to make through his loudspeaker.

Instead of a "HA! HA! HA!" it was a "Ho! Ho! Ho!" And once he started laughing, he couldn't stop. On and on he went, until his stomach started to ache.

"Why, Baron!" Nutmeg exclaimed. "You don't need the schoolmice to teach you how to laugh. Just look — you're doing it all by yourself!"

Chapter Ten

The other mice heard the noise and gathered around the fairground gate, watching Baron Toymouse in astonishment.

"Is he all right?" Tumtum asked nervously.

"Oh, quite all right," Nutmeg replied. She handed the Baron her handkerchief, for he had started to splutter.

Finally, the Baron stopped laughing. Then he wiped his eyes and addressed everyone very solemnly.

"These toys are for sharing," he said, sweeping his arm around the room. "And from now on I will run this nursery as a park, where every mouse in the village can come and play. It will be the finest mouse playground in the whole world. And it will open tomorrow, on Christmas Day!"

Everyone cheered and slapped the Baron on his enormous back.

"I'll bring my son!" said the Brigadier.

"And I will bring my niece!" said the Colonel.

"And I'll bring myself!" said the General, who was longing to take a ride in the Baron's plane.

The Baron looked very pleased.

"Now, hurry, hurry!" he cried. "You must go at once, or Arthur and Lucy won't get their presents in time for Christmas! And remember, you have to deliver an invitation to every schoolmouse in the village, asking them here tomorrow!"

"We will," the officers promised.

They were anxious to get going, for they had a lot of mouse holes to visit. And they didn't want to be home late for Christmas Day.

They all shouted goodbye to the Baron. Then they tucked their pins under their arms, and off they marched—"Left! Right! Left! Right!"—out of the nursery door.

The General and Mrs. Marchmouse stayed behind, for Tumtum had offered them a ride home in Arthur's car.

"Jump in," Tumtum said.

So they all climbed inside, the General and his wife in the back, and Nutmeg in the front. Then Tumtum wriggled in behind the steering wheel, and — *Vroom!*

Off they sped, through the door, and along the corridor, and *Bounce! Bounce! Bounce!* down the stairs.

"Goodbye!" Baron Toymouse shouted.

"Goodbye! Goodbye! Merry Christmas!" the officers cried, as the car clattered past them.

"Merry Christmas!" Nutmeg cried back, waving her handkerchief from the car window.

Tumtum drove with great skill. They raced around the sofa legs in the drawing room, then he pressed his paw down hard on the accelerator, and they shot out through the hole in the glass door.

Then off they went, over the vegetable patch, up the drive, and *r-r-r-r-r-r-r-r*, *rrrr-roar!* down the lane.

The night was icy cold. Nutmeg shivered and pulled her scarf around her face. The car window was still open, and she could feel something damp on her nose. And when she looked into the beam of the headlights, she could see a flurrying cloud of white.

"Oh, everyone — look! It's started snowing!" she cried. "It's going to be a white Christmas!"

"We'd better get home quick — we don't want the car getting stuck," Tumtum said. He hunched over the wheel and drove faster and faster as the snow pattered against the windshield.

Tumtum dropped the General and Mrs. Marchmouse at the gates of the Manor House.

"Good night, and see you tomorrow," the General said. "We shall come on our toboggan!"

"Good night!" Nutmeg cried.

Then Tumtum looped back down the lane and sped toward Rose Cottage.

They drove through the front gate, then raced up the garden path until they reached the back door.

The door was falling apart, just like everything else in Rose Cottage. There was a gap at the bottom where the wood had rotted away — just wide enough for the car to drive through.

Nutmeg clung to her seat while Tumtum skidded across the kitchen floor, then shot into the drawing room.

"There they are!" she cried, seeing Arthur's and Lucy's stockings hanging beside the fire. The children had even left Father Christmas a plate of cookies on the hearth—so they must have been hoping he'd come!

Tumtum and Nutmeg jumped out of the car, then they grabbed the bottoms of the stockings and tugged them to the floor.

Finally, Tumtum drove the car into one stocking, and they pushed the box of magic tricks into the other.

"Just think how pleased the children will be when they find them in the morning," Nutmeg said. "They'll think Father Christmas managed to get in after all!"

"Since we've done Father Christmas's job for him, do you think we could eat his cookies?" Tumtum asked longingly.

"Oh, I'm sure he wouldn't mind," Nutmeg said. She hadn't eaten since breakfast, and she could feel her stomach rumbling, too.

They climbed onto the plate, then sat down on either side of one of the cookies and started breaking off little chunks.

It had been a very long day, and they both ate hungrily. Tumtum sighed contentedly. It was wonderful to feel full and fat again.

"Well, I'm glad that adventure's over," Nutmeg said eventually. "Just think, dear. You might have spent Christmas Day cleaning tin soldiers!"

"I wouldn't have liked that," Tumtum yawned. He had eaten rather a lot, and he was feeling very sleepy.

"I suppose we'd better get back to Nutmouse Hall," he said. But they were so tired it was tempting to stay a little longer.

And by the time the clock chimed midnight, they were both fast asleep on the plate — dreaming of plum puddings, and tin dragons, and mechanical cats that went "Boo!" in the night.

On Christmas Day, Arthur woke up first.

And when he looked out of the window he gave a shout:

"Lucy, look! It's snowed!"

Lucy jumped out of bed and ran to look out. The snow had come thick and fast, and the whole garden was glistening white.

"Do you think Father Christmas managed to get in?" Arthur asked anxiously.

"Come on, let's go and see," Lucy said.

They pulled on their robes and ran downstairs.

Nutmeg, who had been curled up against a cookie, woke with a start.

She sat up and gave Tumtum a shake.

"Quick! Get up, dear!" she said urgently. "The children are coming!"

They leaped off the plate and shot under the bookcase. A moment later the children ran in.

"Look—he *has* come!" Lucy cried when she saw their two stockings bulging on the floor.

Lucy pulled out her box of magic tricks and peered at it in delight. It was full of mysterious-looking things. She took out the little bottles of potions and studied the labels. There was a

pink potion called "Shrinking Liquid: For Making People Smaller," and a bottle of green potion called "Swelling Liquid: For Making People Grow," and a bottle of jet-black potion called "Vanishing Liquid: For Making Annoying People Disappear."

She couldn't wait to try them all on Arthur.

And Arthur had never seen such a splendid car. It had lights and an engine, just like he had asked for, and when he honked the horn it was so loud it made Lucy jump.

"This is the best car ever!" he said, racing it around the sofa. "Wasn't it clever of Father Christmas to find a way in!"

"I knew he would," Lucy said. Then she noticed the plate of cookies. "He can't have liked them much—he's only eaten half of one!" she said.

But half a cookie was an awful lot for Tumtum and Nutmeg. And had the children turned around just then, they would have seen two rather fat mice waddling out the door.

78

A Seaside
Surprise

Chapter One

Tumtum was sitting at the kitchen table with his head slumped in his paws. "*The seaside?* But we can't possibly go to the seaside," he groaned. "We might be chased by a crab!"

Tumtum had heard very frightening things about the seaside. He didn't like the sound of it at all.

But Nutmeg's mind was already made up. "If Arthur and Lucy are going to the seaside, then we are going, too," she said firmly.

"When are they leaving?" Tumtum asked miserably.

"Tomorrow morning," Nutmeg replied. "I must say, I wish they'd given us a little more warning!"

She had only learned of the trip a few minutes ago, when she had poked her head into the kitchen and heard the children discussing it.

"They're going to stay with their uncle Jeremy," she went on. "And would you believe it, Mr. Mildew is going to let them travel by train *all on their own*! We'll have to stow away in one of their backpacks and keep an eye on them."

Tumtum scowled. The last time he'd stowed away in a backpack it had been very uncomfortable. "How long are they going for?" he asked.

"A week," Nutmeg replied.

"*A week!*" Tumtum cried. A week is a long time in a mouse's life. "Who will look after Nutmouse Hall?" he asked.

"I'll ask Mrs. Marchmouse to come and keep an eye on things," Nutmeg said busily. "Now don't look so glum, dear. The sea air will do us good. And a week will go by in a flash. You will love it once you're there!"

"No I won't," Tumtum muttered. The very thought of the seaside made him shudder. He knew he wouldn't like it one bit.

"I hope this doesn't turn into another adventure," he said glumly.

"Of course it won't, dear," Nutmeg replied.

But Tumtum wasn't so sure.

The following day, late in the afternoon, there was a sudden *Toot! Toot!*, then a loud clatter of brakes as the express train drew into Mousewall-on-Sea.

"Hooray! We're here!" Arthur shouted, leaping from his seat. They had been cooped up in the train compartment for hours, and he was longing to get out.

Lucy sat up, looking a little startled. She had been asleep, and her toes were full of pins and needles.

"Can you see him?" Arthur asked, pulling down their bags from the luggage rack.

Lucy pressed her nose to the window and peered into the throng of faces on the platform.

"There he is!" she cried.

She had only met Uncle Jeremy once before, and that had been ages ago, before she turned seven. But she spotted him at once, for he looked quite different from anyone else. He was round and comfortable, with a red nose, and a waistcoat stretched so tightly over his stomach that the buttons looked as if they were about to go *Ping!*

"Here's your bag," Arthur said, heaving down Lucy's back-pack. "Wow! What did you pack in it? It weighs a ton!"

"Well, at least I didn't bring four pairs of shoes like you did," Lucy replied.

But Arthur was right—when she pulled her backpack onto her back, it did feel very heavy. It was odd, for she was sure she hadn't packed much.

But as soon as they got off the train she forgot all about it, because there was so much else to think about.

The children had never been to Mousewall before, and it all felt very strange. Uncle Jeremy drove them home through a maze of little wriggling lanes, with hedges so high you couldn't see over them.

Then finally they came to the top of a very tall hill—and when they looked down, they could see the sea, spread out like a big sheet of silver.

Uncle Jeremy stopped the car for a moment so they could admire the view.

In front of them was a thin road, twisting down to a

crescent-shaped bay circled by green cliffs. In the middle of the bay there was a little beach, and a cluster of cottages.

"That would be a nice place to live," Arthur said enviously.

"Well, I'm glad you think so," Uncle Jeremy replied.

"Is that where *you* live?" the children cried. "Oh, how wonderful!"

Uncle Jeremy looked very pleased. Then he started the car again and drove all the way down the hill until they came to a white cottage with fishing nets piled in front of it and a sign above the door that read:

The children piled out of the car and looked around in delight. The cottage was on a narrow lane, and just on the other side of the lane was a tall stone wall, with steps leading down to the beach. They had known that Uncle Jeremy's house was by the sea — but they hadn't expected it to be as close as this.

And Smugglers' Keep was such a mysterious name for a

88

cottage, they felt something exciting was sure to happen to them there.

"Come in and I'll show you your room," Uncle Jeremy said, hauling their bags out of the car. He opened the front door, and the children ran inside to explore.

Smugglers' Keep was very old. The furniture was dark and dusty, and the floorboards creaked. The children's bedroom was at the top of the house, and it had tall wooden beds and a pointed window looking out to sea.

"Well, you sort yourselves out, then we can have dinner," Uncle Jeremy said, dropping their bags on the floor. "Mrs. Blythe's made a fish pie."

Uncle Jeremy had already told the children about Mrs. Blythe. She was his housekeeper, and she did everything for him — or at least everything that Uncle Jeremy felt he couldn't do himself, such as cooking and cleaning, and making beds.

"We'll be down in a minute," Arthur said. He hadn't eaten anything since lunch, and his stomach was rumbling.

"I'll take this bed," he said, claiming the one closest to the window.

"All right," Lucy said, thinking it wasn't worth arguing about. She tossed her backpack onto the other bed, the one clos-est to the door — and as it landed, she heard something squeak. "Goodness, doesn't this room feel funny," she said. "Everything creaks!"

Arthur ran to open the window, and that creaked, too.

Then — "Wow!" Lucy said. "Look at that!" For tucked beside the wardrobe was a big dollhouse, with pale pink walls and peppermint shutters. Lucy thought it was the prettiest dollhouse she had ever seen. And yet when she peeked inside, it was quite empty. There was not a stick of furniture, not even a bed. "I wonder whose it is," she said. "It's odd that it doesn't have anything in it."

"Oh, I don't know. Maybe it belonged to Uncle Jeremy when he was a little boy," Arthur said. He wasn't interested in dollhouses. He wanted to explore the beach.

But Lucy thought it was rather mysterious. "If only we were here longer, I could make some furniture for it," she said. She was still gazing at it as she pulled open the strings of her backpack and started tugging the contents out onto the bed.

"Oh, come on," Arthur said impatiently. "We can unpack later. Let's eat first."

"Hang on. It won't take long," Lucy said.

"But I'm starving!" Arthur groaned, hovering at the doorway.

"Oh, all right," Lucy sighed. Then she dropped the backpack back on her bed and ran after him down the stairs. And it was just as well she didn't finish unpacking, or she would have found something most unexpected hidden in her clothes.

Chapter Two

Tumtum and Nutmeg clambered out of Lucy's backpack and onto the bed, pulling their luggage behind them.

And what a lot of luggage there was!

Four suitcases, two deck chairs, a folding table, a beach umbrella, their swimsuits, Tumtum's fishing net, Nutmeg's sewing basket, and two picnic baskets—one for savories and one for sweets.

No wonder Lucy's backpack felt heavy!

The Nutmouses were very stiff. They had been traveling all day long, squashed up inside one of Lucy's slippers.

"I'm aching all over," Tumtum groaned.

Nutmeg was aching, too. But once they had stretched their legs on the bed they began to feel better.

"Come on, we'd better find somewhere to hide before the children come back from dinner," Nutmeg said.

They stood gazing around the room, searching for a suitable hideaway.

"What about that dollhouse?" Nutmeg exclaimed. "Doesn't it look grand? It would make a splendid home for us!" The windows had been left open, and she could see inside the rooms: "We wouldn't be getting in anyone's way — it looks quite abandoned," she said. "It's not even furnished!"

Tumtum shook his head. "It would be much too obvious — Lucy's sure to look inside it, and imagine what a fright she'll get if she finds that we've moved in. Let's try under there," he suggested, pointing to a big dresser in the corner. It was raised on four legs, so there would be just enough room for them to stand up underneath it.

"We'll be well hidden — the children will never see us," he said.

Nutmeg agreed, and that was how the adventure began.

They dropped all their luggage over the edge of the bed,

then they climbed down the blankets and dragged everything under the dresser.

But it was not at all agreeable. It was dark and gloomy, and the floor was covered in cobwebs.

"Surely we could find a nicer hiding place than this," Nutmeg said, scrunching her nose. "Why don't we try inside one of the bedside tables?"

Tumtum shook his head. "If we camp in a bedside table, the children will be sure to find us," he said. "This is much safer. And once we've set up our picnic table it will look more homey."

He dug out the lantern so they could see what they were doing. But when he shone it around, they both stared in surprise. For in the wall behind them there was a little round mouse hole, with a green door and a shiny brass bell. And above the door was a sign that read:

LORD SEAMOUSE OF
SEAVIEW HOLLOW

LORD SEAMOUSE OF SEAVIEW HOLLOW.

"Gracious!" they exclaimed. They hadn't expected to find another mouse living here.

"Perhaps we'd better go and hide somewhere else," Nutmeg said anxiously — for Lord Seamouse sounded like a rather frightening name.

"Don't be silly, dear," Tumtum replied. "Now that we're here, we must ring the bell."

"Oh, please don't!" Nutmeg cried. But before they could decide whether to ring it or not to ring it, the door opened and a fat white mouse in a green waistcoat appeared. This was Lord Seamouse and, goodness, he did look lordly! He was wearing a watch on a gold chain, and a pair of satin slippers.

Lord Seamouse hadn't seen another mouse for weeks, so he was rather startled. "Are you here to collect the rent?" he asked anxiously.

"No, no, nothing like that," Tumtum reassured him. "We're here to look after Arthur and Lucy."

Lord Seamouse looked very baffled — for, of course, he had never heard of Arthur and Lucy. So once the Nutmouses had introduced themselves, they had to tell him the whole story—about how they were the children's secret guardians, and how they had traveled all the way from Rose Cottage in the bottom of Lucy's backpack.

Lord Seamouse nodded and looked very impressed.

"So we hope you won't mind if we camp under your dresser for a few days," Nutmeg said finally.

"You'll do no such thing!" Lord Seamouse said in a booming voice. "You will both stay in Seaview Hollow with me! Now come on in. You're just in time for dinner!"

Tumtum and Nutmeg looked delighted.

Lord Seamouse helped them drag everything inside. He was astonished by how much luggage they had. But when he saw Tumtum's family crest on the suitcases, he supposed they must be rather grand.

When the last case had been dragged in, Lord Seamouse gave them a tour of his mouse hole. It wasn't at all like the sort of mouse hole you would expect a lord to live in. The rooms were small and poky, and the furniture was all mismatched. The beds were made out of matchboxes, and the dining room table was a yogurt cup.

When he had shown them around, Lord Seamouse laid a tray with tea and ginger cake, and scones with clotted cream and jam on top. Then they all sat down beside the drawing-room fire, in armchairs made out of seashells.

"Have you always lived here?" Nutmeg began.

"In this hollow? Oh, gracious no!" Lord Seamouse spluttered. "In fact, it's a rather long story —"

And he was about to tell it, but then all of a sudden the doorbell rang.

Lord Seamouse jumped.

"Two sets of visitors in one afternoon!" he said in astonishment. This really was unusual. He hurried through to the hall, wondering who it could be.

The Nutmouses waited by the fire. Presently they heard footsteps in the hall, and a voice that sounded oddly familiar. And when Lord Seamouse showed the new arrival into the drawing room, they could hardly believe their eyes!

"Well, dear me!" Tumtum stammered. "If it isn't General Marchmouse!"

Chapter Three

The Nutmouses were astonished. They had seen the General yesterday, and told him about their trip. But he hadn't mentioned anything about coming too.

"How did you get here?" Tumtum asked.

"In the same backpack as you!" the General replied shiftily. "I was hiding in one of Lucy's shoes. I crept onboard last night. I've never been to the seaside, so when I heard you were going, I decided to come too!"

"Does Mrs. Marchmouse know you're here?" Nutmeg asked. Poor Mrs. Marchmouse. The General was always running away from home, and it made her terribly nervous.

"She'll know by now," the General said, looking a little guilty. "I left her a note on the breakfast table."

Tumtum and Nutmeg sighed. They should have known better than to tell the General about their trip—for he could never resist an adventure. And whenever he turned up, trouble seemed to follow. Which is why the Nutmouses were not altogether pleased to see him.

But Lord Seamouse didn't know how troublesome the General could be. He was delighted to have another visitor. "I'll go and make some more tea," he said. "And I believe I've got some jam tarts left, unless that nuisance of a cockroach has eaten them all."

Lord Seamouse hurried off to the kitchen. But as he reappeared, there was a huge crash that made the whole mouse hole tremor. The paintings shook on the walls, and the cups rattled in their saucers.

"Help!" Lord Seamouse cried, ducking behind the tray.

"There's no need to worry. It's only the children coming back," Nutmeg said calmly.

The children crashed and banged all the time in Rose Cottage, so she and Tumtum were quite used to it by now.

But Lord Seamouse looked horrified. There hadn't been

any children at Smugglers' Keep in his lifetime, so he didn't know how noisy they could be.

"They won't give you any trouble," Nutmeg assured him. "And if the weather's fine they'll spend most of their time out-doors."

Lord Seamouse sat down in his seashell, feeling rather shaken.

"What do human children *do?*" he asked curiously.

"Oh, lots of things," Tumtum replied. "They play soccer and hide-and-seek, and they bake cakes and they make model airplanes, and they read books—"

"*They can read?*" Lord Seamouse asked in astonishment.

"Oh, yes," Nutmeg replied. "And they can write, too!"

Lord Seamouse stroked his whiskers. He looked deep in thought.

"Do you think they could read a map?" he asked eventually.

"Of course," Tumtum said. "They are a highly intelligent breed."

"Then they might be just what I need!" Lord Seamouse said excitedly.

"How do you mean?" Nutmeg asked.

Lord Seamouse got up and walked over to his desk. He opened the drawer and took out a brown envelope labeled

Then he pulled a piece of crumpled paper from it and passed it around so everyone could see.

"It's a map!" the General exclaimed. He always loved a good map to read. And this one looked very intriguing.

It showed a beach covered in sand castles, with a line of big craggy rocks down one side. The biggest rock had its name written beside it — Stargazer — and it was marked with a black cross.

"What does it all mean?" he asked excitedly.

"It is a treasure map!" Lord Seamouse said in a hushed tone. He leaned over and traced it with his paw. "The beach is just here, in front of the cottage. And the cross marks where the treasure is hidden — in a cave inside Stargazer!"

"What treasure?" the General asked.

"Why, my treasure, of course!" Lord Seamouse replied. "The Lost Treasures of Seaview Manor!"

"Seaview Manor?" everyone asked, looking very confused.

"Yes, Seaview Manor," Lord Seamouse said impatiently. "You must have noticed it when you were traveling across the bedroom floor. It's the finest manor house in all of Mouse-wall!"

"Oh, you mean the dollhouse!" Nutmeg said. "Yes, we did admire it. But it looked abandoned. Does no one live there?"

Lord Seamouse sighed. He could see that he would have to explain everything from the very beginning.

"The last person to live in Seaview Manor was me!" he said. "And oh, you should have seen the house then! I lived just as a Lord should. There were tapestries and chandeliers and four-poster beds ... and a piano and a Ping-Pong table ... and the drawing-room chairs were upholstered in chintz! Every room was appointed to the highest standard. And if you could only have seen the dinners I hosted in the dining room! Do you know, my dears — the table could seat twelve!"

Lord Seamouse wiped a tear from his eye. Thinking about his old home always made him sad.

"Whose dollhouse was it?" the General asked.

"It had belonged to Uncle Jeremy when he was a little boy," Lord Seamouse said. "He used to keep it downstairs in the drawing room. But it got in the way — so one day he dumped it up here. And that's when I moved in!"

"Did Uncle Jeremy never see you?" Nutmeg asked.

"Oh, no," Lord Seamouse replied. "Uncle Jeremy seldom comes up here. And besides, he's much too old to be playing with dollhouses now."

"So what happened to all the furniture?" Tumtum wondered. It was clear Lord Seamouse hadn't brought any of it to Seaview Hollow. Everything here was very shabby.

"It was stolen — every last stick!" Lord Seamouse said. "I went out for a stroll one day, and when I came home, Seaview Manor had been stripped bare. They took everything — the wardrobes, the beds, the soup bowls — they even took my stuffed beetles!"

"Who?" everyone cried.

"Beach mice," Lord Seamouse said distastefully. "Bad apples, who sleep rough in sand castles — Mousewall's full of them. Anyway, word had gone around that I had a lot of fancy things, and one day a gang of them came up here and stole everything. My cleaning mouse, old Mrs. Moptail, was polishing the silver when they arrived, and they tied her up! When I got home, she was the only thing left!"

Everyone looked very shocked.

"What a terrible story," Nutmeg said.

"It certainly was," Lord Seamouse replied. "Mrs. Moptail was in a dreadful state. And I couldn't even make her a cup of tea — because the rotters had taken my kettle! Well, I couldn't stay in Seaview Manor without any furniture. It felt much too bleak. So I moved in here, to Seaview Hollow. And of course it's perfectly cozy and all that.... But ... well, it's not quite the same."

The others nodded in sympathy. They could see why the Hollow must seem like quite a letdown after life at the Manor.

"Was Uncle Jeremy upset when everything disappeared?" Nutmeg asked.

"Oh, he never bothers with the dollhouse. He hasn't even noticed that the furniture's gone," Lord Seamouse replied. "But of course the thieves took his teeth, too!"

"*His teeth?!*" everyone cried.

"Yes, his teeth," Lord Seamouse said. "All his milk teeth — you know, the ones that fell out when he was a boy. I found them in Seaview Manor soon after I moved in. They were hidden under a bed!"

"Was there a full set?" Tumtum asked.

"Oh, yes. A full set!" Lord Seamouse replied proudly.

Everyone gasped, for human teeth are very valuable things. If a wicked mouse creeps up in the night and plucks one from under a child's pillow, he can trade it with the tooth fairies for a gold coin. So just think how much a whole set would be worth!

"How did they end up under the bed?" Tumtum asked.

"One of the tooth fairies must have hidden them there when Uncle Jeremy was a little boy," Lord Seamouse replied. "Each time she collected another tooth from under Uncle Jeremy's pillow, she would have added it to the pile, until eventually there was a full set. They were very dusty when I found them. You could tell no one had touched them for years."

"But why would the tooth fairy have left them there so long?" Nutmeg asked.

"Oh, you know what fairies are like," Lord Seamouse said snootily. "Very silly creatures—always forgetting things. The poor fairy probably couldn't remember where she'd hidden them!"

"What did you do with them?" Nutmeg asked.

"I polished them up and displayed them in my drawing-room cabinet," Lord Seamouse replied. "All my visitors admired them—and they became quite a talking point. That was my mistake, of course. Because soon every mouse in the village seemed to know that there was a full set of milk teeth in Seaview Manor—so I suppose it's hardly surprising I was robbed."

"Couldn't you catch the thieves?" Tumtum asked.

"Oh, we tried," Lord Seamouse said. "Every policemouse in Mousewall was on The Case of the Missing Teeth. And our local *Mouse Gazette* ran a big story on it—there was even a reward on the rascals' heads. But they were very clever. They didn't leave a single clue."

"So, what about this map?" the General asked impatiently.

"That's the exciting part," Lord Seamouse said. "The other day, the postmouse arrived here with a letter. And to my surprise, it was from one of the thieves himself — a fellow called Black Jaw. And very strange it was, too. He said he was the only one of his gang left. The rest of them came to a sticky end, or so he told me. Apparently they were stealing candy from the village shop, and they got trapped in a tub of sherbet!"

"Ooh! How horrible," Nutmeg said.

"What happened to Black Jaw?" Tumtum asked.

"He got away," Lord Seamouse said. "And without the others to egg him on, he began to mend his ways. He stopped thieving and started cleaning mouse holes for a living. But he didn't get much work because all the mice around these parts knew him as a crook. So he decided to move farther up the coast and start his life all over again.

"But he felt so bad about robbing Seaview Manor that before he went, he sent me this map, showing me where the gang had hidden everything they stole. And he said that the teeth were still there, too."

"How thrilling," the General cried. "Then we must go at once and get everything back!"

"Oh, that would be out of the question," Lord Seamouse said. "Stargazer's the tallest rock on the whole beach—as tall as two grown men! If you stand in front of it, you can see it's full of caves. The treasure is hidden in one of them, but the rock face is much too steep to climb."

"Then how did Black Jaw's gang get there?" Tumtum asked.

"They went up the Secret Tunnel inside the rock," Lord Seamouse replied. "But the tunnel begins in the Grotto. And to reach the Grotto you have to cross the Rock Pool. And if you try to cross the Rock Pool, you'll be gobbled alive by the Crabby Crab! Well, Black Jaw's gang had a big boat, so they could get past him all right. But I've only got a raft, so I'd never survive!"

Nutmeg turned pale. She didn't like the sound of the Crabby Crab one bit. But the General was fearless: "Who cares about your silly crab! I shall find the cave!" he cried.

Lord Seamouse laughed. "I wouldn't try it, old boy—you'll come to a sticky end!"

The General scowled. He didn't like being laughed at.

"Well, if you don't let me try, then you'll never get your treasure back," he said grumpily.

"Oh, but I will—now that these two human children have appeared!" Lord Seamouse beamed. "Don't you see? *They* can climb up the rock and dig the treasure out. It would be no trouble for those giants. They could carry the entire contents of Seaview Manor home in a pillowcase!"

Lord Seamouse's face shone. At long last, he might see all his lovely things again.

Nutmeg looked anxious. "Are you sure they wouldn't come to harm?"

"Of course I am," Lord Seamouse replied. "The rock wouldn't seem very big to them. And the Crabby Crab won't give them any trouble—it will flee to the bottom of the pool when it sees two humans coming!"

"I think it's a capital plan," Tumtum said. "Just think how thrilled they'll be!"

Nutmeg agreed—the children would love a treasure hunt. "I'll leave the map on Lucy's bedside table tonight, and I'll write them a letter, explaining exactly what to do," she said.

Lord Seamouse hurried to his desk and found her a piece of paper.

Everyone was very excited. But General Marchmouse had suddenly become quiet. Crabby Crabs! Caves! Secret Tunnels! Why, this was just the sort of adventure he craved. And he didn't see why Arthur and Lucy should have all the fun. He shot a sideways glance at the map, which Lord Seamouse had left on his chair. *Forget the children*, the General thought slyly. *I'll find the cave first! And I'll bring back the teeth! And oh, just think what a hero I will be!*

Chapter Four

Later that evening, when the children were asleep, Tumtum and Nutmeg crept out to leave the map on Lucy's bedside table.

When they got back to Seaview Hollow, Lord Seamouse made some cocoa. Then they sat up talking in the drawing room. But General Marchmouse didn't say much. He had decided that as soon as the others went to bed, he would sneak out and steal the map back! Then he would set off for the beach at dawn and find the treasure all by himself!

He fidgeted in his armchair, wishing they would all turn in. But on and on they chattered. The General found their talk very dull. His long day had tired him. He became drowsier and drowsier, and his eyelids felt like lead.

By the time Nutmeg offered him another mug of cocoa, he was fast asleep.

She gave him a prod, but he didn't stir.

"We might as well let him sleep here. There's no point moving him," Lord Seamouse said.

Tumtum pulled off the General's shoes, and Nutmeg tucked a warm blanket around him. Then they all went off to bed, leaving the General snoring in his armchair.

When the General woke up in the morning, he felt very foggy. He heard the clock chime seven, and he rubbed his eyes, trying to remember what he was supposed to do.

Then he gave a gulp. "*THE MAP!*" he cried.

He threw off his blanket and sprang from the chair. Then he let himself out of the mouse hole, and crept into the children's room.

But when he peeked out from under the dresser, he let out a howl. He was too late! Arthur and Lucy were already awake.

They had found the map and Nutmeg's letter. And now they were sitting staring at them on Lucy's bed.

The General seethed. If only he hadn't fallen asleep!

Lucy was reading the letter through the magnifying glass attached to Arthur's penknife.

"Oh, hurry! What does it say?" Arthur said impatiently.

Lucy held it up to her face and read it out loud:

Dear Arthur and Lucy,

You will have noticed that there is a big dollhouse in your bedroom — but you would never guess what a mysterious tale it has to tell. For it isn't really a dollhouse. It is the home of a Very Important Creature called Lord Seamouse. But one day, all of Lord Seamouse's beautiful treasures were stolen by a gang of thieves. He was quite heartbroken — and since he no longer had a bed to sleep in, he had to move out.

But now, my dears, you might be able to help. Attached to this letter you will find a map of the beach in front of Smugglers' Keep. Be sure to take good care of it, for this map is very special. It is a Treasure Map, showing where all the treasures from the dollhouse are hidden. If you follow the map, you will be able to find them and put everything back in place.

Love,

Nutmeg

P.S. Say nothing of this to Uncle Jeremy. We don't want any grown-ups interfering in our adventure.

The children were astonished.

"What sort of creature do you think Lord Seamouse is?" Lucy asked.

"I don't know," Arthur said. "And if he's half as secretive as Nutmeg, then we won't be seeing much of him!"

Lucy frowned. It was true. Nutmeg never appeared. But she was longing to know more about this mysterious-sounding Lord Seamouse. "With a name like that, maybe he's a mouse," she said.

"Don't be silly," Arthur said. "Mice don't have titles. Now come on! Let's hurry up and find this treasure!"

They quickly got dressed and had breakfast, and then they set out.

"Stay in front of the house, where I can see you, and don't go climbing any rocks," Mrs. Blythe fussed.

"We won't," they replied, hurtling out the front door.

They crossed the lane and ran down the steps to the beach. It was a very small beach, about the size of a tennis court, and this morning there was no one else there. There were just some rowboats propped against the wall and three canoes lying upside down in the sand.

On one side of the beach there was a little harbor full of fishing boats. And on the other side there was a line of tall black rocks snaking down to the sea.

The children shaded their eyes, for it was already hot, and the sun was making the sand glare.

Arthur pulled out the map, and then they tried to work out what was what. But the drawing was so small that the lines just looked like squiggles.

Lucy had to have a long look through the magnifying glass before she could make any sense of it.

"I've got it!" she said finally. Then she held the map so Arthur could see and pointed each thing out:

"There's the sea . . . and that's the harbor wall . . . and these shapes must be those big rocks over there. And the cross is on the biggest rock. So that must be where the treasure's hidden! Come on! We'll find it in no time!"

"Hang on," Arthur said. "What are these?" He was pointing to some tiny marks on the side of the map, next to where the cross was drawn.

Lucy peered through the magnifying glass again. The marks were so small that at first they just looked like dots and dashes. But when she looked harder, she could see that they were letters.

She studied each one very carefully and read them aloud. "S—T—A—R . . . Stargazer!" she said. "That must be the name of the rock where the treasure's hidden!"

Stargazer. It sounded very mysterious.

"Come on. Let's go and find it!" Arthur said.

They ran over to the rocks and stood looking at them from the sand. In the front, the rocks were smooth and flat, like giant pebbles. But the rocks behind were much bigger. Some were as tall as a front door.

"That must be Stargazer," Arthur said, pointing to the tallest rock of all.

It was very different from the other rocks. It was black and ragged and shaped like a witch's hat. And there was a big rock pool in front of it—but not the kind anyone would want to paddle in. The water was dark and thick with seaweed.

The children could see from where they were standing that Stargazer was full of little nooks and crannies. The treasure could be hidden in any one of them.

They scrambled up onto the boulders and started exploring around the bottom of the rock. They searched and searched, digging their hands into each little groove—but all they found was an old golf ball.

Then Arthur climbed a little way up the rock, hanging on by the juts and grooves, and poked around in the caves higher up. The smallest cave was the size of a jam jar, and the largest was as big as a laundry basket. To a mouse, they would have seemed quite spacious. But they did not take Arthur long to

explore. He groped his hands around each cave in turn — but there was no treasure.

Soon the whole morning was gone.

"Come on," Lucy said eventually. "We'd better get back — it's nearly lunchtime."

"The treasure must be hidden in one of those little caves right at the top," Arthur said, jumping back down.

Lucy frowned. "Then how will we reach it?" she said.

"Look! I can climb up onto that," Arthur said, pointing to a place where the rock bulged to form a small platform. "Then I'll be able to reach across."

"All right," Lucy said. "Let's come back this afternoon and try."

Arthur tucked the map into his pocket, then they ran back to the cottage.

Mrs. Blythe opened the door. She was holding a pair of oven mitts, and her face was red from cooking. "Goodness, look at the state of you!" she exclaimed, seeing Arthur's blackened hands. "You'll need a good wash before lunch. Whatever have you been doing?"

"We were just playing around the rock pool," Lucy said guiltily, remembering that they had promised not to climb on the rocks.

But Mrs. Blythe did not seem very interested.

"Hurry up, dears, lunch is ready," she said, bustling back into the kitchen. The children ran to the bathroom to wash.

"Don't let Uncle Jeremy see the map!" Lucy whispered.

"Of course I won't," Arthur said, patting his pocket. "I'm not that stupid!"

But unbeknownst to him, there was danger lurking.

As the children walked into the dining room, General Marchmouse — dressed in his green summer uniform — was hiding under the sideboard, following their every move.

He watched as Arthur sat down in the chair closest to him, and he saw him reach a hand into his pocket to check that the map was still there.

Aha! the General thought.

He waited until Mrs. Blythe had gone back to the kitchen. Then he sprang out from his hiding place and shot up Arthur's leg like a bullet.

Arthur felt something tickling him and darted a hand to his pocket—but the General had already snatched the map in his paws and leaped to the floor.

Uncle Jeremy, who was sitting at the opposite side of the table, was suddenly aware of something flashing across the carpet toward the door.

He put down his fork and frowned. It looked like a mouse. In fact, he could have sworn it was a mouse. But it was *green*!

He assumed his eyes must be playing tricks on him—and he took off his spectacles and gave them a good polish.

The General moved so fast that the children, who were facing the other way, did not see him at all.

But Arthur was rummaging in his pocket. Then he turned to Lucy, looking very pale.

"The map!" he whispered. "*It's gone!*"

Chapter Five

It was only when lunch was over and they had run back upstairs to their bedroom that Arthur was able to tell Lucy what had happened.

"I felt something tickling my leg, then when I looked in my pocket, the map wasn't there!" he said. "Something took it!"

"But what?" Lucy said. "It must have been something very small."

Arthur shuddered. "It felt like a spider," he said. "It moved very fast, that's for sure. And it obviously knew what it was looking for—it went straight for my pocket, then shot off like a bullet!"

"A spider? But that's impossible," Lucy replied. "Spiders can't read maps!"

Arthur shrugged. "Maybe it was a mouse, then."

Lucy shook her head in bewilderment. "But mice can't read maps either."

The children felt very confused. Someone had taken the map, that was for sure. But whoever could it have been?

"We'd better write Nutmeg a letter, warning her that the map's been stolen," Arthur said. "If there's someone else after the treasure, then she might know who they are."

Lucy agreed, so they found a piece of paper and wrote a hurried note:

Dear Nutmeg,

Arthur was looking after the map while we were having lunch, but then something ran up his leg and dived into his pocket and stole it. He didn't see what it was, but it felt like a mouse. And now they might be looking for the treasure, too. But we'll make sure we find it first.

Love,

Arthur and Lucy

They folded the letter in half and propped it against the lamp on Lucy's bedside table, where they had found Nutmeg's letter that morning.

"She'll know what to do," Lucy said, feeling reassured.

"I hope so," Arthur said. "But now let's go back to the beach and start looking again. We've still got a lot of caves to explore."

They ran downstairs. But Uncle Jeremy was waiting for them in the hall. "How would you like to come fishing?" he asked brightly. He was holding some rods and a canvas bag stuffed with tins and reels.

The children flinched. Normally they would have loved to go fishing. But this afternoon they had treasure to find.

"Come on, jump in the car," Uncle Jeremy said, grabbing his straw hat. "We might catch something for dinner."

"We'll have to go with him," Lucy whispered, fearing Uncle Jeremy would be hurt if they refused.

"But what if someone else finds the treasure while we're gone?" Arthur asked.

"They won't," Lucy said firmly. "We'll only be away a few hours. No one's going to find the cave as quickly as all that."

But unbeknownst to the children, General Marchmouse had already set off. And now he was clambering down the steps to the beach, with the map clutched tightly in his paw.

"I shall find the teeth first! And just let anyone try to stop me!" he muttered.

He jumped off the bottom step into the sand and stared around him in astonishment. He had never been on a beach before, and it was a thrilling sight. There were huge strands of seaweed, coiled like snakes, and exotic shells colored pink and blue and green. And there was ridge after ridge of golden sand, stretching as far as the eye could see.

The beach may have seemed small to Arthur and Lucy. But to General Marchmouse it was like a desert.

He took out his map and noted in which direction the rocks must be. And when he looked through his field glasses, he could just see them, black and hazy in the distance.

He tucked the map back into his breast pocket and set off toward them. It was a long way, and the sun was very hot. Soon his throat felt parched, and he could feel the sweat trickling down his shirt.

On and on he tramped, until eventually a dark shadow fell on him. And when he looked up, there they were — a range of huge black mountains, towering out of the sand!

He took out the map again, trying to work out which one was Stargazer.

"It must be that one!" he cried excitedly, gazing up at the biggest mountain of them all. And when he saw how steep and craggy it was, his heart soared. The General loved danger — and he was sure to find it here.

At the base of the rock there was a smooth, flat boulder, about a foot high. He climbed onto it, pulling himself up by a strand of seaweed. When he got to the top, he saw the Rock Pool — and then even the brave General Marchmouse felt a pang of fear.

It stretched before him like a vast lake, and it looked horribly cold and slimy! On the far side of the pool, he could see the opening of a small black cave. He knew it must be the Grotto

that Lord Seamouse had told them about, where the Secret Tunnel began.

But how would he reach it? It was much too far to swim.

I need a boat! he thought. And he started searching around the rock for something he could use.

Then suddenly he saw a Popsicle stick lodged under a pebble. *That will make a fine surfboard!* he thought. He tugged it free and lowered it into the pool. Then he slithered down onto it, so that he was lying flat on his tummy, and pushed himself off from the bank.

He waited until the stick had stopped wobbling. Then he dipped his paws into the pool and started to paddle.

The water was icy cold, and he couldn't see below the surface. He didn't like it one bit, but he kept his eyes fixed on the cave and paddled on as fast as he could.

At least there's no sign of the Crabby Crab, he thought. *Lord Seamouse was probably making that up!*

But when he was halfway across the pool, he felt a sudden tremor beneath the water, then all at once a huge pink beast reared up in front of him.

The General lay on his Popsicle stick, quivering like jelly. It was the pinkest, most hideous creature he had ever seen. It had claws like nutcrackers, and a huge pink head with monstrous pink eyes goggling straight at him.

The Crabby Crab! The General trembled.

The Crabby Crab fixed him with a very crabby stare. Then it reached out a long pink claw and grabbed him around the stomach.

"Put me down!" the General squealed. But the Crabby Crab just squeezed him tighter. The General saw his life flash before him. He thought he was going to be gobbled up in one bite.

The Crabby Crab was drawing him closer and closer, and its crabby mouth was gaping.

My pistol! the General thought desperately. But he had left it at home. Then suddenly he remembered his slingshot. With a thumping heart, he fished it from his pocket and loaded it with his last peppermint.

He had catapulted many a beetle before, but never a beast as big as this. He aimed very carefully, for he knew he could not afford to miss.

TWANG! went the elastic.

Then *THWACK!* went the peppermint, as it hit the Crabby Crab smack in the eye.

The beast reared backward, giving a roar of pain. He loosened his claw and let the General fall—*plop!*—into the pool.

The General kicked furiously, gasping for air as the seaweed and the black water swirled all around him. Then he felt something clonk against his shoulder—and when he looked up he saw the Popsicle stick floating beside him. He hauled himself onto it and started thrashing furiously back across the pool, not daring to turn around. When he reached the bank, he hauled himself out of the water. Then he raced to the edge of the rock and slid back down the seaweed into the sand.

And then he ran and ran and ran—faster than he had ever run before.

Puff! Pant! Stumble! Wheeze!

On and on he fled until Stargazer and the Rock Pool and the Crabby Crab were all far behind him. Then finally he collapsed, panting, in the sand.

Everything was suddenly very quiet. All the General could hear was his heart going thump.

His backpack was soaking wet, and his clothes were covered in slime.

"The map!" he yelped, fearing he had lost it. But when he felt in his pocket, it was still there.

He fished it out carefully, for the paper was soggy, and held it in the breeze to dry.

It was late in the afternoon, and he still had a long walk home to Seaview Hollow. He felt very wretched. He was cold and hungry, and he knew Tumtum would be furious with him when he got back.

He staggered to his feet and started trudging miserably through the sand.

But the beach seemed to go on forever.

I must be nearly there, he thought, when he had been walking for a very long time.

But there was still no sign of the steps in the wall. He looked through his field glasses, but all he could see was sand. Never-ending sand.

"Oh, dear," he said. "I'm lost!"

Chapter Six

Tumtum and Nutmeg and Lord Seamouse had spent a pleasant day sunbathing on the drawing-room windowsill.

It was nearly four o'clock when they finally packed up their picnic basket and made their way back to Seaview Hollow.

"I wonder why the General suddenly darted off like that, before we'd even started our picnic," Nutmeg said as they climbed upstairs.

"Oh, I don't know," Tumtum said sleepily. "He probably went off exploring. You know the General. He can never sit still."

But Nutmeg still felt uneasy. It was unlike the General to miss lunch.

And as they were walking home across the children's bedroom floor, she noticed the letter on Lucy's bedside table.

"Oh, look, Tumtum!" she said eagerly. "Quick, let's see what it says! Perhaps they've found the treasure already!"

Lord Seamouse waited on the floor while Tumtum and Nutmeg climbed up the bedcover and hopped onto the bedside table. They unfolded the letter under the lamp—but when they saw what it said, their faces fell.

For the Nutmouses guessed at once that it was the General who had stolen the map from Arthur's pocket.

"I should have known he's been up to no good!" Tumtum said angrily.

They dragged the letter back down to the floor. And when they told Lord Seamouse what had happened, he looked very alarmed.

"I hope the General's not still out on the beach," he said. "The tide will be coming in soon!"

Nutmeg turned pale. "We must go and look for him!" she cried.

"The beach is huge. We'd never find him," Lord Seamouse said. "And there's no point in us all getting into trouble."

Tumtum was worried, too. But he agreed with Lord Seamouse that it would be foolish to go out looking for the General now.

"Just you wait, dear—he's probably on his way home already," he said, giving Nutmeg's paw a squeeze. "And you know how greedy he is. He's sure to be back in time for dinner!"

Just then they heard a thunder of feet on the stairs. It was Arthur and Lucy, back from their fishing trip.

"Come on, quick!" Tumtum said—and they all darted under the dresser.

Lord Seamouse let them back into Seaview Hollow, then they waited anxiously for the General to reappear.

But by dinnertime there was still no sign of him. Nutmeg had made a seafood pie, yet somehow no one had much appetite.

They sat in an unhappy silence, listening to the tick-tock of the clock, and to the breeze whistling under the mouse hole door. Nutmeg pulled her shawl tighter. The thought of the General all alone on the beach made her shiver. Oh, if only he would come home!

They were quite right to be worried. For General Marchmouse was having a terrible time.

He was stumbling through the sand, with no idea which way he was going.

Goodness, he was tired!

Gradually it got dark, and the moonlight started playing tricks on him. At one point, he thought he could see a strawberry ice cream bar in the distance. He hurried hungrily toward it — but when he got there it was just a pink shell. He trudged on a little farther. Then he saw a sight that made his heart leap. Just ahead of him, parked in the sand, was a beautiful silver car. *Hooray!* he thought. *I can drive home.* But when he reached it, he let out a howl of dismay. It was just a lemonade can!

Eventually, when he thought he couldn't walk another step, he saw a castle up ahead. He staggered toward it. And this time his eyes weren't deceiving him. It was quite real—and what a splendid castle it was! It had a moat and a drawbridge and tall yellow turrets, all built out of sand.

And there was light spilling from the arrow slits.

That's good, he thought. *There must be someone at home!*

Whoever lived here was sure to know the way back to Smugglers' Keep.

He walked across the drawbridge, which was made from a plastic spoon. Then he poked his head through the archway.

There was a campfire burning in the corner, and a delicious smell of roasting flies. There was no chimney, so the room was full of smoke. The General couldn't see much.

"Is there anyone at home?" he called.

No reply.

"I say, it's General Marchmouse here! Is there anyone at home?" he called again, in a more commanding voice.

Then he heard a horrible laugh. Laughs aren't usually horrible, but this one was. It was harsh and cackling, and it cut through the air like a whip.

Then out of the smoke there appeared the nastiest-looking mouse the General had ever seen.

He was tall and gray and whippet-thin, and he was dressed in a black cape and tall scarlet boots. And he had bloodshot eyes and yellow fangs. And — most unusually — purple claws.

He gave the General a long icy stare. And yet when he spoke his voice sounded as smooth as molasses.

"I am Purple Claw," he purred. "*Doooo* come in. You're just in time for dinner."

The General didn't like the look of him one bit. But he was so hungry, the thought of a roasted fly lured him in.

Purple Claw didn't have much furniture. There were just two pebbles to sit on, and a purple shell for a bed.

"Sit down," Purple Claw said, pointing to one of the pebbles.

Then he took the frying pan from the fire and tossed the General two roasted flies, keeping four for himself.

"It's very kind of you to share your dinner with me," the General said nervously.

There were no plates or cutlery, so they ate with their paws.

It was so smoky, the General could barely see Purple Claw's face. He could just hear his jaws smacking.

"What a nice castle you have," the General said brightly, gobbling his food. "Have you lived here long?"

"Only a few hours," Purple Claw drawled. "And I shall move out after dinner."

"Why are you leaving so soon?" asked the General, who did not know that sand castles get washed away by the tide. Purple Claw did not reply.

"What brings *you* here?" he asked instead, changing the subject.

"I was just taking a little stroll," the General said cagily. Something told him that it would be best not to tell Purple Claw about the treasure.

"The beach is a very dangerous place for a *stroll*," Purple Claw sneered.

"Well, seeing as it was such a nice day, and seeing as I'm only in Mousewall for a week, I thought a stroll would be rather pleasant," the General said nervously.

The conversation was not going at all as conversations should, and he suddenly felt very ill at ease. He had finished his flies, so he decided to ask for directions back to Smugglers' Keep and get straight on his way.

"If you'll excuse me, I really must be getting home, or my friends will be worrying about me," he said, getting up from his pebble. "Perhaps you could be kind enough to direct me back to Smugglers' Keep—"

"*Smugglers' Keep?*" Purple Claw barked, sounding very interested.

"Yes...er...that's where I'm staying, you see—with Lord Seamouse," the General replied. "And...er..." He knew he was saying too much. But he was so confused, he had started to babble. "And...er...now I really must get back. Don't worry if you don't know the way, I'm sure I'll find it. At least I've still got the map—"

"*THE MAP?* What map?" Purple Claw hissed.

"Oh, just a map of the beach, nothing...er...nothing special," the General stammered. "Now if you'll excuse me, it really is time I went."

146

He stumbled across the room, trying to find the way out. But the smoke was so thick that everything was a haze.

Then suddenly he saw Purple Claw standing just in front of him. He was smiling a fiendish smile, and his eyes were blazing.

"Give me the map," Purple Claw said very slowly, reaching out his paw.

"*NO, I WON'T!*" the General squealed.

"*YES—YOU—WILL!*" Purple Claw snarled. Then all at once the General saw a pistol being pointed at him.

He heard a peal of hideous laughter, then everything went black.

Chapter Seven

The General woke up feeling very groggy. He looked around in a daze, trying to work out what had happened.

He was lying on the floor of the sand castle, and his clothes were soggy. The fire had gone out, and there was moonlight piercing through the arrow slits.

Suddenly he remembered Purple Claw pointing a pistol at him — and when he felt in his pocket, the map was gone.

"The thief!" he cried. "Just wait till I catch him!"

He tried to get up, but he could feel a sharp pain in his lungs, and his head was throbbing.

Then he noticed a funny smell. He twitched his nose, trying to identify it. But it was so strong it made his nostrils burn. And he could see clouds of green vapor wafting in the air.

The smell hit him again — and all at once, he realized what it was.

"PEPPERMINT GAS!" he yelped. "He's sprayed me with peppermint gas!"

The General staggered to his feet, desperate to get out. He had seen peppermint gas used in the Royal Mouse Army, and he knew it was a terrible thing.

It is a deadly weapon that can knock a mouse out cold, and make his nose turn blue, and his whiskers crinkle. And if a mouse inhales enough of it, his stomach knots, and his kidneys twist, and his lungs sizzle. . . .

So no wonder the General felt alarmed.

He stumbled through the archway and stood outside, taking big gulps of air. Slowly, his head stopped throbbing, and he knew he had escaped in time.

Even so, he felt very wretched.

The mission had gone horribly wrong. He had hoped to

return to Seaview Hollow a hero, carrying Uncle Jeremy's teeth. But instead he had lost the map. Just think how cross Tumtum and Lord Seamouse would be when they found out! And that's if he ever managed to find his way home!

He stood miserably on the drawbridge, gazing at the vast moonlit beach. He had no idea which way to go.

But then suddenly he saw a light high up in the distance — and to his relief he realized that it was the lantern in front of Smugglers' Keep. And below it, he could just make out the pale outline of the steps in the harbor wall.

Well, at least I'm not lost anymore. That's one good thing, he thought as he stepped onto the drawbridge.

But then he heard a sudden roar — and all at once a jet of ice-cold water scooped him up and flung him back through the archway.

There was water everywhere. Black water, swirling all around him. Around and around it went, tossing and rolling him across the castle floor.

Then it suddenly gushed away.

The General scrabbled to his feet and ran to an arrow slit. And when he looked down the beach, he saw a sight that froze the blood in his tail—a huge foaming wave, snarling toward him.

In an instant, it reared up over his head and smashed into the castle turret. He ran blindly across the room and stumbled outside onto the drawbridge. The castle was tumbling to the ground, and the moat was rising. And the next moment the bridge was sucked from under him, and the General was plunged into the water.

He thrashed his paws frantically, but the current was battering him in all directions.

"Help!" he cried.

Then a big wave swelled up beneath him and spat him out onto the sand.

He lay there trembling. He felt too weak to move. But he could hear the next wave coming toward him.

With a final burst of strength, he dragged himself to his feet, and staggered toward the harbor wall. Then he started

frantically clambering up the big stone steps as the sea crashed behind him.

Finally, he reached the lane. He was out of danger now. But there was still a voice in his head crying, *Run! Run! Run!* And run he did, as fast as his trembling legs would carry him — into the cottage, and through the kitchen, and across the hall, and up the stairs. . . .

He didn't stop until he reached Seaview Hollow.

The Nutmouses and Lord Seamouse were sitting in the drawing room when they heard his exhausted *Rat! Tat! Tat!* on the front door.

"He's back!" they shouted, racing into the hall.

It was long past midnight, but they had all been much too worried to go to bed. And they were so relieved to see the General again that at first they forgot to be upset.

"You're soaking wet!" Nutmeg cried. "Now sit down by the fire, and we'll find you some nice dry clothes to wear."

Everyone made a great fuss over him. Lord Seamouse lent him a pair of pajamas, and Tumtum lent him a pair of slippers, and Nutmeg hung up his suit to dry and warmed up a big slice of pie for his supper.

They could see he was half-starved. So they waited until he had finished eating before the questions began.

"Where have you been?" Tumtum asked finally.

The General scowled. Normally he loved boasting about his adventures. But he felt that what had happened today did not show him off in a very good light.

"We know you stole the map," Tumtum said when he didn't answer. "Really, General, what on earth were you thinking? You must know there's no point in trying to find the treasure yourself. You could never carry it all back from the cave."

"I just wanted to find Uncle Jeremy's teeth," the General replied sulkily. "The children can go and get all the rest. But *I* want to find the famous missing teeth, and bring them home all by myself, and get my picture in the *Mouse Times*. And I will! So there!"

"Oh, no you won't!" Tumtum retorted. "We will return the map to Lucy's bedside table tonight. This is their treasure hunt, and you've no right to interfere!"

The General crossed his arms grumpily.

Everyone thought he was being very silly.

"We might as well take the map back now, then we can all go to bed," Nutmeg said.

"Good idea," Tumtum agreed, getting up from his chair. "Come on then, General — give it back."

The General squirmed and turned very red.

There was a tense silence. Everyone was thinking the same thing.

"You didn't . . . you didn't *lose it*, did you?" Lord Seamouse asked eventually.

The General grunted.

"Oh, how could you be such a fool?" Tumtum cried. "You knew it was the only copy in the whole world!"

"Oh, leave me alone. I didn't lose it!" the General shouted. "IT WAS STOLEN!"

"*Stolen?*" Lord Seamouse gulped.

"Yes, STOLEN!" the General said indignantly. "I met a horrible villain on the beach, who lured me into his beastly sand castle, then poisoned me with peppermint gas, then stole the map while I was out for the count! But just you wait. I'll teach that rascal to mess with *me*!"

Lord Seamouse had started to quake.

"Wh . . . wh . . . what was his name?" he stammered.

"Purple Claw," the General replied. "Very silly name if you ask me — silly name for a — "

But then there was a loud crash. It was Lord Seamouse, fainting on the floor.

Chapter Eight

Lord Seamouse was soon revived with a cold cloth and a glass of water.

But he was in a terrible state.

"Purple Claw! Oh me, oh my! Oh no, oh no!" he kept saying.

Everyone tried to calm him, but he wasn't making any sense.

Tumtum and Nutmeg wondered who Purple Claw could possibly be that his name had such a terrible effect.

"Was he very frightening?" they asked the General.

"*I* certainly wasn't frightened of him," the General lied. "I don't know what Lord Seamouse is making such a fuss about."

"Oh, but you don't know anything!" Lord Seamouse spluttered. "Purple Claw is the biggest villain in all of Mousewall. Everyone trembles when they hear his name! He lurks in sand castles, ambushing any mouse who comes near. He jumps out and says, 'BOO!' and steals your picnic basket and your fishing nets. He'll even steal your swimming trunks, if he likes the look of them! And oh, woe betide you if you make an enemy of him! He'll appear at your mouse hole in the middle of the night and drag you from your bed, kicking and shrieking . . . then he'll take you to the beach and throw you to the sea horses!"

"What an awful-sounding fellow," Tumtum said. Nutmeg was so alarmed that as a precaution she went and bolted the front door.

"I've seen him lurking around in Uncle Jeremy's garden, watching me come and go," Lord Seamouse quivered. "He must have suspected that I knew where the teeth were hidden. He'd give his front fangs for them. He knows how much gold he'd get from the tooth fairies for a full set of milk teeth! If he got his claws on them he'd become so rich there would be no end to his powers. He'd build himself castles and battleships, and he'd employ his own private army to rob and steal for him, and every mouse in Mousewall would live in dread."

"Then we must make sure the children find the treasure first!" Nutmeg cried.

"But we can't stop Purple Claw now!" Lord Seamouse trembled. "Why, if he finds out we've been plotting against him, he'll . . . he'll . . ."

"Oh, nonsense!" the General said sharply, wishing Lord Seamouse would be a little more brave. "We can't let that old tramp frighten us off. Besides, he'll never be able to get across the Rock Pool without a boat."

"Oh, Purple Claw can get anywhere," Lord Seamouse replied. "He's got SPECIAL POWERS!"

"Well, he's still only a mouse. He'll be no match for Arthur and Lucy," Nutmeg said firmly. "They'll scare him off all right."

Lord Seamouse looked a little reassured by this. He supposed even Purple Claw might be frightened when he saw the size of *them*.

"I will write the children a letter, asking them to go back

to Stargazer first thing in the morning," Nutmeg said. "They already know which rock the treasure's hidden in, so they shouldn't need the map."

"But what if they can't find the right cave?" Tumtum asked. "They've already spent a whole morning looking for it. And Stargazer's huge—it must contain a thousand nooks and crannies."

"Tell them to search near the peak," Lord Seamouse said. "Whoever stole the treasure's sure to have hidden it as high up in the rock as possible. When the tide comes in, the caves lower down the rock face are completely submerged—no mouse in his right mind would have hidden it there."

"Good idea," Nutmeg said. "That should narrow down their search!"

Lord Seamouse found her a piece of paper, and she quickly scribbled the children a message, telling them exactly what to do.

"Arthur and Lucy won't let Purple Claw get the better of *them*!" she said.

When the children woke up the next morning, they found Nutmeg's letter:

Dear Arthur and Lucy,

Your map has been stolen by a wicked villain called Purple Claw, who is determined to find the treasure before you do and use it to make himself rich and powerful. But we must not let him! You must return to Stargazer at once, my dears, and resume your search. I am told the treasure is most likely to be hidden close to the rock's peak, where the tide cannot reach it. But be careful how you go, for Purple Claw will be lurking. Try not to arouse his suspicions. Now hurry, my dears. For time is running out.

Love,

Nutmeg

The children read it with astonishment.

"*Purple Claw*," Arthur said quietly, thinking it a very strange name. "Who do you think he is?"

"Well, if he's anything like his name suggests, then he'll have purple claws," Lucy replied. "Maybe he's a rat!"

Arthur shuddered. Mice were one thing, but rats were quite another. Once a gang of them had kidnapped Nutmeg and kept her prisoner on the farm pond. If it weren't for Arthur and Lucy, she might never have escaped. From everything he had heard, rats were a very nasty breed.

He looked at his watch. It was nearly eight o'clock.

"Come on! Whoever Purple Claw is, we'd better hurry if we're going to get there before he does," he said. They quickly got dressed, then as soon as they had finished breakfast they ran outside.

But when they saw the beach they both groaned. The sea was lapping against the rocks, and the pool was completely hidden.

"Well, we can't look for the treasure now," Arthur said.

"Is the tide coming in or going out?" Lucy asked.

"Uncle Jeremy will know," Arthur said. They ran back inside to ask him.

"Let's see," Uncle Jeremy said, looking at his watch. "Yes, it'll be on its way out now."

"How long till it goes down below the rocks?" the children asked impatiently.

Uncle Jeremy looked at them curiously. They had a whole week to play on the beach, so he wondered why they were in such a hurry.

"I'd give it an hour or two," he said. "Now don't look so glum. The beach will still be there." Then he disappeared behind his newspaper again.

The children paced restlessly around the cottage, rushing to the window every five minutes to look out. It seemed as if the tide would never go out. But eventually it did.

"Come on!" Lucy said, and they rushed into the hall.

"Don't forget your backpack," Mrs. Blythe said, appearing from the kitchen. "I've packed a water bottle for you, and some cookies."

"Thank you!" Lucy said, taking the bag and tugging it onto

her shoulder. Then they ran outside and raced down to the beach. Arthur was carrying his toy speedboat. "Why did you bring that?" Lucy asked.

"So we can pretend we're playing with it," he replied. "Nutmeg said we should try not to make Purple Claw suspicious. Well, if he sees us playing with our boat in the Rock Pool, he'll think we're just messing around, instead of looking for treasure."

"Good idea," Lucy said—for they certainly didn't want Purple Claw getting wind of what they were up to.

When they reached the rocks, they scrambled up to the pool, and Lucy slung the backpack onto the ground.

"According to Nutmeg, the cave must be right up there," Arthur said, pointing to the top of Stargazer.

"*Don't point*," Lucy whispered. "He might be watching!"

There was no one on the beach, and there was not a mouse or rat to be seen. But Lucy had an uncomfortable feeling, as though they were being followed.

She made Arthur feel anxious, too.

"Come on, let's pretend we're playing with the boat," he

said. He knelt down beside the Rock Pool and placed it on the water. Then Lucy knelt beside him and pretended to be playing, too. But she kept glancing over her shoulder. It was a horrible feeling—but whoever Purple Claw was, she felt sure he could see them.

And indeed he could. For Purple Claw was crouched behind a pebble on the far side of the pool, watching their every move. He was dressed in a black wet suit, and he was holding his sailboard.

Purple Claw wasn't fooled by the boat. From the moment Arthur pointed up to the rock, he guessed that the children were after the treasure, too.

"Curse them!" he hissed.

He could see the Grotto on the far side of the pool, and he knew from the map that it was where he must get to. He had been about to slip into the water and start windsurfing toward it—but then the children had appeared.

He crouched very still, watching Arthur and Lucy with

167

gritted fangs. If he was going to reach the treasure first, he would have to get rid of them. And there were all sorts of wicked thoughts whirring through his mind.

But Purple Claw was not the only mouse present.

Though he did not know it, Tumtum and Nutmeg and Lord Seamouse and General Marchmouse were also there — hiding inside the pocket of Lucy's backpack.

It had been Nutmeg who had insisted that they come. She knew Purple Claw could be no match for the children. But she wanted to be there just in case.

But Purple Claw was so well hidden behind his pebble, the other mice hadn't spotted him.

The pocket was quite a squash. They were all standing tight together, with their heads poked over the side. The General was scouring the rocks with his field glasses. He felt sure Purple Claw must be there somewhere. He had brought his slingshot along and was looking forward to pelting him with an ink cartridge.

"Can you see him?" Lord Seamouse asked, not for the first time.

"No. Still no sign of him," the General said—but then suddenly he caught sight of Arthur's boat, and he felt his heart throb.

It was a deep crimson, with a fat wooden steering wheel, and a motor as big as a matchbox.

Oh, what a golden opportunity this was!

The General had almost given up hope of finding the teeth by himself—for he didn't dare make another attempt to cross the pool. But in a boat such as that he would be quite safe. He could zoom into the Grotto so fast the Crabby Crab would hardly see him coming!

It was a heavenly plan! But he knew he must act fast.

He lowered his field glasses and gave the others a sly look. Then he suddenly sprang out of the backpack pocket and started hurtling off across the rock.

"Come back! They'll see you!" Tumtum shouted.

But the General didn't care.

He made straight for the edge of the pool, where the children were sitting. Then he tore over Arthur's lap and leaped down into the speedboat.

"What was that?" Arthur cried, looking around in astonishment.

The General flung himself onto the driver's seat and turned the control switch to ON. There was a sudden *Vroom!* and the boat sped out across the water.

"Stop!" the children cried, trying to grab it. But it was already out of reach. Faster and faster it went, its propellers thrashing through the slime.

The children watched in astonishment. And as the boat veered around to enter the cave, they saw a mouse in a green suit whooping at the wheel.

Chapter Nine

The boat sped into the Grotto. Then it gave a last *Vroom!* of its motor and disappeared from sight.

Purple Claw, who was watching from behind his pebble, stamped his feet with rage.

He had left the General to drown in a sand castle. However did he escape?

Purple Claw knew he must move fast, or the General would reach the teeth first. He would have to windsurf after him at once — and it didn't matter if the children saw him!

He slunk out from his hiding place and crept to the edge of the pool. But just as he was lowering his sailboard into the water, he saw something moving in the slime. Then all at once the tip of a pincer appeared . . . and then out peeked a big goggling eye!

It was the Crabby Crab! And when he caught sight of the children, he sank back to the bottom of the pool. But he had given Purple Claw a terrible fright.

"Eek!" he cried, reeling backward. He didn't dare wind-surf across the pool now. Not with a pink monster lurking underwater!

But then how was he to reach the Grotto?

He needed a plan. And his eyes went black as he started plotting.

Tumtum and Nutmeg and Lord Seamouse, who were watching from the backpack pocket, had also seen the General disappear.

And they were furious.

"How dare he!" Tumtum raged. "He promised he wouldn't interfere again!"

"Oh, promises, promises! The General's never thought much of those," Nutmeg groaned.

"Well, we can't stop him now," Lord Seamouse said. "We'll just have to hope that Purple Claw doesn't follow him."

"Oh, I do wish Arthur and Lucy would buck up and find the treasure! Then we could all go home!" Nutmeg cried.

But the children were still standing beside the Rock Pool—staring at the little black hole into which the boat had disappeared.

"That mouse must have been Purple Claw!" Lucy said. "Oh, we are stupid! Nutmeg warned us to look out for him— and now we've gone and let him steal our boat!"

"We must get it back!" Arthur cried.

But the only way to reach the Grotto was to wade into the pool. And the water looked so dark and slimy that neither child dared to.

Besides, the entrance to the Grotto was very small—if the boat had gone in far, they would never be able to fish it out.

"Come on!" Lucy said. "As long as we can find the cave, we'll still get to the treasure first!"

"All right," Arthur said, scrambling to his feet. "I'll climb up and look for it."

He ran around the side of the pool and started pulling himself up the rock face, until he could reach the top. He hugged the rock with one hand, and started groping around with the other.

Presently, his hand found a small opening in the rock. "There's a cave here!" he shouted. "This might be it!"

Lucy watched impatiently as he moved his hand around inside — but all he found was an old water bottle. He tossed it down to her in disgust. Then he felt all around the rest of the rock. But there were no more openings. And the top of the rock was covered in a thick blanket of moss.

Eventually, he gave up and jumped back down.

"Well, now we've looked everywhere," he said glumly. "The treasure must not be here after all."

The children looked at the rocks in dismay. If the treasure wasn't in Stargazer, then wherever could it be?

Chapter Ten

I t was pitch-black in the Grotto. All the General could see was the pale beam of his flashlight, slipping over the walls.

He turned down his engine and inched the boat forward, searching for the Secret Tunnel.

The deeper he went, the colder the cave became. He shivered and rubbed his paws on the steering wheel to keep warm. Then all of a sudden there was a pool of green light — and to his surprise, he saw a lantern hanging from the wall.

Below the lantern, there was a small opening in the rock. And beside it, carved into the rock in big rough letters, was written:

THE SECRET TUNNEL

"Hurrah!" the General cried.

He chugged toward it and tied his boat to a thin sliver of rock jutting from the wall. Then he clambered up into the passage.

The tunnel twisted and turned, and got fatter and thinner, and colder and damper — and in places it was so steep that the General had to clamber along on all fours.

Eventually, he came to a big cave, the size of a shoe box. On the other side of the cave, there was a flight of stairs cut into the rock. He climbed up — but when he got to the top, there was just a brick wall.

"Drat," he said, thinking he had reached a dead end.

But just as he was about to turn back down the stairs, he noticed a thin ribbon of light on the floor.

And when he shone his flashlight over the wall, he saw to his astonishment that there was a door cut into the rock. It didn't have a bell. There was just a rusty iron latch and a bolt that had been left drawn.

He lifted the latch and gave the door a push. It felt very stiff, as if it hadn't been used for a long time. And when it opened, there was a flood of golden light.

The General stepped inside and gave a loud gulp. He was in a cave lit by a chandelier and filled with the most sumptuous treasures. There were paintings and tapestries, and silver candlesticks, and four-poster beds with velvet curtains. And there was a big enamel bath, and a toilet with a pink seat! And there were chests full of fancy jugs and plates, and a dining room table with a china swan on top!

General Marchmouse was astonished — he had never seen such luxuries.

No wonder Lord Seamouse was so anxious to get everything back.

But the General wasn't interested in the furniture. It was Uncle Jeremy's teeth he had come for. He started rummaging through the chests, pulling out the jugs and plates and cups and candlesticks, determined to find them.

Then he noticed a rusty candy tin in the corner of the cave, with the words FRUIT PASTILLES written on it. He scuttled over to it and pried off the lid . . . and there they were — a whole set of fat white milk teeth, shining like pearls.

The General gasped. Just think how much the tooth fairies would pay for teeth such as these! It was hardly surprising Purple Claw was so keen to get his paws on them.

But he was too late now!

The General quickly pressed the lid back on the tin and heaved it up onto his shoulders.

He wanted to get away as quickly as possible and show off his amazing discovery. He couldn't wait to tell the *Mouse Times* that he had solved The Case of the Missing Teeth — when every other mouse in Mousewall had failed!

But as he was staggering to the door, he suddenly saw some bright beads of light coming from the back of the cave — and

when he looked more closely he saw that there was a dark green curtain hanging there.

He put down the tin and walked over to it, wondering what was on the other side. And when he touched the curtain, he found to his surprise that it was made of moss. It was so thick he had to push his way through it with both paws — and when he came out on the other side he let out a cry.

For suddenly he was outside again! He was standing at the mouth of the cave, right at the top of Stargazer — and below him the rock fell away in a steep cliff. He could see the Rock Pool far below him, and the gulls wheeling in the air above.

He was so high up he felt quite giddy.

He crouched down on all fours, for he was frightened of losing his balance. Then he took up his field glasses and peered down the cliff. Arthur and Lucy had gone. But he wondered if Tumtum and the others had stayed behind.

He looked all around the beach, as far as he could see. But there was no sign of them. He supposed they must have gone back to Smugglers' Keep with the children.

Then he noticed a big sand castle, just in front of the rocks, with a red flag flying from the turret. He twisted his lens to bring it into focus and then his heart gave a sudden jolt.

Purple Claw was standing in the watchtower with a telescope — peering straight at him.

Chapter Eleven

The General stumbled back through the curtain, shaken to the core. He knew Purple Claw had seen him. And now he would be lying in wait, ready to ambush him as he made his way home with the teeth.

The situation was hopeless. The General would never be able to get past him.

He paced anxiously around the cave, wondering what to do.

Then all of a sudden he heard a rap on the door.

The General froze. It must be Purple Claw! But how could he have been so quick?

The General grabbed a chair and pressed himself against the wall, waiting to clobber him as he came in.

As the door opened, the General heard some muffled voices on the other side. *He must have accomplices,* he thought anxiously.

He tightened his grip on the chair, ready to bash the first mouse who appeared. But when he saw who it was, he gave a cry of relief.

"Tumtum! Nutmeg! Lord Seamouse! What are *you* doing here?" he gasped.

There was a lot of explaining to be done. But when Tumtum and Nutmeg saw all the treasure, they were too astonished to speak.

Seaview Manor must have been the grandest dollhouse in the world!

Lord Seamouse became quite emotional. He wandered around the cave in a delighted trance, gazing at his beloved possessions. He had feared he might never see them again. But here they all were — and everything intact! Some of the furniture looked a little dusty. But it was nothing that a good polishing wouldn't fix.

"Just think!" he said tearfully. "Soon everything will be

returned to Seaview Manor, and I'll be able to move back into my old home again!"

The General looked sulky—for despite his grand adventure, no one was making a fuss of him. "I got here first!" he said stubbornly.

Tumtum turned to him savagely. He had been so overawed by the sight of the treasure that for a moment he had forgotten how angry he was. But he had remembered now. "I know you did, and I've a good mind to box your ears!" he said. "We need Arthur and Lucy to carry this treasure back to Smugglers' Keep—we'll never get it home on our own. But then you try and mess everything up! First you steal the map, then the children's boat. You should be ashamed of yourself! But if you think you can spoil everything, you can think again. We've come to take you home!"

The General huffed. "How did you come after me without a boat?" he asked. "You're lucky the Crabby Crab didn't catch you."

"It was thanks to the crab that we got here," Tumtum replied. "After you went into the Grotto, we climbed out of the backpack and came down to the pool to see if we could call you back. Then the Crabby Crab poked his head out of the water. And the four of us got talking, and it turned out that he's not crabby at all. Most delightful fellow, in fact. And when we explained why we were there, he told us all to hop onto his back — then he carried us to the Secret Tunnel."

"Huh," the General said. "I wish he'd been as helpful with me."

"And he said that if Purple Claw tries to cross the pool, he'll gobble him up!" Nutmeg added. "So we don't have to worry about him following us up here."

Tumtum frowned. He was looking at the walls of the cave. "How will the children ever find the cave from the outside?" he asked. "Look, it's completely enclosed!"

"No, it's not," the General replied. He walked to the back of the cave and pulled aside a corner of the moss, so they could see outside.

Everyone was astonished. The cave was very well hidden. No wonder the children hadn't found it yet.

"Keep still," the General warned them. "When I went out

earlier, Purple Claw saw me. He was lurking in that sand castle down there."

"Oh, no!" Tumtum said. "I hope he doesn't give us any trouble on the way home."

"It's all right, I know another way we can go," Lord Seamouse said. "We can cut up beside the rocks, then creep along the bottom of the harbor wall. That way we won't have to pass the sand castle."

"Good," Tumtum said. "Now come on, we'd better get going. We've still got a long walk back to Smugglers' Keep."

They all made for the door — except the General, who was hovering by the tin of teeth.

The children could fetch the rest of the treasure — but he didn't see why he should leave the teeth behind. He could carry them home himself — they weren't too heavy. Then he could tell the *Mouse Times* that it was he who had found them, and imagine what a fuss there would be!

"Come on, General — hurry up," Tumtum said impatiently.

"I'm bringing these with me!" the General said stubbornly, hoisting the tin onto his back.

"Don't be ridiculous," Tumtum said. "If Purple Claw sees us carrying a treasure chest across the beach, he's sure to come and ambush us!"

"Oh, forget Purple Claw!" the General snapped.

It was blowing up into quite an argument. But then all of a sudden there was a huge roar outside the cave.

"It's a bird!" Nutmeg cried.

"That's not a bird," the General said. "*That's a plane!*"

"Duck!" Tumtum cried.

At that moment, a fierce blast of wind swept open the curtain. And as the mice looked out in horror, they saw a toy airplane swooping toward the cave.

They flattened themselves against the wall, for it was coming straight at them. It was so close they could see into the cockpit. And what a surprise they got! For there was Purple Claw, grinning horribly as he clung to the joystick.

Chapter Twelve

Purple Claw's plane plowed into the cave. It landed with a ferocious bang. One of the engines exploded, and there was smoke billowing down the wings.

The plane had taken a terrible battering. Everyone wondered if Purple Claw had survived — but then the roof of the cockpit flipped open, and out he stepped.

He looked very calm, despite his crash landing.

"How do you do?" he purred, peeling off his goggles.

"You are outnumbered, Purple Claw! Now get out of here or I'll make you sorry!" the General shouted in his most menacing voice.

He stepped forward, baring his fists — but Purple Claw

just sneered. Then he reached into his cape and whipped out his peppermint gun.

"Face the wall, with your paws behind your backs!" he barked.

And they all did, including General Marchmouse — for even the bravest mouse does what he's told when someone's pointing a peppermint gun at him.

Then Purple Claw tied them up with cotton, pulling it so tight it burned their skin.

"You rogue! You won't get away with this!" the General fumed.

"Oh, yes I will!" Purple Claw said, giving him a push. When all the prisoners were bound, he started upturning the chests and cupboards, searching for the teeth. And when he finally found the candy tin and pried open the lid, his heart gave a surge.

"Ha! Ha! Just think how much gold the tooth fairies will give me for all of these!" he cried. "Oh, what a rich and power-ful villain I will be!"

In his mind's eye, he saw the life he would lead — a life of banquets, and fine clothes, and feather beds. "I'll buy a yacht!" he shrieked, clapping his paws in delight. "And I'll build myself a palace in the dunes, and I'll employ an army of field mice to do all my robbing and thieving!"

The idea was so glorious it made him shiver. But this was no time for dreaming. He knew he must get out fast, in case those tiresome children came back.

He stuffed his gun back into his holster and heaved the tin onto his shoulders.

"Goodbye, my dears!" he said mockingly — then he staggered through the door, and slammed it shut behind him.

A moment later, the mice heard the bolts being clicked shut on the other side.

"He's locked us in!" the General fumed.

They were trapped!

"Come on, let's get these ropes off. We can't do anything with our paws tied," Tumtum said. But Purple Claw had secured the knots very tightly.

"I'll try and gnaw through them," Nutmeg said. She leaned down and started nibbling at Tumtum's ropes.

194

Nibble! Nibble! Nibble!

Chomp! Chomp! Chomp!

On and on she gnawed, until her jaws ached.

The others watched tensely. Eventually, Tumtum's ropes fell free.

The others cheered. Then Tumtum quickly undid their paws, too. They all rubbed their wrists, for the cotton had left red welts on their skin.

"Come on! Let's try and smash open the door," the General cried. "Purple Claw will still be in the tunnel — we can catch up with him and spring on him from behind!"

Everyone agreed that the scoundrel must be stopped.

But though they all kicked and battered the door together, and rammed it with a dresser, it wouldn't budge.

"We'll never escape!" Nutmeg said tearfully.

They felt helpless. With every second that passed, Purple Claw would be creeping farther down the Secret Tunnel toward Arthur's boat.

"We must stop him!" the General cried.

Everyone racked their brains, wondering what to do.

"I wish we could get a message to the children," Tumtum said. "Then we could tell them to come back to the Rock Pool and catch Purple Claw red-handed when he comes out of the Grotto!"

"Oh, that would serve him right!" the General said.

It was a wonderful idea. But of course it would never work, for how could they get a message to the children when they were stuck up here in the cave?

"If only Purple Claw's plane weren't smashed to bits, then one of us could fly back to Smugglers' Keep!" Nutmeg said.

The General gave a start. He had been so busy battering the door, he hadn't thought about the plane.

He turned and looked at it very carefully. Was it *really* as badly damaged as all that?

It was still belching smoke, and the wings were dented. But the propellers had survived, and the cockpit looked intact. He peered inside and saw that the fuel gauge was showing half-full.

The General's whiskers started to quiver. "She'll fly for me!" he said.

Chapter Thirteen

The Nutmouses and Lord Seamouse were still standing by the door, discussing what to do.

"I say, where's the General?" Tumtum asked suddenly.

Then all at once there was a deafening roar. And when they looked around they saw the airplane tearing across the cave, with General Marchmouse shrieking in the cockpit. There was smoke choking the wings, and the engine was spluttering.

Everyone ran for cover.

"*BRAKE!*" Tumtum shouted.

"*STOP!*" Nutmeg screeched.

But there was no stopping him now.

"Clear the skies!" the General cried. "I'm taking off!" Then he tore into the curtain.

Everything went black as the moss slapped over the windshield. Then the plane slipped over the edge of the cave and plunged into the air.

"I'm flying!" the General shouted, giggling with glee.

But when he looked out of the window he gave a loud yelp. He wasn't flying. He was sinking like a stone! And all he could see were vast black rocks, spinning all around him.

He turned to the control panel and started yanking the knobs. But the plane kept falling.

"Help!" he cried.

He saw a green lever beside the joystick, labeled THROTTLE.

He grabbed it with both paws and wrenched it toward him.

There was a violent shudder, and he felt his seat vibrate. Then the engine went *VROOOOOM!* and suddenly the nose of the plane flipped up and the craft soared skyward.

The General gave a whoop. He was going higher and higher, wheeling through the air like a gull.

"I am Lord of the Skies!" he cried, thumping his paw on the roof.

When he looked down, he could see Tumtum and Nutmeg and Lord Seamouse, watching him in terror from the mouth of the cave.

And how he showed off!

He looped the loop, and he did a figure eight, then a figure nine. . . .

But then suddenly he saw an enormous shape looming toward him. It had white wings and an orange nose.

The General gulped. It was a seagull, coming straight at him!

He yanked the joystick, trying to steer out of its path. But it was too late . . . and a moment later the bird thwacked into the cockpit.

The plane batted backward like a football and started plummeting from the sky.

The General peered through the window in terror.

He could see the beach below him, spinning nearer and nearer . . . then there was a sudden *Smack! Bang!* as the plane crashed into a sand castle.

The General lay in the cockpit, too dazed to move. Everything was swimming. All he could see were stars.

Then he smelled burning. "Fire!" he cried.

He wrenched open the door and dragged himself out. Then he started racing frantically through the sand. He saw a shell in front of him and ducked behind it. And the next moment there was a huge blast, as the plane exploded in a torrent of flames.

The General cowered there for a moment, feeling very shaken. He fished a handkerchief from his pocket and wiped his brow. What an adventurous day it had been! And it wasn't over yet.

He still had to get a message to the children. And if they were to come back in time to catch Purple Claw, then he didn't have a moment to lose.

He dragged himself to his feet and stumbled on through the sand. Then he clambered up the steps, shot across the lane, and wriggled into Smugglers' Keep under the garden door.

He crept through the kitchen, into the hall. He could hear voices coming from the dining room, and when he peeked around the door, he saw Arthur and Lucy sitting at the table, finishing their lunch.

He quickly searched around the hall for something to write on. Then he saw a small table against the far wall, with a telephone cord dangling down to the floor. He scrambled up it, and when he reached the tabletop he found a pad of white paper and a jar of pens.

He pushed the jar over, then he pulled out a pen and wrestled off the cap. He tore a piece of paper from the pad and, holding the pen with both arms, he scrawled out his orders:

ATTN: Arthur and Lucy Mildew

Your uncle's childhood teeth have been stolen by Purple Claw, and he intends to sell them to the tooth fairies and make himself rich and powerful. But if you do exactly as I say, you can stop him.

You must proceed at once to the Rock Pool and wait for Purple Claw to emerge from the cave in your speedboat. As soon as you see the boat coming, swoop down and grab it. Then the teeth will be yours! As for Purple Claw — he will

be yours, too. Do what you wish with him. A cage with
thick bars might be the best answer.

Yours,

General Marchmouse

When he finished writing, the General folded the note in half
and clenched it between his teeth. Then he scrambled down to
the floor and crept into the dining room.

Mrs. Blythe had appeared now and was standing beside
Lucy, spooning a second helping of rice pudding into her
bowl.

The General was in such a hurry, he did not even wait until
she had gone. He shot straight across the carpet and scurried up
the tablecloth. Then he leaped over the sugar dish and dropped
his letter beside Lucy's bowl.

Mrs. Blythe shrieked and dropped her spoon with a clang.

"It's that green mouse again!" Arthur cried as the General
darted back to the floor. Arthur leaped from his chair, trying to
see where he went. But the General was gone.

Chapter Fourteen

Mrs. Blythe shrieked and squawked and made a terrible fuss.

Uncle Jeremy had been in the drawing room, answering a telephone call. When he heard the noise, he came hurrying into the dining room.

"What's going on?" he asked.

"It was a mouse!" she cried. "A mouse wearing *clothes*!"

"Don't be silly, Mrs. Blythe. Mice don't wear clothes," Uncle Jeremy said.

"Well, I'm telling you, this one did!" Mrs. Blythe replied. "Ask the children. They saw it!"

"Did you?" Uncle Jeremy asked, looking somewhat bewildered.

Arthur and Lucy went red. They *had* seen it—and now Lucy was clutching the General's note in her hand. But they didn't want to have to start explaining everything to the grown-ups.

"There was a mouse," Lucy said. "It ran across the table. Then it ran off again."

"And was it wearing clothes?" Uncle Jeremy asked, clearly thinking this was all very silly.

"I . . . er . . . I didn't notice," Lucy mumbled.

"Neither did I," Arthur said.

"But it *was*," Mrs. Blythe cried. "I tell you, it *was!*"

Uncle Jeremy gave Mrs. Blythe a kindly look. He wondered if it was the hot weather that had made her so muddled.

"Well, I can assure you, if I see it I'll send it running!" he said. But Mrs. Blythe did not look at all reassured.

As soon as lunch was finished, Arthur and Lucy ran upstairs to read their letter. The General's writing was much bigger than Nutmeg's, so they did not need a magnifying glass. Arthur sat next to Lucy on the bed while she read it out loud.

"So there we were thinking the mouse in the green suit was Purple Claw — but he's on *our* side," Arthur said.

"And he's a general!" Lucy said in astonishment.

The children had never heard of the Royal Mouse Army, so they hadn't known that mice could be generals.

"Do you think Purple Claw is a mouse, too?" Arthur asked — for the General's letter hadn't made it clear.

"He must be, if he's small enough to fit in the boat," Lucy said.

"Come on!" Arthur said, jumping up from the bed. "Whoever he is, we'd better go and catch him!"

"We'll need something to put him in," Lucy said. "We should take a cookie tin."

"We needn't bother. We can put him in a fishing net," Arthur said.

"All right," Lucy agreed. So they ran downstairs and grabbed their fishing nets from the hall, then they raced back to the beach.

Tumtum and Nutmeg and Lord Seamouse, meanwhile, were looking out anxiously from the mouth of the cave.

They had seen the General's plane crash into the sand castle. And they had seen him clamber out of the wreckage and start scurrying back toward Smugglers' Keep.

And when they saw Arthur and Lucy charging back across the sand, they knew he must have gotten his message to them and warned them about what was going on.

"Oh, hooray. They're back!" Nutmeg cried.

"And look! They've brought nets to catch Purple Claw!" Lord Seamouse said gleefully.

They couldn't wait to see what happened.

"Hurry, hurry!" they all cried, fearing the children might not reach the pool in time.

The mice watched as Arthur and Lucy scrambled up onto the rocks and stood waiting by the pool with their nets.

But they waited and waited—and still Purple Claw didn't come. The Rock Pool was quite still. There was nothing stirring, not even a dragonfly.

"Do you think he's escaped already?" Lucy asked.

"He can't have, otherwise the boat would be here," Arthur said.

The mice craned over the ledge, trying to see what was going on. It seemed like ages since Purple Claw had left the cave. He should have reached the Grotto by now, and found the boat. They wondered why he was taking so long.

Then suddenly Tumtum noticed Lucy looking their way.

"Keep still!" he hissed.

But it was too late. Lucy had seen them moving out of the corner of her eye, and now she was staring straight at them.

"Look!" she cried. Then Arthur saw them, too — three little mice high up on the rock, dressed in smart summer clothes! The children stared at them in astonishment.

"However did they get up there?" Lucy said.

"They must have come out from behind that green plant," Arthur replied. "Maybe that's where the cave is. I'm going to take a look."

"Oh, do be careful," Lucy said. "They might bite!"

But Arthur was already scrabbling up the rock.

"Quick, get inside!" Tumtum said. And the mice all ran back through the curtain, into the cave.

Arthur pulled himself up slowly, gripping the rock with

both hands. When he was about three feet off the ground he found a little groove to tuck his feet into. He hugged one arm around the rock to keep his balance. Then he reached out his other arm and pushed his hand into the moss.

It was thick and slimy. But when he pressed his hand all the way through, he could feel a hole on the other side.

"There's a cave here all right!" he shouted down to Lucy.

"Is there anything in it?" she asked excitedly.

"Hang on," he said, pushing his hand in deeper.

He stretched out his fingers and groped them around the floor.

Tumtum and Nutmeg and Lord Seamouse pressed themselves against the wall, watching in terror as the huge, pink hand edged closer and closer.

"Quick! Let's hide over there in the wardrobe!" Tumtum whispered.

They jumped over Arthur's thumb and ran toward the wardrobe. They all squeezed in together, then Tumtum pulled the doors shut tight.

"Shhh!" he whispered. "Keep very still!"

Arthur pushed his hand deeper into the cave. Then suddenly his fingers bumped into something solid. It felt like wood. He grabbed it in his fist and pulled it out.

"It's a wardrobe!" he cried.

"Let's see!" Lucy said.

He passed it down to her, and she looked at it in astonishment. It was the prettiest piece of dollhouse furniture she had ever seen. The wood was painted pale blue, and the knobs were made of tiny pearls.

It felt very heavy. *I wonder if there's anything inside*, she thought. She tried to pull the door open, but it was stuck—Tumtum had locked it from the inside.

Arthur turned and reached back into the cave.

"There's tons more!" he said excitedly, fishing out a four-poster bed.

Lucy put the wardrobe on the ground. And as she reached up to take the bed from Arthur, the door squeaked open, and Lord Seamouse and Tumtum and Nutmeg crept out.

"Come on! Let's hide here," Tumtum said, leading them behind a fat gray pebble.

They crouched behind it, watching as Arthur pulled more and more treasures from the cave.

"Here's a bathtub...and a piano....And look! There's even a Ping-Pong table!" he cried.

He passed each item down to Lucy, who packed them into her fishing net. She could hardly wait to take everything back to Seaview Manor. Just think how grand it would look!

"That's everything," Arthur said eventually, when he had explored the last inch of the cave.

"What about the mice?" Lucy asked. "I wonder where they disappeared to."

"They must have escaped somehow. I bet the rock's full of secret tunnels," Arthur said.

He clambered down, then the children stood by the Rock Pool, admiring all the lovely things they had found.

But there was still no sign of the boat.

"Where can it be?" Arthur said.

"*Shhh!*" Lucy said suddenly. "What's that?"

The children both crouched by the pool and listened. They could hear a rumbling noise coming from somewhere deep

inside the rock. It was very faint at first. But then it got louder and louder, until it became a *Vroom! Vroom! Vroom!*

"It's the boat! He's coming!" Arthur whispered. He grabbed his fishing net, then they both waited by the pool, keeping very still.

"He won't get past *us*!" Lucy said.

Chapter Fifteen

Purple Claw shivered with glee—for his job was almost done.

He had trudged all the way down the Secret Tunnel, lugging the tin of teeth on his shoulders. The tin was very heavy, and it had been a terrible slog. So imagine his joy when he finally reached the Grotto and found the speedboat tied to the wall.

He heaved the tin down into the boat, then he jumped in and undid the mooring rope.

He shone his flashlight over the control panel and pressed the knob marked ON. The engine spluttered, then the boat slowly pulled out onto the water. Purple Claw chugged forward until he could see the big arch of sunlight at the mouth of the Grotto.

He pointed the boat toward it. Then he tightened his grip on the steering wheel and slammed his paw on the accelerator.

The engine roared. The propeller spun. Then the boat reared up its nose and shot forward like a bullet.

"No one can stop me now!" Purple Claw cried as he sped into the sunlight.

The boat skimmed across the Rock Pool — *Buong! Buong! Buong!* — with seaweed splattering the windshield.

"The teeth are mine!" Purple Claw shrieked. "*Mine! Mine! Mine!*"

Then suddenly a black shadow fell over him, and the boat flew into the air.

Purple Claw frantically wrenched the steering wheel — left! right! left! — thinking he had been hit by a wave. But the boat kept rising.

He stood up and peered out over the windshield. Then he let out a terrible squeal.

He was caught in a fishing net!

"Help!" he cried as Arthur and Lucy gazed at him in astonishment. They thought he looked very odd.

And it wasn't just his long black cape or his scarlet boots or his purple claws that were so unusual. He was simply the mangiest, meanest, most horrible-looking mouse they had ever seen.

"Are Uncle Jeremy's teeth in there?" Lucy said, peering into the net.

"They must be in that candy tin," Arthur replied.

But Purple Claw looked so nasty, neither child dared to reach down a hand to take it out.

Tumtum and Nutmeg and Lord Seamouse watched in delight from behind the pebble.

"What will they do with him?" they all asked. The children were wondering just the same thing.

He didn't look like the sort of mouse one would wish to keep as a pet. And yet he was a villain, so now that they had caught him, they couldn't just let him go.

"Let's take him back to Smugglers' Keep and tip him out into the bathtub," Arthur suggested.

"All right," Lucy said. "Uncle Jeremy will know what to do with him. Now that we've found the treasure, we can tell him the whole story."

They hurried back across the beach.

Lucy carried the fishing net full of the dollhouse furniture, and Arthur carried the net containing Purple Claw, who squealed and scrabbled all the way home.

Tumtum and Nutmeg and Lord Seamouse followed them. But they couldn't run as fast as the children, so they trailed some way behind.

When the children got back to Smugglers' Keep, Mrs. Blythe let them in.

"What have you caught in there?" she asked Arthur, seeing the net swinging back and forth on the end of his rod.

"Er, nothing . . ." Arthur mumbled, edging past her.

The children didn't want Mrs. Blythe to see Purple Claw, or she would be sure to start shrieking again.

"Just something we fished out of the Rock Pool," Lucy explained hurriedly. Then they fled upstairs before she could ask any more questions.

They ran into the bathroom and locked themselves in. Then Arthur shook the fishing net over the bathtub, and Purple Claw and the boat and the candy tin all tipped out.

Purple Claw tore up and down, trying to escape. But the bathtub was very deep, and he couldn't get out.

He saw Arthur and Lucy peering down at him and shook his fists with rage.

"You hateful little creatures!" he cried. "I'll get my revenge for this! Just you wait and see!" But all the children heard was a squeal.

When Purple Claw was at the other end of the bathtub, Lucy quickly reached down and picked up the tin.

Purple Claw gave a howl and rushed toward her, gnashing his fangs.

But the bathtub was so slippery he skidded over and landed flat on his back, kicking his scarlet boots in the air.

Lucy pried off the lid of the tin, and the children looked inside. And there they were, Uncle Jeremy's milk teeth! It was very strange to see them.

"They must be ancient," Arthur said. "Nearly as old as he is!"

"We'd better give them back to him," Lucy said.

Arthur agreed, so they put the lid back on the tin, then

they left Purple Claw alone in the bathtub and went downstairs.

Uncle Jeremy was in the drawing room, reading a book about fishing. He could see from the children's faces that something very serious had taken place.

And when they showed him the tin of teeth, and told him that they all belonged to him, from when he was a little boy, he was very surprised.

He had given his teeth to the tooth fairies. He hadn't expected to see them again.

"Wherever did you find them?" he said.

So the children told him the whole story, beginning with the map on the bedside table and ending with Purple Claw's dramatic capture in the Rock Pool.

The only thing they left out was Nutmeg, for she had warned them ages ago that they must never tell anyone about her. But after she had left the map on the bedside table, she didn't come into the story much. So they managed not to mention her.

Uncle Jeremy listened very carefully. He didn't interrupt,

and he didn't make them feel silly, as some grown-ups might have done.

"So what did you do with Purple Claw?" he asked.

"He's upstairs in the bathtub!" Lucy said. "Come and have a look at him. He's the strangest mouse you've ever seen."

Uncle Jeremy didn't really believe that the children had found a mouse with purple claws, of course. But he didn't want to spoil their fun. So he levered himself out of his chair and followed them upstairs.

Lucy opened the bathroom door and let him go in first. And when he saw Purple Claw, he got such a fright that he fell over backward and crashed into the towel rail.

It was just as the children had said . . . an angry, mangy little mouse, with a black cape and scarlet boots . . . and *purple claws*!

Uncle Jeremy pulled off his spectacles and gave them a good polish. Then he stood up and peered into the bathtub again.

"Astonishing. Quite astonishing. Quite, quite astonishing," he said over and over again. He had seen mice before in Smugglers' Keep — but never one like this.

"What will we do with him?" the children asked.

"We must treat him like a king!" Uncle Jeremy replied. "Go down to the kitchen and ask Mrs. Blythe for the finest mouse-size morsels she can find. Bring him cheese and chocolates and the tastiest slivers of ham!"

"What's he done to deserve that?" Arthur asked.

"Don't you see? He is a rare breed!" cried Uncle Jeremy, hopping around the room in excitement. "He's the rarest of the rare! The strangest of the strange! We must put him in a cage and lend him to the zoo, so that everyone can come and look at him! He will be the most famous mouse in the whole wide world!"

Chapter Sixteen

The rest of the day passed in a daze.

Uncle Jeremy telephoned all the local newspapers, and soon the house was full of reporters and photographers jostling to see Purple Claw in the bathtub.

Everyone wanted to ask the children questions:

"So this mouse who appeared on the dining room table was wearing a suit, you say?"

"And he left you a letter?"

"And he signed himself 'General Marchmouse'?"

Somehow, the children sensed that the reporters didn't believe a word they were saying.

"They think we've made it up," Arthur whispered, when he and Lucy finally had a moment on their own.

"But they've seen Purple Claw, so they must know it's true," Lucy said.

"They probably think we dressed him up in doll clothes and painted his toes purple ourselves," Arthur said. "You know what grown-ups are like."

It was true, grown-ups could be awful. The children started wishing they would all go away.

But then suddenly one of the reporters pointed across the room and let out a shriek.

The other reporters all turned to look. And to their astonishment, they saw a little mouse marching up and down the mantelpiece, swishing a match at them.

And he was dressed in a green suit!

"Don't believe a word those children tell you!" he roared. "*I* discovered the treasure first, and if it hadn't been for me, no one would ever have found it! It was all thanks to *me*! The great General Marchmouse!"

No one could understand a word he was saying, of course — for, to the human ear, his voice was just a squeal. But the General didn't care. He strutted up and down, boasting and saluting, and causing a terrible stir.

The reporters became very excited and started shouting into their cell phones, saying, "Hold the front page!"

Everyone was egging the General on.

The General was a natural show-off, and he didn't need much encouragement. He puffed his chest, and preened his whiskers, and posed beside a china shepherdess.

The children stood at the back of the room, watching with delight.

"They'll have to believe us now!" Arthur said.

The General showed off for a whole hour solid. Then one of the reporters cried, "Come on! Let's catch him, and take him back to the office in a cardboard box!"

"Just you try!" the General roared. Then he leaped to the floor and shot out the door, moving so fast no one could stop him.

Tumtum and Nutmeg and Lord Seamouse had just returned from the beach. As they came inside, they heard a terrible commotion.

They crept across the kitchen floor and peeked into the hall—just in time to see General Marchmouse fleeing from the drawing room.

"General, over here!" Tumtum cried.

The General shot toward them, then they all darted into the vegetable rack, and hid inside the leaves of a cabbage.

The reporters saw the General fleeing into the kitchen and stormed after him.

"He must be here somewhere!" they cried, and started going through the cupboards. The mice trembled in the cabbage, terrified they would search the vegetable rack, too. But when Mrs. Blythe saw what a mess the people were making, she shooed them away with a dishcloth.

Finally, the kitchen was quiet. But the mice were very shaken. "Whatever's been going on?" Nutmeg whispered.

"Some reporters came to see me," the General replied grandly. "They tried to kidnap me—I think they wanted to make me into a television star!"

"What a terrible thought!" Tumtum shuddered.

They could still hear people clonking around in the hall.

"Let's stay here until everyone's gone," Nutmeg said. "We don't want to take any more chances."

They stayed hidden inside the cabbage for a long time. They waited until the last reporter had left the house and until they had heard Uncle Jeremy and the children going upstairs to bed. Only when everything was absolutely quiet and still did they dare to clamber out of the vegetable rack and start creeping back to Seaview Hollow.

But as they were passing the drawing-room door, they heard someone snoring. And when they peeked inside, they saw the most astonishing sight. In the middle of the floor was a big cage, with a domed roof and gold bars. It was the most magnificent cage they had ever seen, almost like a palace.

They crept over to it and looked through the bars. And there was Purple Claw — and didn't he look grand! He was fast asleep in a bed of duck feathers, with a swollen stomach, and a dreamy smile on his face. Beside his bed was a pretty china trough, full of raisins and pastry crumbs, and chips and chocolates, and slivers of the finest, smelliest cheese.

"These bars look very flimsy. I wouldn't be surprised if he manages to break out!" Lord Seamouse said anxiously.

"Oh, he won't do that," Tumtum said. "Just see how fat and content he looks. He's the luckiest villain there ever was!"

Tumtum was quite right, of course. Purple Claw had done very well for himself—much better than he deserved. The next morning, he would be taken to the zoo, where he would live happily ever after in his gilded cage, being photographed and fussed over, and given delicious things to eat.

And he would never bother another mouse again.

Finally, the Nutmouses, the General, and Lord Seamouse clambered upstairs and crept into Arthur and Lucy's bedroom. The children were asleep, but the mice crossed the floor on tiptoe all the same. But when they saw the dollhouse, they all let out a cheer.

For it was a splendid sight. The children had put all the furniture back in place, and arranged all the tapestries and the paintings and the lamps and the candlesticks. They had even made Lord Seamouse's bed!

"Gracious," Tumtum said enviously. "It's even grander than Nutmouse Hall!"

When he saw all his pretty things again, Lord Seamouse shed a tear. His dark days in Seaview Hollow were over. Now he would be Lord Seamouse of Seaview Manor again!

"Come in, come in!" he said, longing to show the other mice around. He took them upstairs and downstairs, and everyone was very impressed.

"Oh, look!" Lord Seamouse cried when they came to the drawing room. For there were Uncle Jeremy's teeth, neatly arranged in the glass cabinet! How delightful to see them back on display!

And when they reached the dining room, they found that the children had left them a dish of breadcrumbs and a whole chunk of chocolate for their dinner.

Lord Seamouse sliced the chocolate with his carving knife and served everything on his best silver plates, then they all sat talking and eating late into the night.

And there we must leave them, for their adventures are over, at least for now. And Arthur's and Lucy's adventures are over, too. But there was so much chatter and laughter coming from the dollhouse that, as the night wore on, Lucy started to toss and turn. And it was a funny thing, but as she was drifting back to sleep, she could have sworn she saw someone peeking at her from the bedside table. Someone with a long, nutmeg tail.

A Circus Adventure

Chapter One

General Marchmouse swaggered past the compost heap with a song rising in his heart. He had a good reason to be cheerful. Today was his birthday! And mice do not have nearly as many birthdays as humans do, so it was a very special day indeed.

He had been busy since breakfast time, opening his cards and unwrapping his presents. And now it was almost dark, but the best part of the day was about to come. Tonight his friends Tumtum and Nutmeg had invited him for a birthday feast at Nutmouse Hall.

Perhaps Nutmeg will have made a spider pie, he thought hungrily. *And a toad-in-the-hole, and a cockroach soufflé, and a strawberry tart with toasted almonds on top!*

He hurried on under the rosebushes and clambered down onto the garden path with his invitation clutched tightly in his paw.

It was a cold November evening, and he was glad to see Rose Cottage looming ahead of him. The curtains were still open, and all the windows were brightly lit.

"Hmm. It looks like the Mildews are at home," the General muttered. "I shall have to be very careful—Tumtum will be furious if I give his hiding place away."

He crept up to the back door and cautiously peeked his nose into the kitchen. There was a spider dozing on the doormat and a cockroach scuttling beneath the sink. But there were no humans in sight, so he set off across the floor.

The General had visited Nutmouse Hall many times before, so he knew just where to go. It was a very grand house, built inside the Mildews' broom cupboard. And its front gates were hidden beneath the old wooden dresser that stood against the kitchen wall.

The General marched along, taking his usual route—past

the laundry basket, under the vegetable rack, then around the leg of the kitchen table.

He hadn't far to go, and when he peered under the dresser he could see a tiny beam of light coming from the Nutmouses' gates. He hurried on, wading through a puddle of ketchup on the floor.

But then suddenly he stepped into a dark shadow. And when he looked up, he saw the most amazing sight.

It was a circus top!

The General froze and stared at it in astonishment. It was the finest circus top he had ever seen. It had red and green stripes, and it was bigger than a beach ball. And standing beside it was a little ticket booth with velvet curtains.

The General felt his heart quicken. *It must be a mouse circus,* he thought, *for everything is mouse-sized! Well, whomever it belongs to is a very lucky mouse, that's for sure!*

He stood there a moment, gazing admiringly at the big top. He wondered if there were dancing dragonflies or racing slugs inside, or any of the other wonderful things that mouse circuses usually have.

He decided to go in and see. But when he reached the entrance, he saw a silver card lying on the ground with HAPPY BIRTHDAY! written on the front.

"Well, I never!" the General said. "This circus must be a present for me!"

The birthday card was bigger than he was. He hoisted it open, wondering who it was from. But when he read what it said, his face fell:

Dear Arthur,

This is a toy circus that I was given long ago, when I was not much older than you are now. I hope you enjoy playing with it as much as I did.

Happy birthday!

Love,

Uncle Jeremy

"Humph!" the General grunted. "It must be Arthur's birthday, too. So it's not a mouse circus after all — it's just a toy!"

The General thought it was very unfair that the circus was

given to Arthur instead of him. "It's far too good for a boy!" he said crossly.

He was still curious to see what was inside the big top. So he pulled open the flap and marched through. And how he stared! For he had entered an enormous stadium, with bleachers around the edges and a trapeze up above. And in front of him was a huge crowd of circus animals — wooden horses and leopards, and tigers and elephants, all parading around the ring.

The General rubbed his eyes in astonishment. The toys looked so real that for a moment he thought they were! And standing in their midst was a toy ringmaster, dressed in a shiny gold suit and holding a whip.

"Lucky fellow," the General muttered. How he wished that he had a gold suit, too! What fun it would be to stand in the middle of the ring, making the toy animals dance to the crack of his whip!

He stood there a long time, filled with envy. From somewhere in Rose Cottage, he heard the clock chime seven. He knew Tumtum and Nutmeg would be wondering where he was, for he was late for dinner. And yet he couldn't tear himself away.

"Surely there can't be any harm in just trying the suit on and having one tiny crack of the whip," he reasoned. "It is my birthday, after all!"

He glanced guiltily over his shoulder. And after a very short period of hesitation, he marched up to the ringmaster and tore off his clothes. He tugged off his own Royal Mouse Army uniform and tossed it to the floor. Then he pulled on the gold pants. They were very tight, and he had to take a big breath in order to do up the button. But the jacket was a perfect fit.

He strutted around the ring, feeling very grand.

Then he snatched the whip and gave it a hard crack. "March!" he shouted. The lions shook, and an elephant toppled to the floor. The General whooped with joy — oh, what fun!

On he went, swaggering and shouting and cracking his whip, until soon all the wooden animals had been thrashed to the ground. No creature was spared. An elephant's trunk had snapped off, and one of the lions had lost a leg.

The General wondered what to do next.

244

Then he had a daring plan. *I know!* he thought. *I shall go and catch that cockroach I saw scuttling off under the sink, and train it to swing on the trapeze!*

He tucked the whip under his arm and ran out of the tent. But as he stepped into the kitchen, he got a terrible fright. Arthur and Lucy had appeared!

The General gulped and dived back inside the big top. He had been making such a racket that he hadn't heard them come downstairs. He knew he was in terrible trouble, for Arthur would be furious if he found him here and saw what he had done to his toys.

He stood trembling, listening to the children's feet thudding on the floor.

"Just look at my circus!" Arthur said proudly. "Isn't it the most splendid circus you've ever seen? I don't mind one bit that Pa forgot my birthday—this circus makes up for everything. Come on, let's play with it again."

"All right," Lucy replied. "Wait a second while I put the milk bottles out."

The General's mind whirled with terror. He heard Lucy open the garden door and clank the bottles outside. Then he saw the children's shadows looming over the tent. . . . There was nowhere he could hide. . . .

There was a rustle, then the flap of the tent slowly opened, and a huge pink hand reached inside. . . .

"Oh, what am I to do?" The General quivered. "It's my birthday, and I'm going to be caught!"

But the General was a very lucky mouse. At that very moment, as if by a miracle, the telephone rang.

"I bet that's Uncle Jeremy, calling to check that the circus has arrived!" Arthur said.

The children both jumped up and ran into the hall.

The General seized his chance to escape. In his haste, he forgot all about his army uniform and dashed out of the big top wearing the ringmaster's clothes. Then he turned and hurtled toward Nutmouse Hall.

But as he skirted around the big top, he saw something so completely and utterly dazzling that for a moment it made him forget all his fear. It was a bus. But not just any bus. This was, quite simply, the most splendid, most beautiful bus the General had *ever* seen.

It was gleaming red, with shining headlights and silver wheels the size of saucers. He crept up to it and pressed his nose to the window. There was a fat leather steering wheel and a gear-shift as big as a lollipop. And in the back of the bus were two little camp beds, with green quilts and crisp white pillows.

The General's whiskers twitched. The bus looked very inviting. He could hear Arthur and Lucy talking on the tele-phone in the hall. He knew they might come back into the kitchen at any moment, but he was too excited to care.

Then suddenly he felt a cold breeze cutting his ankles, and when he looked around he saw to his surprise that the children had left the garden door open.

He stood there for a moment, peering out. The sky had turned black, but he could see the pale outline of the garden path, twisting away in the moonlight.

And suddenly it was as if the moon and the stars were all calling out to him, whispering his name.

With a pounding heart, he pulled open the bus door and clambered into the driver's seat, tossing the whip beside him. Then he gripped the steering wheel and turned on the ignition.

The engine gave a shudder, then a *Vroom!*, and the bus shot forward across the floor.

"Faster! Faster!" the General cried, slamming his paw on the accelerator. He swerved under the kitchen table and tore toward the open door.

The spider fled out of his way, and a fly splattered on the windshield. The General clung tightly to his seat as the bus bounced over the doorstep and crashed onto the garden path. Then, with a shriek of joy, he turned on his headlights and sped into the night.

Chapter Two

N utmeg had been bustling and baking all afternoon, preparing the General's birthday feast. While she worked, Tumtum dozed in the library. When everything was nearly ready, he woke up and wandered into the kitchen, hoping to find a spoon to lick.

"Isn't it a coincidence that Arthur and the General have a birthday on the very same day?" he said, dipping his paw into one of Nutmeg's mixing bowls.

"Yes, isn't it!" Nutmeg agreed. "Oh, Tumtum, wouldn't it be lovely if Arthur could come to the birthday feast, too? For you can be quite sure there won't be any feasting going on in Rose Cottage. I peeked out into the kitchen earlier this afternoon, and there was no sign of a cake!"

"Are the children having something special for dinner?" Tumtum asked.

"I don't think so," Nutmeg replied sadly. "When I last looked in their pantry, there was nothing except a stale loaf of bread and a few cans of spaghetti!"

Tumtum wrinkled his nose. He had tried canned spaghetti once, and he had not liked it at all. It was a great mystery to him how humans could bear to eat it.

"Yes, it's a pity the children are too big to fit through our mouse hole," he agreed. "Just think how astonished they would be if they could see all the delicious things *we* have to eat!"

"Well, at least Arthur got a nice birthday present," Nutmeg said, briskly rolling her pastry.

"Did he?" Tumtum asked in surprise. "Was it from his father?"

"Oh, gracious no! You know Mr. Mildew never remembers about presents!" Nutmeg said. "It was from his uncle Jeremy — you remember, the nice old man we all went to stay with last summer at the seaside. He sent Arthur a beautiful toy circus. I saw it when I went out this afternoon. Arthur's thrilled with it!

I heard him telling Lucy that it was the best present he'd ever had!"

"Oh, I am glad," Tumtum said, looking very relieved. It would have spoiled his appetite to think of Arthur having no presents as well as no cake. Now he could enjoy tonight's feast all the more.

"Now let's see," Nutmeg said, rubbing her paws on her apron. "I've still got to toast the croutons for the soup, and ice the cake, and glaze the ham, and chill the jellies, and warm the cockroach pastries. . . . Then I think we're all done!"

"Goodness, what a lot of food we have," Tumtum said. "It's enough to feed a hundred mice! But there will only be three of us, now that Mrs. Marchmouse can't come."

"Yes, it is a rotten shame that she can't be here." Nutmeg sighed. "But it's good of her to go and look after her sister. Apparently she's been very ill. Mrs. Marchmouse will probably have to stay with her for some time."

"Well, I shouldn't think the General will mind being left on his own," Tumtum said. "You know what he's like. He loves getting into trouble, but it's not so easy when Mrs. Marchmouse is watching over him!"

They both smiled. Mrs. Marchmouse had been keeping a very strict eye on the General lately. He would probably consider a few days on his own a nice birthday treat!

"Well, I shall make jolly sure that he doesn't get into trouble when he comes here," Nutmeg said firmly. "I'm not having him dragging us into another adventure."

"Certainly not," Tumtum agreed. "We've had quite enough of those! But you needn't worry, dear. The General will be much too busy eating to get into any of his silly games. The only trouble he's likely to encounter tonight is indigestion!"

Nutmeg laughed. And as she put the finishing touches on all her little dishes, she felt full of warm feelings. Tumtum was right; there would be no adventures tonight. Just lots of fun and hearty eating!

"Gracious!" Nutmeg exclaimed, looking at the clock. "He'll be here soon, and we still haven't set the table!"

They both hurried through to the dining room and lit all the candles. Then they set the table with their best silver and china, for today was a very special occasion.

When everything was ready, they went upstairs to change.

Nutmeg put on her pink dress, and Tumtum wore his velvet jacket. Then they settled down in the library and waited for the General to arrive.

But the time ticked by, and the clock struck seven — and still he did not appear.

"I hope he hasn't gotten lost," Nutmeg fussed.

"Oh, no, he knows the way much too well for that," Tumtum said. "He's probably just stopped to chase a beetle. You know what he's like!"

But Nutmeg still felt anxious, for it was unlike the General to be late for a meal.

They sat there a little longer, nervously sipping their cider. Then suddenly they heard raised voices coming from the Mildews' kitchen. It was Arthur and Lucy, and they sounded very upset.

Tumtum and Nutmeg could hear everything that was being said:

"But where is it? It can't just have disappeared!" Arthur cried.

"Well, I haven't taken it," Lucy said.

"Well, you must have. No one else has been in here today!" Arthur shouted.

"No, I haven't!" Lucy said angrily. "Why would I want to take your silly circus bus? I tell you, I haven't touched it."

"But you *must* have taken it," Arthur wailed. "I left it here, by the big top. It can't just have vanished into thin air!"

Tumtum and Nutmeg looked at each other in surprise. They knew all the comings and goings at Rose Cottage, and no one else had been there today.

"Arthur must be mistaken—no one could have taken the bus," Nutmeg said. "It was certainly there when I went out earlier, and I know no one's been in the kitchen since then...except—"

The mice looked at each other and gulped, for the same thought had entered both of their minds.

"Oh, Tumtum," Nutmeg whispered. "You don't think...I mean...you don't really think...Oh, *surely* the General couldn't have been wicked enough to steal Arthur's bus!"

Chapter Three

Tumtum and Nutmeg hurried out of Nutmouse Hall and tiptoed to the edge of the dresser. They could see Arthur kneeling on the floor, his head lowered as he peered into the big top.

"Someone's knocked over all the animals," he cried. "And look—they've broken my wooden elephant *and* one of the lions!"

He turned around and glared at his sister. "It must have been you!" he shouted, holding up the broken toys for her to see. "Who else could have done it?"

"Oh, don't be so awful!" Lucy said. "I told you, I didn't touch your circus. I haven't been near it all day!"

Arthur scowled, but he knew Lucy was telling the truth. She wouldn't have broken his toys; she wasn't that sort of sister.

"Let me look inside," Lucy said, for she could see that he was very upset. Arthur shuffled aside, then Lucy knelt down in front of the big top and peeked in. And she noticed something very strange. "Look!" she said. "Someone's taken the ringmaster's clothes!" She reached in through the flap and pulled the toy out of the tent. And when Arthur saw it, he was amazed. When he had last played with it, it had been wearing a gold suit. But while he had been gone, someone had stripped it to its underpants!

"Hang on, here are some other clothes!" Lucy exclaimed, spotting the tiny garments that the General had tossed to the floor. She took them out and spread them on the palm of her hand. The children looked at them curiously.

"Look at these tiny gold medals on the jacket!" Arthur said. "It looks like an army uniform. But it's not like the uniform a toy soldier would have. It seems too . . . well . . . *too real!*"

Lucy agreed. "And isn't this strange?" she said, pointing to the jacket. "There are hairs on the collar! This can't belong to a toy. It must belong to something real.

"Whoever he is, he must have crept in here when we

weren't looking, then changed into the ringmaster's clothes and driven away in the bus!"

It was very mysterious, and the children were longing to find out who the secretive creature could be.

"Come on, let's find him!" Arthur said, running to grab the flashlight from the shelf by the back door. "The bus was still here when the telephone rang, so he can't have gotten far!"

Tumtum and Nutmeg were still crouched under the dresser. They watched in dismay as the children raced outside.

"So it *was* him!" Nutmeg said. "And to think that he even stole the ringmaster's clothes! Oh, Tumtum! How could he be so silly? And now the children are sure to catch him and lock him up in a cage!"

"It serves him right if they do," Tumtum said crossly. "Really, what a fool he's been!"

"But just think how upset Mrs. Marchmouse will be when she comes home and we have to tell her the General's been taken prisoner," Nutmeg said anxiously. "You remember how upset she was when it happened before!"

The children had once caught the General playing in their

dollhouse, and they had taken him to school with them and put him in a cage full of pet gerbils. Mrs. Marchmouse had been quite distraught.

Tumtum sighed. Nutmeg was right. It would be too awful if Mrs. Marchmouse had to go through that again.

"Let's see what's going on outside," he said. They crept over to the back door and peered anxiously into the garden.

The children were rushing back and forth, shining the flashlight around the lawn and the flower beds.

"Come on! It must be here somewhere," Arthur cried.

They searched high and low—around the vegetable patch, and the trash cans, and the compost heap—but the bus wasn't there.

"I don't understand it!" Arthur groaned. "It can't just have vanished!"

"We'll have to come and look again in the morning," Lucy said. "It's too dark now."

"All right," Arthur agreed, for they had been outside for ages, and his hands were turning numb with cold. "But I can tell you, whoever the thief is, he'll be sorry when I catch him!"

Tumtum and Nutmeg crouched behind the shoehorn

while Arthur and Lucy walked back inside. They did not speak until the children had gone upstairs.

"Well, I'm glad they didn't catch him — Arthur sounded very cross!" Nutmeg said. "But where can he be? They've looked everywhere!"

"He must have gone farther away," Tumtum said. "I'll bet he's driven down into the meadow! You know how much he loves to go exploring there, and just think how much fun he could have in a bus!"

"Oh, dear. You're probably right," Nutmeg said anxiously. She hadn't thought of the meadow. It was below Rose Cottage, and it was a wild place. It had a stream running along the bottom, and at night it was full of foxes. Tumtum was right; the General loved to go adventuring there. "I hope he doesn't come to any harm," she said.

"Oh, he'll be all right inside his bus," Tumtum said. "And you can be quite sure he'll turn up as soon as he's hungry. He won't want to miss his birthday feast! Now come on, we'd better get home in case the children come downstairs again."

They hurried across the kitchen and let themselves back into Nutmouse Hall. They were no longer in the mood for cider, so Nutmeg made a pot of tea, and they sat by the fire waiting for the General's familiar *Rap! Tap! Tap!* at the door.

But the hours slipped by, and when ten o'clock struck, he still hadn't appeared.

Nutmeg was very nervous.

"I wouldn't worry, dear; he's probably just decided to camp out for the night," Tumtum said, trying to raise her spirits. "You know how he can't resist an adventure. We might as well get some sleep. Then we can set out and look for him first thing in the morning."

Nutmeg agreed, for after her long day of baking and worrying, she felt very tired. And the house felt sad now that their party had been spoiled. So they blew out the candles in the dining room and climbed upstairs to bed.

But if they could only have seen what their friend was up to, they would not have slept a wink that night. For the General had fallen into a much bigger adventure than he had planned.

Chapter Four

The General drove so fast, it was no wonder the children hadn't been able to find him. After leaving the cottage, he had sped straight to the far end of the garden, then slipped underneath the tall farm gate and into the meadow.

He heard an owl hoot and a fox howl, but he felt not a drop of fear. The night was his, and his whole body ached for adventure.

The General knew just where he wanted to go. He would drive all the way to the bottom of the meadow, and spend the night camping by the stream! The grass was taller than the hood of his bus, and the ground was rough. But he pressed on the accelerator and bumped along as fast as he could, plotting his adventure.

Tonight he would ambush a snoozing snail and roast it for his dinner. And tomorrow morning he would catch a beetle and make it trot around in circles to the crack of his whip. Then he'd put up a sign reading GENERAL MARCHMOUSE'S CIRCUS TROUPE, and invite all the field mice to his first show!

In his mind's eye, he saw himself strutting about in his gold suit, showing off to the crowd, and he felt a lovely warm glow in his stomach.

What a glorious adventure it promised to be! But adventures cannot be planned, as the General was about to discover.

For just as he was coming within sight of the stream, he saw something that surprised him. A little way off along the bank, there was a cluster of bright lights! On spotting them, he quickly stopped and turned off his engine, wondering what they could be.

At first he could not see much, for the moon had been swallowed by a cloud. But a few moments later, the cloud shifted in the breeze, and a pale moonlight dribbled along the bank. The General sat back in his seat with a jolt. For he could see now where the lights were coming from, and he could hardly believe it.

For lo and behold if it wasn't another circus—with wooden wagons and a big top! But this circus was no toy. It was a *real live* mouse circus—there could be no mistake of that. There was a snail roasting on a campfire, and the General could see four mice silhouetted against the tent.

He was *furious*. He couldn't set up his own circus now, this one would steal all his business! There hadn't been a mouse circus visiting the village for ages. It was just his luck that one should show up now.

"I can't let them spoil my adventure. I shall go and tell them to go away!" he decided. He grabbed his whip and slid out of the bus. Then off he marched toward the big top, grinding his teeth.

The circus was surrounded by a low fence lit with lanterns. And at the entrance there was a wooden booth with a sign reading TICKETS. The General knocked on the window, but there was no reply. He knocked again. Then he barged through the gate.

He looked around in surprise, for the circus did not look nearly as impressive from the inside. The ground was strewn with garbage, and the big top was full of holes.

The circus mice were sitting around their campfire, eating roasted flies. When they saw the General, they all looked up and snarled.

The General was taken aback, for the mice looked very sinister.

There were four of them, and they all had fiery red fur and piercing black eyes. One was as thin as a pipe cleaner and had a scar on his nose. He was dressed in a scarlet suit, and was clearly the ringmaster. The other three were dressed as clowns, but somehow they did not look at all like clowns should. Their mouths turned down at the corners, and their teeth were dark yellow.

"What are you doing here?" the ringmaster snarled.

"I've come to tell you to go away!" the General replied rudely. "There is only room for one circus in this meadow and..."

But before he could get any further, one of the clowns walked up and gave him a shove. "Who do you think you are, telling us to go away?" he shouted. "*You* go away! Go on, *SHOO!*"

The others had gathered behind him, looking at the General with menace. But the General was too angry to feel afraid.

"How dare you! Do you know who I am?" he raged. "I am the Great General Marchmouse, and I warn you now, if you mess with me, you will have the Royal Mouse Army to answer to!"

His words had a powerful effect. General Marchmouse was a very famous name in the mouse community, and the circus mice were clearly impressed.

The ringmaster gave a cold smile. And when he spoke, his voice sounded quite different. It was almost oily.

"General Marchmouse! This is an honor, to be sure! You will surely stay and share our roasted flies?"

The General was flustered by the ringmaster's sudden change in tone. It struck him that his voice sounded suspicious. But the flies did smell very delicious, and it seemed ages since he had last eaten.

"Well, all right then, I suppose I might have time for a quick bite," he said, eyeing the frying pan greedily.

One of the clowns handed him a plate, and they all sat down around the campfire. Once he had started munching his fly, the General felt quite chatty.

"Tell me, are you new to these parts?" he asked. "It's been ages since we last had a circus in the meadow."

"We just arrived here this afternoon," one of the clowns replied. "And we're doing our first show tomorrow. All the field mice are coming. We're nearly sold out."

"Gracious!" the General said. "And where are you all from?"

"Oh, we're town mice!" the ringmaster replied proudly. "I used to belong to the Red Top Circus, but I left a few months ago to set up on my own."

The General looked at him in amazement. "The Red...Y-you mean to say... *Good grief!*"

The General was very impressed. For the Red Top Circus was the most popular circus in the whole land. Mice traveled for miles to see its giant beetles and its dancing dragonflies and its racing slugs. As a schoolmouse, the General had saved his pocket money to see its shows, and had known the names of all the performers.

"By Jove!" he said.

The ringmaster looked very pleased. "Mr. Goldtail's my

name," he said, "and these are my clowns—Mr. Merry, Mr. Mirth, and Mr. Moody."

Everyone shook paws. "Goldtail, *Goldtail* . . ." the General said, racking his brains.

"Yes, yes . . . I'm sure I remember your name!" he said uncertainly. "Were you one of the trapeze mice, who shot out of a cannon?"

"That's right," Mr. Goldtail said quickly. "I was one of the best trapeze mice they'd ever had. But I wanted to be a ring-master, so I left Red Top and set up on my own!"

"Good for you," the General said admiringly. He had been very upset to find another circus in the meadow, but he felt quite differently now that he knew whose circus it was.

"I must buy a ticket to your show," he said, digging in his pocket. But then his face fell. "Oh, drat," he muttered. "My wallet's in my jacket pocket—and I left my jacket in Arthur's big top!"

"Arthur's big top?" Mr. Goldtail asked anxiously. "But I haven't heard of Arthur's Circus. I thought *we* were the only mouse circus in these parts."

"Oh, Arthur's circus isn't a mouse circus!" the General laughed. And then he told them the whole story of his adventure so far—about how he had stumbled across a toy circus at Rose Cottage, and stolen the ringmaster's gold suit, and sped off in the shiny red bus before anyone could catch him.

Mr. Goldtail and his clowns laughed admiringly. They thought his story was very funny. No one showed any disapproval.

"There's my bus, over there," the General said proudly, pointing through the grass. "She's as fast as a hare—and she's got a horn and camp beds and real leather seats!"

The others all turned to look. And when they saw the big red bus gleaming in the moonlight, their jaws dropped in surprise.

No one said a word. The General could see they were very envious.

"Where's your bus?" he asked, feeling rather superior.

"We don't have one," Mr. Goldtail replied. "We had a squirrel to tow our wagons, but she escaped when we were unloading this morning, and now she's disappeared up one of these trees. It's a wretched nuisance. We want to move on after tomorrow's show, but we won't be able to go anywhere until we've caught another squirrel, and that could take weeks!"

The General pondered this. Then, because he was feeling starstruck, he made a rather careless suggestion.

"I say," he said. "Why don't you borrow my bus? I'm sure Arthur wouldn't mind if you took it for a day or two—and in return you can give me a ticket for tomorrow's show!"

The mice all looked very pleased.

"That's most kind of you," Mr. Goldtail said, his eyes glinting greedily.

Then he made a suggestion, too. "Why don't you join our circus?" he said. "You would make a fine beetle tamer, dressed in that splendid gold suit. And the crowds would come from far and wide if they heard that General Marchmouse was in the ring!"

"Oh, yes, what a smashing idea!" cried Mr. Mirth. "You could be our star attraction!"

The General looked overwhelmed. It was such a thrilling proposal, his brain had started to spin.

"It's a fine life in the circus," Mr. Merry said temptingly. "We'll give you your own wagon, with a dressing table, and chocolate truffles to eat!"

The General's heart quickened. He adored chocolate truffles.

"And you get lots of money!" purred Mr. Moody.

"And lots of fan mail!" Mr. Mirth added silkily.

Suddenly, all the mice were talking at once, egging him on. Their voices swam around him as if in a dream. The General closed his eyes and saw it all. . . . The crowds and the chocolates and the fan mail — and, above all, the open road, paved with adventure.

"Oh, yes!" he said. "I'LL COME!"

Chapter Five

Nutmeg woke up first the next day. She sat up in bed and pulled her shawl around her. It was cold and bright. The sun was shining through the broom cupboard window, and she could hear a bird rustling in the bushes.

She turned to Tumtum and gave him a shake. "Wake up, dear!" she said. "It's time to go and find the General!"

Tumtum groaned, and buried his head under the covers. He didn't like waking up.

Nutmeg left him in bed and hurried downstairs to the kitchen. A little while later, Tumtum smelled sausages cooking.

Perhaps I will get up after all, he thought.

He got dressed, and soon he and Nutmeg were having a big breakfast in the kitchen. After the sausages there were oatmeal and pancakes and eggs, then a fruit salad with cherries on top. And they washed it all down with milk and hot coffee.

Tumtum felt much more cheerful once his stomach was full. "Well, we'd better be setting off," he said, wiping the crumbs from his whiskers.

It was still early when they crept out into the kitchen of Rose Cottage, and no one had stirred. The toy circus towered before them, with the big top flapping in the draft.

"If we find the General quickly, then we can have the bus back in place by the time the children come down for breakfast," Nutmeg said. "Just think how pleased they'll be to get it back!"

"Hmm, it should be possible," Tumtum said. "It's a Saturday, so the children will probably get up late. Come on, let's hurry up and find him. He can't have gone far in a small bus like that."

They ran across the kitchen and wriggled outside under the garden door. Then they crossed the path and scrambled onto the lawn. They were glad they had put on their rain boots, for the grass came up to their waists, and it was soaked with dew.

They were in little doubt as to which way the General had gone. They trekked straight down to the farm gate at the bottom of the garden and hurried into the meadow.

"Look how tall this gate is," Nutmeg said, peering up at it. "The bottom bar's so high up off the ground, there would have been plenty of room for the bus to go underneath."

They walked on, but soon the grass became taller than their heads, and they couldn't see where they were going. It was like being in a forest.

"Oh, dear. We'll never be able to find the bus now!" Nutmeg said anxiously.

But then Tumtum saw a big rock poking out of the grass just ahead of them. "Let's climb up on that. Then we'll be able to see where we are," he said.

They traipsed over to it and scrambled to the top. And when they stood up, they could see the whole meadow sloping away in front of them. Tumtum perched his field glasses on his nose and carefully scanned all over the meadow.

"Can you see anything?" Nutmeg asked impatiently.

"There he is!" Tumtum cried finally. "There's something

red down by the stream! I can't see it clearly; it's too far away. But it must be him!"

He passed Nutmeg the field glasses, so she could have a look, too. "Yes, that must be the bus," she said, seeing something red and shiny through the lenses. "But what's that big thing next to it? It looks like a balloon — or a huge tent. How very strange!"

Tumtum looked again, and now he saw it, too. Nutmeg was right; it did look like a tent. But it was too far off for him to be sure. "How very odd," he said. "Come on, let's go and investigate."

They scrambled down from the rock and started making their way through the grass. It was a long way, and it was nearly two hours later before they reached the other side of the meadow.

The grass around the stream was much shorter, and they could see all the way along the bank. And when they spotted Mr. Goldtail's big top, with Arthur's bus drawn up beside it, they stopped still in surprise.

"It must be a touring mouse circus!" Nutmeg said. "But I didn't know there was a circus coming to the meadow.

How strange that they didn't put up posters around the village!"

"Hmm. And it looks like the General's gotten mixed up with them somehow," Tumtum said anxiously. "Come on, we'd better go and see what's going on."

They hurried toward the gate. And when they reached the ticket booth, they found a notice pinned to the window:

MASTER GOLDTAIL'S MOUSE CIRCUS

COME AND SEE

GENERAL MARCHMOUSE

FIGHT A

GIANT BEETLE!!!!

1 p.m. Today

"He's joined the circus!" Nutmeg gasped.

"Well, he'll un-join it soon enough when I've spoken to him!" Tumtum said crossly. "What nonsense! I'll soon talk some sense into him!"

He peered in the window of the ticket booth. There was no one inside, so they walked in through the gate.

The circus seemed deserted; there was no one in sight and not a sound coming from the wagons. And everything looked very shabby. Nutmeg was surprised. She had been taken to mouse circuses as a child, but they had all been spick-and-span—not a bit like this.

Something about the place made her feel uneasy.

"Let's look in those wagons over there and see if we can find who's in charge," Tumtum said.

The wagons were lined up on the far side of the big top. They walked over to them nervously, wondering what they would find. The first wagon had a filthy wool blanket hanging across the bars.

"I say, is anyone at home?" Tumtum asked.

There was a sudden scraping noise. Then the corner of the

284

blanket moved aside — and a big beetle glared out. When he saw Tumtum and Nutmeg, he gave an angry hiss. He did not look at all pleased to be disturbed.

They hastily apologized and tried the next wagon. But this one was just full of cardboard boxes, sealed with brown tape and labeled FRAGILE.

"I wonder what's in them," Nutmeg said curiously.

"Oh, probably just some circus junk," Tumtum said. He wasn't interested in the boxes. He just wanted to find the General and go home.

The third wagon had a plush red curtain drawn across it and a sign reading DO NOT DISTURB.

"Perhaps this is the ringmaster's wagon," Tumtum said. "It says not to disturb him, but I'm afraid we shall have to!"

Nutmeg watched anxiously as Tumtum stepped forward and called out through the curtain.

"I say, sorry to disturb you, but it's Mr. and Mrs. Nutmouse here, from Nutmouse Hall and — er, well . . . we just wondered if we could have a quick word with you, if you would be so kind. You see, we're looking for our friend. . . ."

There was no reply, and they wondered if the wagon might be empty.

But then they heard someone yawn. It was a long, bored yawn. The sort of yawn that says, "Oh, go away and leave me alone!"

The next moment, the curtain swished aside, and General Marchmouse appeared!

The Nutmouses looked at him in astonishment. This was not the General Marchmouse of yesterday, or the day before that. He had *completely* changed.

He was dressed in the gold suit that he had stolen from Arthur's ringmaster, and he had a daisy in his buttonhole — and his whiskers had been dyed bright pink! He did not look at all as a General should.

"Darlings, hello," he purred, in a voice most unlike his usual one. "Have you come to see the show?"

"Certainly not!" sputtered Tumtum, who found the General's new style most unsettling. "What are you playing at, General? How *dare* you steal Arthur's bus? You must drive us all back to Rose Cottage at once and return it to the kitchen!"

"I'm afraid that is out of the question," the General replied, peering down at them snootily. "I have lent the bus to Mr. Gold-tail, seeing as he has no transportation of his own. And we shall need it this afternoon, when we drive off on the next stage of our tour!"

Tumtum and Nutmeg looked at him in horror. "*A TOUR!* Don't be so ridiculous!" Tumtum said. "Just think what Mrs. Marchmouse would say if she discovered you were touring with a circus! You are coming home with us!"

The General gave Tumtum a withering stare. His eyes had a stubborn gleam. "Oh, no, I am not!" he said tauntingly. "Can't you see, old boy? I'm a circus mouse now!"

Chapter Six

Tumtum and Nutmeg did all they could to make the General be reasonable.

"Just think what the *Mouse Times* will say when they learn that the Great General Marchmouse has joined a circus!" Tumtum cried. "You'll be a laughingstock! The Royal Mouse Army will disown you!"

Tumtum felt sure this would bring the General to his senses, for he took great pride in his army career. But the General didn't seem to care.

"It's no good. I'm not coming," he said stubbornly. "My army days are over. I've become a star of the circus, and I shall become the biggest star there ever was! Now leave me alone. I've got to curl my whiskers for the show!"

And with that he stood up and swished shut the curtains.

Tumtum and Nutmeg were aghast.

"How *could* he be so beastly?" Nutmeg cried. "Can you imagine what Mrs. Marchmouse will say when we tell her he's run away to become a beetle tamer? Oh, if only we hadn't invited him to dinner at Nutmouse Hall, then none of this would have happened!"

"I've an idea," Tumtum whispered. "Let's drive the bus back to Rose Cottage ourselves! We've every right to take it—it belongs to Arthur, after all. And without a bus, the circus can't drive away. They'll be stuck here in the meadow until they find another means of transportation. So in the meantime at least we shall know where the General is."

"Oh, what a splendid idea!" Nutmeg said. "We can let the General have his fun until Mrs. Marchmouse comes home. Then *she* can come down here and talk some sense into him."

Feeling very cheered by this plan, they slipped away from the General's wagon and crept over to the bus.

But when Tumtum tried the driver's door, it was locked. And the passenger door was locked, too.

"Oh no!" he said.

"Let's see if the back door's open," Nutmeg suggested—but then there was a sudden shout.

"Hey! What are you doing?" a furious voice cried. "Get away from that bus!"

They both jumped. And when they turned around, they saw a clown marching toward them. It was Mr. Merry. His teeth were bared, and he looked very frightening.

"I, er...We were just, um...just admiring it," she said nervously.

"Well, now that you've admired it, you can leave," the clown replied.

"That's no way to speak to—" Tumtum began, but then Nutmeg gave him a sharp nudge. She sensed it was time to go.

"Shh," she whispered, "don't get into a fight." She tugged Tumtum by the sleeve, and they hurried out through the gate while the clown stood glaring at them.

"Let's get away," Nutmeg whispered. They walked on along the bank, until the circus was out of sight. Then they sat down on a rock, feeling very shaken.

"Isn't it strange that a clown should be so awful?" Nutmeg said. "In fact, there's something strange about the whole circus. I didn't see any dragonflies or racing slugs, or the sort of things that a mouse circus usually has. There was just one tired old beetle. And their big top was in tatters, but the General's wagon had a fancy chandelier. And did you notice those silver plates on his dressing table? I wonder where they got them from."

"Hmm, it did seem odd," Tumtum agreed. "I'd like to go back and have another snoop. But I don't want to run into that clown again. I reckon we've already made him suspicious. There's no point asking him for the bus back—I shouldn't think he'd care one bit if we told him it really belongs to Arthur!"

Nutmeg agreed. "But we *must* get it back before this afternoon," she said, "or else the circus will drive away, taking the General with them! You can see that he's completely under their spell! And just think what a fool he's going to look if word of this gets out. His reputation as a famous war hero will be ruined!"

The situation did look bleak. They both sat on the rock, wondering what to do.

Then Tumtum had an idea. "I know!" he said, jumping up

excitedly. "Let's go home and write the children a letter, telling them that their bus has been stolen by another circus. We can explain exactly where it is—they know the stream very well. Then they can come down here this afternoon and fetch it back!"

"Oh, what an excellent plan!" Nutmeg agreed.

But then she looked worried. "What about the General?" she said anxiously. "I mean, if the children find him sitting there curling his whiskers in his wagon, they might take him prisoner!"

"Oh, I wouldn't worry," Tumtum said. "The children are so big, the General and the circus mice will spot them from a mile away, and they'll hide in the grass. And without a bus, the circus won't seem nearly so appealing to the General. He'll give up the whole thing and head for home before anyone finds out about his silly adventure!"

"Oh, I do hope so," Nutmeg said.

There was no time to lose, so they hurried straight back across the meadow. But it took even longer this time, for they were going uphill. It was nearly midday by the time they reached the farm gate leading into the garden of Rose Cottage.

"At last!" Tumtum said, and at the sight of home his

stomach gave a loud rumble. It seemed ages since breakfast, and he could hardly wait to have a nice big lunch in his cozy kitchen.

But as they were walking under the gate, a field mouse appeared. The Nutmouses recognized him at once. It was little Timmy Twigmouse, whose father delivered the firewood to Nutmouse Hall. He appeared to be in a great hurry, and he looked very excited.

"Have you heard the news?" he asked breathlessly. "There's a circus in the meadow, and would you believe it? . . . *General Marchmouse has joined them!*"

Tumtum and Nutmeg gulped. News of the General's adventure had spread faster than they'd thought.

"When did you hear about this?" Tumtum asked anxiously.

"Oh, everyone's talking about it," Timmy Twigmouse squeaked. "The General's the star act. He's going to FIGHT A GIANT BEETLE! Just think! It's going to be quite a show!

"I say!" he said, noticing that they did not look very enthu-siastic. "You are going, aren't you?"

"Oh, I don't think we have time today," Tumtum said vaguely, not wishing to reveal what they knew. "We'll go tomorrow."

"But you can't go tomorrow!" Timmy Twigmouse cried. "They won't be here tomorrow. Haven't you heard? This is a *ONE TIME SHOW*! And when it's finished they're packing up and moving on to London. Their next performance will be in Covent Garden!"

Tumtum and Nutmeg looked at him in horror.

"London!" Nutmeg gasped. "Are you sure?"

Timmy Twigmouse nodded. "That's right," he said. "My father went over this morning to deliver them some firewood, and they told him all about it. They're going to drive all the way there in their new circus bus! This is the only chance to see them, so I'm going down there right now to make sure I get tickets!"

Timmy Twigmouse said goodbye and hurried on his way. But Tumtum and Nutmeg stood by the gate, frozen with shock. They had never been to London, but they had heard all sorts of horrible things about it.

"*London!*" Nutmeg said faintly. "But London's miles away! And the mice there are all bandits! Oh, Tumtum, we *must* stop him! Just think what kind of trouble he might get into there!"

Chapter Seven

After Tumtum and Nutmeg left, the General sat down at his dressing table, feeling a little deflated. Tumtum's words kept echoing around his head: "Think what Mrs. Marchmouse will say. . . .Think what the *Mouse Times* will say. . . .You will be a disgrace . . . disgrace . . . disgrace . . ."

The General pressed his paws over his ears, trying to block the voice out. "*Forget Tumtum!*" he muttered. "This is too good an adventure to miss. I'm not going to let him spoil my fun!"

He glanced in the mirror, thinking he would cheer up when he saw how handsome he looked. But even his reflection had lost its dazzle.

He stood up with a sigh and took his whip from the bed.

He was sure he would feel all right again once the show had begun.

"Tumtum is quite wrong, a beetle tamer is a very fine thing to be!" he told himself firmly. "When Mrs. Marchmouse hears about it, she'll be proud of me! And she isn't due back from her sister's until Friday. . . . Our tour will probably have finished by then, and I shall be back home, so it's not as though she'll miss me!"

He felt a little happier. And he decided to go and find Mr. Goldtail to discuss his act. The show was due to start in less than two hours, but there had been no dress rehearsal.

But just as he was opening the door of his wagon, Mr. Mirth appeared. He looked very agitated.

"Quick! Get in the bus! We're moving on!" he shouted.

"Moving on? But I don't understand," the General said in astonishment. "What about the show? The fans will be arriving soon."

"Just do what I say—and hurry!" Mr. Mirth snapped. Then he ran on down the line of wagons, slamming the doors, and fastening the towropes.

299

When the General stepped outside, he saw that all the other mice were rushing about, too. Mr. Moody was pouring water on the campfire, and Mr. Merry and Mr. Goldtail were pulling down the big top!

The General could not understand it. "What's happening?" he asked. But everyone was too frantic to explain.

In almost no time, the big top and the ticket booth had been loaded onto the wagons.

"Right, that's everything!" Mr. Goldtail shouted, leaping behind the wheel of the bus. "Come on, everyone. Jump in, QUICK!"

The clowns all leaped on board. Only the General remained on the ground.

"*COME ON!*" Mr. Goldtail shouted, poking his head through the window. "Hurry, or we'll be caught!"

"*Caught?* Caught by whom?" the General asked — but then suddenly he heard a piercing whistle, so loud it made him jump.

Mr. Goldtail revved the engine, and the bus and the wagons lurched forward through the grass. The General did not know

where the whistle was coming from, or whose whistle it was. But he felt sure that something awful was going to happen.

"Wait for me!" he cried, running after them. Mr. Mirth opened the back door and hauled him inside. Then off they sped along the bank, with the six wooden wagons clattering behind them.

Mr. Goldtail drove the bus as fast as it would go. *Bump, bump, bump* they went until they reached the edge of the field. Then they scraped through the bushes and into the woods on the other side.

The clowns kept looking anxiously through the back windshield to see if they were being followed. The mood was very tense. No one said a word.

Deeper and deeper into the woods they drove, crunching through the bracken.

Eventually, when the trees had become so thick it was almost dark, Mr. Goldtail turned off the engine and let out a deep sigh.

"Well, I reckon we've gotten away all right," he said. "They'll never find us now."

"We'll have to be more careful in the future," Mr. Mirth said. "I wonder how they found out about us. I didn't think the word would spread so fast."

"Found out *what?*" the General asked. "Oh, *please* will someone tell me what's going on? Who were we running away from? Who was blowing that whistle so loudly?"

The others looked at him coldly.

"*Who?*" Mr. Goldtail sneered. "Why, the policemice, of course! We can recognize their whistles from a mile away. I don't know how many of them there were — there might have been only one, but there might have been half a dozen of them, running up on us through the grass. We weren't going to take any chances. If we'd hung around a minute longer, we'd have been arrested!"

"*Arrested?*" the General gasped. "But why? What have we done?"

Mr. Mirth gave a nasty cackle. "Haven't you guessed what our game is by now, General? We're not *real* circus mice. The ticket booth and the big top—they're just a cover-up. Our real job is BURGLING MOUSE HOLES. That's what we do for a living!"

"Burgling m—*Good grief!*" the General stammered. He was too horrified to speak.

Mr. Moody jerked a paw over his shoulder. "In those wagons back there, we've got more gold and silver than you've ever set eyes on!" he said proudly. "We're going to sell it when we get to London, and we'll all be rich!"

"London?" the General gasped.

"That's right," Mr. Goldtail said. "It's a long way, but we reckon we'll get there in a month or two—especially now that we've got this nice fast bus to drive. And there'll be lots of rich mouse holes to burgle on the way!"

The General was appalled. He thought he had joined a circus—but instead he had fallen in with a gang of common thieves!

"I am leaving!" he said, flinging open the door. "And when I see those policemice, I shall tell them which way you've gone!"

"Oh, but I wouldn't be so hasty if I were you, General," Mr. Goldtail said mockingly. "You don't want to be caught by the police any more than we do!"

The General snorted. "I'm not a thief!" he fumed. "I've nothing to fear from the police!"

"Ha! What do you think the policemice will say when we tell them that you've stolen a bus?" Mr. Goldtail snarled. "You've supplied the getaway vehicle, General. You're as much a crook as we are! If the police catch us, you'll go to prison, too."

"How ridiculous! I only borrowed the bus—I didn't steal it!" the General said furiously. "Now let me out!"

He turned to the door and started to lower himself to the ground.

"Get back in your seat," Mr. Goldtail growled. His voice was ice cold—and when the General turned around, he saw that he had whipped out a gun.

"Close the door!" Mr. Goldtail barked, pointing the nozzle straight at the General's nose. "This gun's loaded with sherbet."

The General gulped, for as he knew very well, sherbet guns are horrid things. They are loaded with a sickly yellow powder, which comes shooting out in a choking blast. If a mouse is hit by one at close range, the sherbet burns his throat and stings his eyes, and turns his stomach into lemonade.

He squirmed back into his seat and pulled shut the door.

Mr. Goldtail passed the gun to Mr. Mirth and started the engine. The bus gave a sputter and lurched on through the woods.

The General was numb with fear. And it was not just the gun he was afraid of. The others were right — when the police found out he had taken Arthur's bus, they would think he was a thief, too. He would be stripped of his medals and sent to prison. Oh, the shame of it! The shame!

"Don't look so glum, General," Mr. Goldtail sneered. "You do as we say and we'll make sure you're all right. You can help us rob the village shop!"

"*The village shop?*" the General said feebly.

"That's right," Mr. Moody said. "We're going to bust in after closing time, and gorge ourselves on ice-cream cakes and jam tarts! And we'll steal all the tubes of sherbet to use in our guns!"

"And no one will catch us now that we've got this speedy bus to escape in!" Mr. Mirth gloated. "We'll be able to carry out twice the number of robberies now! And it's all thanks to you, General!"

"Three cheers for General Marchmouse!" Mr. Merry cried.

"Hip hip hooray! Hip hip hooray! Hip hip hooray!"

The General sat silently in the backseat. He could hear the joyless cheers echoing around him, as if in a terrible dream. He thought of his wife, and of Tumtum and Nutmeg, and of all the army medals for which he had fought so bravely . . . but they seemed lost in a distant past.

Oh, if only he hadn't borrowed the wretched bus! But he had gone too far to turn back. . . . He was a criminal now.

Chapter Eight

Meanwhile, back at Rose Cottage, Arthur and Lucy had been looking for the bus all morning.

They had searched the garden twice, and they had gone up and down the little lane that ran in front of Rose Cottage, peeking under the parked cars and poking along the ditch.

But by lunchtime there was still no sign of it.

"Whoever stole it has probably driven out of the village by now," Arthur said as they came back inside. "I bet we'll NEVER get it back. And we'll never discover who the mysterious thief was either!"

They sat down to lunch feeling very glum. They did not say anything to their father about the missing bus, for they suspected he would not believe such a peculiar story. And this was *their* adventure — they didn't want any grown-ups interfering before they'd had a chance to solve it themselves.

Mr. Mildew spent the whole meal with his head buried in a book, so it was easy to exclude him from the conversation.

"I think we should write to Nutmeg and tell her everything that's happened. I bet she'll know who took the bus, and perhaps she'll be able to help us find it," Lucy said.

Arthur nodded. Lucy was right. This was just the sort of mystery Nutmeg might be able to solve. So as soon as lunch was finished, they hurried upstairs to write her a letter.

But when they walked into the attic, they got a big surprise — for Nutmeg had already been there. And she had left a letter for *them*! It was in the usual place, propped up against Lucy's hairbrush on top of the chest of drawers.

"Quick, let's see what it says," Arthur said excitedly. "Perhaps she knows about the bus already!"

Lucy grabbed the magnifying glass from her bedside table, then sat down on her bed, holding the letter in her palm.

"Oh, hurry!" Arthur cried impatiently. "What does it say?"

Lucy was squinting. "It looks like Nutmeg wrote it in a

great hurry—her writing's even more squiggly than usual!" she said. Eventually, stumbling a little, she managed to read it out loud:

URGENT

Dear Arthur and Lucy,

Last night your toy bus was borrowed by a mischievous friend of mine. He only took it for a bit of fun, and I'm sure he intended to bring it back. But now the strangest thing has happened, for he has fallen in with a real circus, which is camped down in the meadow! And they plan to run away to London together, traveling in YOUR BUS!

But you can stop them! You should go down to the meadow at once, my dears, and snatch the bus back! You will find it parked on the bank of the stream, beside the willow tree. Now hurry, for they plan to set off this afternoon!

Love,

Nutmeg

The children were astonished.

312

"A friend of Nutmeg! But who could that be?" Arthur asked. "And I wonder who the *real* circus belongs to."

"Perhaps it's a fairy's circus!" Lucy said.

Arthur scrunched his nose. He did not believe in fairies, except for Nutmeg, of course. And she was a Fairy of Sorts, which was different. "Perhaps it's a rat circus!" he said eagerly. "Now *that* would be exciting!"

"Rats! Ugh! I hope not!" Lucy shuddered.

"Well, whoever they are, we've got to stop them before they take my bus to London," Arthur cried, leaping up from the bed.

The children set off at once. They ran across the garden and clambered over the gate, wondering what on earth they were going to find.

But the mystery was not going to be solved as soon as they had hoped, for General Marchmouse's adventure had just taken a turn for the worse.

Chapter Nine

The children ran across the meadow until they were in sight of the stream. Then they stopped still and stood peering along the bank. They could see the willow tree, but the ground around it was hidden by ferns.

They crept up to it on tiptoe, hoping to take the circus by surprise. But when they reached the tree and searched all around it, the circus wasn't there!

"Nutmeg must have gotten the wrong tree," Arthur said impatiently. "Come on, we'll have to look all along the bank. It must be here somewhere!"

He ran back to the stream and started poking around in the bushes. Lucy was about to follow him, but then she noticed something curious on the ground. To one side of the willow tree, there was a patch of grass that had been flattened, as though a bucket had been placed on it. And in the middle of it, she could see something glowing.

She wondered if it might be a bead, or a gold coin. But then she noticed little feathers of smoke rising from it. And when she knelt down and peered at it more closely, she saw that it was a tiny campfire, and lying on the grass next to it was a frying pan the size of a penny.

"Arthur, quick! Come and look at this!" she cried.

She held the pan in the palm of her hand, and they both studied it in astonishment.

"Yuck!" Arthur said. "It's got a burnt fly in it!"

"And look at the fire!" Lucy said, pointing to the tiny pile of smoldering twigs. "This must be where the circus was parked. You can see where the grass has been flattened by the wagons and the big top!"

"So Nutmeg was right. The circus *was* here!" Arthur cried. "But we're too late. It's gone!"

"Well, they wouldn't have gotten far," Lucy said. "Their campfire's still burning, which means they can't have left long ago. But I wonder which way they went."

Then Arthur noticed something else. "Look here!" he said, pointing to two little tracks weaving through the grass. "These must be the marks made by the wheels of the bus and the wagons."

"Well spotted!" Lucy said. "If we follow these, then we're sure to find them!"

They jumped up excitedly and followed the trail along the grass by the bank of the stream. But when they reached the edge of the meadow, the ground turned to bare earth, and the tracks disappeared.

"Oh no!" Arthur said. "Now what do we do?"

"Well, they can't have crossed the stream—there's no bridge," Lucy said. "They must have driven into the woods."

"Then we'll never find them!" Arthur said glumly. "The woods are huge and full of bracken."

"Yes, but if you go straight through it, you come out behind the church," Lucy said. "Well, I bet that's what they've done. If they plan to drive all the way to London, they'll have to get onto a proper road at some stage. So they'll probably cut down through the church, and then on to one of the lanes leading out of the village."

"All right, then," Arthur said. "Let's see if we can find them on the road."

They clambered over the gate at the bottom of the meadow and followed the footpath along the edge of the woods until they reached the churchyard. Then they ran up the path to the gate and let themselves out onto the lane.

To the right, the lane forked in two, with both ways weaving into the open countryside. To the left, it led past the war memorial and the village shop, then back toward Rose Cottage.

"Which way's London?" Arthur said.

"Let's go to the shop and ask Mrs. Paterson," Lucy suggested. "She's sure to know."

They hurried along, for it was nearly teatime, and the shop would be closing soon.

But just as they were crossing the lane toward it, Lucy suddenly stopped. "Arthur, look!" she whispered. "*Over there!*"

She was pointing to the side of the shop, where the trash cans were kept.

"What is it?" Arthur asked.

"It was the bus!" Lucy said. "I just saw it, really I did—it went racing off behind those trash cans! And it was towing lots of wagons!"

Arthur looked doubtful. He wondered why he hadn't seen it, too. But they would know soon enough if Lucy had been imagining things. "Come on, let's go and have a look," he said. "The alley's a dead end. If the bus did go down there, then we're sure to find it!"

They crossed the lane and ran around to the side of the shop. The alley was very short, but it was full of trash cans and cardboard boxes, so it took them a while to dig around. And just as Arthur was peering under a box, Mrs. Paterson appeared. She had just closed the shop for the day, and now she was carrying out the trash.

"Whatever are you doing here?" she asked the children in surprise.

"Er, I was . . . looking for something I'd lost," Arthur mumbled. "A . . . um . . . well, a toy bus actually!"

"Well, I'll let you know if I see it!" Mrs. Paterson said. "Now, shouldn't you be getting home? It will be dark soon. Your father will be worrying about you."

"Yes, we're just going back now," Arthur said, for he didn't want Mrs. Paterson interfering in their secret search. "And, er . . . one thing, Mrs. Paterson. Do you know which way London is?"

"London? Well, it'll be about a hundred miles that way." Mrs. Paterson tutted, pointing up the lane toward the church. "Why, are you planning to run away?"

"Oh, no . . . we were just, well . . . just wondering," Lucy said hurriedly. "Anyway, we'd better be getting home."

The children said goodbye, then they turned and headed back toward Rose Cottage.

"Are you *sure* you saw it?" Arthur said.

"Quite sure! How many times do I have to tell you?" Lucy replied impatiently. "It drove behind the cans, then it just sort of . . . vanished!"

"Well, if you're sure you saw it, then we'll have to go back first thing in the morning and have another look," Arthur said. "It can't just have disappeared. Maybe it's hiding under one of those cardboard boxes, though I'm sure I looked under them all."

"Yes, let's go back after breakfast," Lucy said. "You know, I feel *sure* we're going to find it!"

But the morning was still a long way off. There was the whole night to get through first, and all sorts of mysterious things were going to happen.

Chapter Ten

Tumtum and Nutmeg had spent an anxious afternoon in Nutmouse Hall, waiting for the children to come back from the meadow. But by teatime there was still no sound of them.

"They should have been home *ages* ago—it couldn't have taken them more than half an hour to run down to the stream and back," Nutmeg said anxiously. "Whatever could have happened?"

"Perhaps we had better go and look for them," Tumtum said. But just as they were getting ready to go out, they heard a door slam, and then the thunder of feet in the kitchen.

"Oh, thank goodness! That must be them!" Nutmeg said.

They hurried out of their front gates and crept to the edge

of the dresser. The children had just come in from the garden, and they still had their coats on. But when the Nutmouses saw that they had returned without the bus, they looked at each other in dismay.

"Whatever could have happened?" Nutmeg whispered. "Why didn't they bring it back?"

She and Tumtum crouched beneath the dresser, trying to find out what had gone wrong.

"I wish it would hurry up and be morning so we can go and have another look in that alley," Arthur said.

"It's a good thing tomorrow's Sunday and there's no school," Lucy said. "Let's go and write Nutmeg a letter, telling her everything that's happened."

"All right," Arthur said. "Just think how surprised she'll be when she learns that the circus has already gone!"

The children dashed upstairs, leaving Tumtum and Nutmeg in suspense.

"Gone? Oh, how dreadful!" Nutmeg cried. "But whatever could have happened? The children should have had plenty of time to snatch the bus before the circus set off!"

"Hmm. And I wonder what this alley is that they're referring to," Tumtum said.

"They're sure to explain everything in their letter," Nutmeg said. "We shall just have to wait until we can read it."

They did not dare follow the children up to the attic, for fear of being seen. So they waited impatiently under the dresser, until eventually Arthur and Lucy came downstairs. Lucy went into the pantry to find something for dinner, and Arthur went to put his bicycle in the shed.

The mice seized their chance. They shot out from beneath the dresser and dashed up the baseboard beside the stairs. Then they crossed the first-floor landing and slowly heaved themselves up the steep wooden steps to the attic.

When they finally crept into the children's bedroom, they saw a letter addressed to Nutmeg propped against the mirror on the chest of drawers.

They climbed up to it, clutching onto the tights and socks that had been left tumbling from the drawers.

Then they anxiously unfolded the letter and stood together at the bottom of the page, reading what it said:

Dear Nutmeg,

We went straight down to the meadow, just as you told us to, but the circus had left! Then we thought we saw the bus on our way home, going down the alley at the side of the village shop. We tried to follow it, but we couldn't find it.... And then Mrs. Paterson appeared. But we'll go and look again in the morning.

Love,

Arthur and Lucy

"The morning might be too late!" Nutmeg cried. "We shall have to go to the shop tonight and try to persuade the General to come back! Oh, I do hope the circus is still there!"

"Let's go back to Nutmouse Hall and fetch our coats and a flashlight, then we can set out straight away!" Tumtum said.

They climbed back to the floor and hurried downstairs. The children had moved into the living room, so they were able to dash back under the dresser unseen.

Then they ran back into Nutmouse Hall and put on their

warm clothes. But as they were about to leave the house, there was a sharp *Rap! Tap! Tap!* on the front door.

The Nutmouses looked at each other in surprise. They had not been expecting visitors tonight, and they wondered who it could be.

"Perhaps the General's come back!" Nutmeg said hopefully. But when Tumtum opened the door, they got a big surprise.

It was Mrs. Marchmouse. She looked very upset, and her nose was wet with tears.

"Mrs. Marchmouse!" Nutmeg said in astonishment. "We thought you were away looking after your sister — oh, goodness! Is she all right?"

But before Mrs. Marchmouse could speak, a second mouse appeared.

This time, Tumtum and Nutmeg were even more taken aback. For it was Chief Constable Watchmouse, the head of the local Mouse Police Station! His snowy whiskers glared against his black uniform, and his big red nose was twitching.

He stood twisting his baton in his paws, looking very grave.

"Mr. and Mrs. Nutmouse," he said, addressing them in his most solemn voice. "I believe you might be able to assist me in my investigation."

"Assist you in . . . *Gracious me!* What's all this about?" Tumtum said, looking very alarmed.

The Chief Constable cleared his throat, clearly relishing the drama. But before he could explain his business, Mrs. Marchmouse started sobbing out the story herself.

"Oh, Mr. and Mrs. Nutmouse, you will never guess what's happened!" she cried. "I came home early from my sister's mouse hole, seeing as she was so much better, and seeing as I was missing my poor Marchie so much. . . . But when I let myself in, there was the Chief Constable, *searching the gun cupboard*! And he's got A WARRANT FOR MY HUSBAND'S ARREST!"

Tumtum and Nutmeg were astounded.

"*A warrant for his arrest?* Whatever do you mean?" Tumtum asked.

"They say he's joined a gang of thieves!" Mrs. Marchmouse trembled. "Oh, tell me it's not true, Mr. Nutmouse! Tell me it's not true!"

"Gang of th——" Nutmeg stammered, hardly able to believe her ears.

"But, Chief Constable, there's been a terrible mistake!" Tumtum cried. "The General hasn't joined a gang of thieves. He's joined a circus! We went to see him! Now, I can assure you, this is all a misunderstanding!"

But Mrs. Marchmouse did not seem at all reassured. At the mention of the circus, she started sobbing even harder.

The Chief Constable raised his baton, gesturing for them to stay calm.

"The information I have is that General Marchmouse has joined Mr. Goldtail's circus," he said. "Or should I say, Mr. Goldtail's circus that is *pretending* to be a circus...but that is not really a circus at all!"

"*Not* a circus? But what do you mean, Chief Constable?" Tumtum asked, finding this all very confusing. "They must be a circus—they've got wagons and a big top!"

"Ha!" The Chief Constable snorted. "You weren't the first to be fooled by all that, Mr. Nutmouse! But I tell you, those wagons and that tent are just a front! Mr. Goldtail and his gang don't have one single drop of *real* circus blood in their tails!"

"But then who are they?" Nutmeg asked.

"They are *WANTED CRIMINALS*," the Chief Constable announced dramatically. "Burglars, Mrs. Nutmouse—and very accomplished ones, too. They have robbed more mouse holes than you've eaten cockroach pies! Their wagons aren't used for carrying circus insects—they're used for carrying *stolen goods*!"

Nutmeg clapped her paw to her mouth in horror. No wonder the circus had given her such funny feelings! *They were thieves!* The cardboard boxes that she and Tumtum had seen stacked high in the second wagon must have been full of all the things they had stolen. And the silver in the General's wagon must have been stolen, too! And to think that the poor General had been fooled by them!

"But why aren't they behind bars?" Tumtum asked.

"Because they are VERY CUNNING," the Chief Constable replied. "We call them the Vanishing Circus: here one minute— and gone the next! I'll tell you how they operate, Mr. Nutmouse. They turn up somewhere quite out of the blue—somewhere they've never been before, where the local mice don't know how wicked they are. Then they quickly put up some posters and put on a show. And very poor shows they are, from all I've heard—no trapeze, no dragonflies, just an old beetle! But the field mice all go because it's not every day a mouse circus comes by—and of course, they don't know how rotten the show's going to be until it's begun."

"What about the clowns?" Tumtum interrupted.

"The clowns? Ha! That's the whole business!" The Constable grunted. "Only one of the clowns performs in each show. The other two sneak out when the field mice are in their seats and burgle their mouse holes. And by the time everyone gets home and discovers they've been robbed, the circus has packed up and gone!"

"How shocking!" Nutmeg cried.

"But now the law's catching up with them," the Chief Constable said confidently. "The last set of burglaries they carried out were on Apple Farm, less than two miles away! So the police station down there sent me a telegram, warning me they might be coming this way!"

"Why haven't you arrested them?" Tumtum asked.

"Oh, believe me, I've tried!" the Chief Constable said. "When I heard they'd arrived in the meadow, I rushed straight there. But they must have gotten wind I was coming, because by the time I reached their campsite, they were gone! But I saw one of their posters tacked to a fern stem. And imagine my surprise when I saw that General Marchmouse was topping the bill!"

"But it must be a mistake!" Mrs. Marchmouse wept. "Why would he want to get mixed up with wicked mice like that?"

Tumtum and Nutmeg agreed that there must have been a terrible misunderstanding.

"Now listen here, Chief Constable," Tumtum said firmly. "We spoke to the General this morning, and he had *no idea* that Mr. Goldtail was a crook. He thought the circus was real, just like we did. And he only joined it for a bit of fun. He'll change his mind soon enough when he finds out they're burglars! I wouldn't be surprised if he doesn't try to arrest them himself!"

But the Chief Constable did not look convinced. "I have

been reliably informed, Mr. Nutmouse," he said, "that the circus is traveling in a stolen bus. And last night, near the hour of seven o'clock, the General was seen driving that *very same* bus at considerable speed across the garden of Rose Cottage."

The Chief Constable paused to allow this dramatic information to sink in. "I have three different field mice who will back up that version of events," he added solemnly. "So it would appear, Mr. Nutmouse, that the General has not only teamed up with Mr. Goldtail and his gang . . . he has supplied them with a getaway vehicle!"

Tumtum and Nutmeg looked horrified.

"But Chief Constable," Tumtum said, "this is quite unfair. It's true that the General took the bus last night—but he only meant to *borrow* it. He's borrowed the children's toys before, but he's always given them back! And he wouldn't have driven anywhere near that wicked circus-that-isn't-a-circus if he knew what sort of mice they were."

The Chief Constable hesitated. He agreed that borrowing was not the same as stealing. Even so, the General had a lot to answer for. "He must have found out by now that Mr. Goldtail and his mice are crooks—they've been on the run all after-noon," he said. "So why hasn't he come home?"

"They're probably holding him against his will," Tumtum said. "They'll be frightened that if they let him go, he'll come and tell you where they are!"

"That's right!" Nutmeg cried. "And I bet they think that when they reach London, his famous name will attract more visitors to their shows!"

"Holding him against his will! Oh, how awful!" Mrs. Marchmouse sobbed.

But the Chief Constable was still not convinced. "So long as the General is on the run, he remains under suspicion," he said firmly. "The important thing is to catch these crimi-nals—*then* we can decide whether the General's guilty or not. But I've already searched the village high and low!"

"But *we* know which way they went!" Nutmeg said. "The children left us a letter just before you arrived saying that they saw them disappearing down the alley beside the village shop!"

"The village shop!" the Chief Constable exclaimed. "Well,

well, I can guess why they've gone there. They'll be doing a break-in, I bet!"

Everyone gasped. "But surely they're not planning to rob Mrs. Paterson!" Nutmeg said in horror.

"Oh, yes! Believe me, Mrs. Nutmouse, that is just the sort of mice they are!" the Chief Constable replied. "They'll dig into her ice-cream cakes, and steal all the gold coins from her cash register! COME ON! Let's go and catch them red-handed!"

Chapter Eleven

The Chief Constable was longing to make an arrest. Mr. Goldtail was one of the most wanted mice in the country, so it was thrilling to think he might catch up with him at last.

"I shall set off at once!" he said gleefully, bounding from the door.

"But you'll never catch them on foot!" Tumtum cried. "Not if they're in a bus. You must take Arthur's police car!"

"*A police car!*" the Chief Constable exclaimed. "Does he have one? A *real* one, with a siren and an accelerator and a flashing light?"

"He certainly does," Tumtum replied. "It's parked just outside in the kitchen, behind the vegetable rack. I saw it there this afternoon. And it's a splendid car, I can tell you!

Nutmeg and I had a test-drive in it one night, and it went at a speedy pace!"

"That's right!" Nutmeg said excitedly. "We'll come with you—there's plenty of room for us all to ride inside."

The Chief Constable's face glowed. He had always longed for a police car of his own, but the Mouse Police Force only provided cars for officers in the city.

"Are you sure the children won't mind us using it?" he asked anxiously.

"Of course not," Tumtum replied. "We'll leave them a note, explaining that we've borrowed it in order to arrest Mr. Goldtail and rescue their bus."

Everyone agreed that it was an excellent plan.

"We must hurry," Nutmeg said anxiously. "Mr. Mildew usually takes out the trash after dinner. When he opens the back door, that will be our chance to drive the car outside!"

Nutmeg went to the living room and hastily scribbled a letter to the children. Then she tucked it into her pocket, and they all hurried out of Nutmouse Hall.

The kitchen was deserted, but a light had been left on, and

when they peered across the floor, they could just see the police car's yellow hood shining behind the vegetable rack.

They crept over to it. And when the Chief Constable saw the car, he gave a whoop of joy. It was longer and sleeker than he could ever have dreamed, and it had a big blue siren and a walkie-talkie!

He tugged open the driver's door, his heart pounding with glee. But then he suddenly jumped back in fear.

"Who's *that*?" he cried, pointing to the backseat.

Tumtum and Nutmeg laughed. "Oh, that's nothing to be frightened of! It's only Arthur's toy policeman!" Nutmeg said.

"Goodness me!" the Chief Constable said, for it looked very fierce. It was sitting in the backseat, holding a pistol and a pair of handcuffs.

"He's a mechanical toy," Tumtum explained. "When you wind him up, he marches along, shouting, 'Stop where you are, you are under arrest!' He makes Lucy's dolls tremble!"

"Let's take him with us," the Chief Constable said eagerly. "He'll frighten the wits out of Mr. Goldtail!"

"Where shall I leave the letter?" Nutmeg wondered. "I haven't time to go and deliver it to the attic."

"I've an idea," Tumtum said. "Why don't you leave it on Arthur's shoe? He always wears that pair that's on the doormat, so he'll find your letter when he puts them on in the morning."

"Good idea," Nutmeg said, and she ran over and tucked the letter into one of the laces.

Then they all jumped into the car. Tumtum sat in the front next to the Chief Constable, and Nutmeg and Mrs. Marchmouse squeezed into the back, on either side of the toy policeman.

They waited silently. They didn't dare start the engine, for fear someone might hear.

Eventually, they felt the whole floor tremble, and when they peered up out of the windshield they saw Mr. Mildew striding in. They all sat still as stone, watching as he switched on the kettle, then started clattering some dishes in the sink.

Finally, he stooped down and heaved the bag out of the trash can, then turned and pulled open the back door.

"Get ready!" Tumtum whispered.

The Chief Constable turned the key in the ignition, and the engine gave a gentle purr. They waited until Mr. Mildew had walked a little way down the path, toward where the cans were kept.

"Quick, go now!" Tumtum cried.

The Chief Constable pressed forward the gearshift, and the car glided across the kitchen floor and slipped out the doorway.

Mr. Mildew was heaving the trash into the cans as the toy car slunk past behind him. The Chief Constable drove slowly, fearful of making too much noise. It was only when they had swung out onto the lane that he dared to slam down his paw on the accelerator. Then off they sped, with their siren wailing.

The village was dark and still. The fires had been lit, and the televisions were flickering behind the curtains. No one saw the little toy car racing toward the village shop.

Nutmeg and Mrs. Marchmouse clung to their seats, for the Chief Constable was driving at a terrifying speed. When they reached the shop, he screeched to a halt. The shutters had been

pulled down, and there was a CLOSED sign hanging from the door. But there was light coming from the windows of Mrs. Paterson's flat, on the floor above.

"Try down the alley," Tumtum said. "That's where the children thought they saw them."

The Chief Constable sped around the side of the building. They came into a forest of crates and trash cans towering ghoulishly in the moonlight.

They raced all around them, skidding and looping. Round and round they went, searching behind every box and bin. But the circus bus wasn't there.

The Chief Constable thumped the steering wheel in frustration. "We're too late!" he said furiously. "The scoundrels have gone!"

"They must have left for London already!" Mrs. Marchmouse sobbed. "Now I'll never get my husband back!"

Everyone felt wretched. But then Tumtum noticed something in the headlights. "Hey, look over there!" he said, pointing ahead. "That's where they must have broken in!"

They all peered through the windshield. The Chief

Constable had stopped the car at the bottom of the alley, beside a tall thin door leading into the back of the shop. The door was made of glass, and the bottom pane had been broken. The hole was just big enough for a mouse to get through!

"Ha! So they *did* break in, as I suspected!" the Chief Constable said. "I shall go inside and take some paw prints."

He grabbed his baton, then turned to Tumtum and flung him a pair of handcuffs. "Come on, Mr. Nutmouse!" he said. "Let's see if they're still in there."

Tumtum clambered nervously out of the car.

"Don't forget the toy policeman!" Nutmeg said, for she was frightened Mr. Goldtail and his gang might become aggressive.

Nutmeg and Mrs. Marchmouse jumped out and helped Tumtum drag the policeman from the backseat.

They turned him around, so that he was facing the door of the shop, then Tumtum took hold of the silver knob on his back, and wound it in three full circles.

The policeman shuffled his feet, and his eyes rolled in their sockets. "Stick 'em up!" he cried. "You are under arrest!"

Then the policeman marched forward and smashed through the broken pane of glass. Tumtum and the Chief Constable scrambled after him, while Nutmeg and Mrs. Marchmouse waited anxiously outside.

It was very dark inside the shop. All they could see was the outline of a pile of pie shells towering in front of them. They crept around it, with the policeman marching alongside. Then they took out their flashlights and shone them around.

They both gasped, for the shop was in a terrible mess. On every shelf, packets of chips and cookies and jam tarts had been gnawed open, and candy wrappers nibbled to shreds.

"Come out, wherever you are!" the Chief Constable shouted. "Do you hear me? This is the police! Come out with your paws above your heads!"

"Stick 'em up! Stick 'em up!" the toy policeman squeaked.

There was not a sound.

"Drat, they've gone!" the Chief Constable muttered. "But what's happened to the General?" Tumtum said. "He would *never* have assisted in a burglary like this. They must be holding him against—"

But then the Chief Constable shone his flashlight onto the cake shelf and gave his whistle a piercing shrill. And when Tumtum spun around he saw the most astonishing sight.

It was General Marchmouse, trapped in a chocolate éclair! What a sight he looked. He was sunk in it up to his waist, and it was so sticky he couldn't get out. His gold suit was covered in goop, and he looked very embarrassed.

"Come down with your paws above your head! You are under arrest!" the Chief Constable shouted.

"I can't come down. Can't you see? I'm stuck!" the General replied, blushing furiously. "Now get me out of here. The burglary had nothing to do with me. All I did was borrow a silly old bus!"

Hearing the commotion, Nutmeg and Mrs. Marchmouse came running in from outside. When she saw her husband, Mrs. Marchmouse let out a cry of relief—but she could see there was a lot of explaining to be done.

"Get me down! My boots are soaking!" the General cried.

The Chief Constable stared at him coldly. The evidence looked very bad. But Tumtum still could not believe that the

General had been part of the burglary. "Let's get him down and hear what he has to say," he said.

He and the Chief Constable climbed up to the shelf, and between the two of them they pulled General Marchmouse out of the éclair and helped him down to the floor.

"All right, then, General," the Chief Constable said gruffly. "What's been going on here?"

"Oh, Constable!" the General spluttered. "I have been treated most hideously! Mr. Goldtail and his gang forced me to come in here and help them load up their wagons with stolen sherbet. And they said that if I didn't do as I was told, they'd throw me in the licorice jar and leave me there for Mrs. Paterson to find in the morning!

"But I escaped by bravely burrowing inside a chocolate sponge cake. They hunted high and low for me, but I was too well hidden. You should have heard them cursing! Anyway, I waited until they'd left, then I crawled out, and I was going to come straight to the station to tell you what had happened. But I didn't have a flashlight, and I stepped into an éclair! And now everyone will think I'm a criminal, just because I borrowed

Arthur's bus. . . . But I only borrowed it for a bit of fun! And it was my birthday, after all."

Tumtum and Nutmeg and Mrs. Marchmouse could tell at once that the General was telling the truth. He had been very wrong to take Arthur's bus. But it was clear that he had been horribly fooled by Mr. Goldtail and his gang.

They looked anxiously at the Chief Constable, wondering if he would believe the General, too.

The Chief Constable was scratching his chin.

"You have been a very foolish mouse, General," he said eventually. "And next time you borrow a toy bus without permission I might not be so lenient! But for now you have been punished enough."

Tumtum and Nutmeg and Mrs. Marchmouse looked very relieved. "Oh, thank goodness!" Mrs. Marchmouse cried.

The General was relieved, too, for he had feared he might go to prison. But he was too proud to let his gratitude show.

"Well, I'm glad that's cleared up, then," he said briskly. "Now I can go home and have a hot bath."

"Not so fast, General," the Chief Constable said. "We need

to catch this gang, and FAST. Have you any idea where they went?"

"Well, yes. Now that you mention it, I think I do," the General replied.

"*Where?*" the Chief Constable asked excitedly.

Everyone looked at the General, waiting to hear what he would say. There was an atmosphere of great suspense. The General knew his news would come as a big surprise, and he started to feel rather important.

He brushed a blob of cream off his jacket, trying to look more as a general should.

"I regret to inform you," he announced dramatically, "but Mr. Goldtail and his gang have GONE TO BURGLE NUT-MOUSE HALL!"

Chapter Twelve

Everyone gasped.

"They've gone to burgle Nutmouse Hall!" Nutmeg cried.
"Oh, my, oh, my! They'll take our family portraits and our silver
candlesticks and our crystal glasses and our gold goblets and our
stuffed ladybugs and our rare books and . . . and . . ."

She clapped a paw to her mouth, too upset to continue.
Tumtum was distraught, too.

"Have they gone straight there?" he asked anxiously.

"No, the burglary's planned for tomorrow evening," the
General replied. "They've got to do some repair work first on
their wagons."

"Where are they tonight?" the Chief Constable asked.

"I couldn't really tell," the General replied. "Mr. Goldtail just said, 'Okay, let's camp under the rhubarb again!' and the clowns all agreed with him. Well, I didn't want to ask which rhubarb he meant, for fear of making him suspicious. Mr. Goldtail doesn't like being asked questions, you know."

"*The rhubarb*," the Chief Constable said thoughtfully. "Hmm, that could be anywhere. The village is full of rhubarb — just about every garden has got some. We could search all night and still not find them. We shall have to wait until tomorrow and catch them when they go to Nutmouse Hall."

Everyone agreed that it was pointless looking for the thieves now. "You must all spend the night with us," Nutmeg said firmly. "We've plenty of food, and we can all drive home together in Arthur's car."

"An excellent plan," the Chief Constable said. He lived alone, and he would be glad to have some company after such an eventful day. The General was pleased, too, for he didn't want to walk home knowing that Mr. Goldtail and his gang were still lurking in the village.

Nutmeg was glad to have guests to fuss over. It would take her mind off all the worrying things that had been taking place.

"Now, let's see," she said. "I shall put the General and Mrs. Marchmouse in the green bedroom . . . and the Chief Constable can have the blue bedroom, next to ours and . . ."

THUD!

Everyone froze.

THUD!

It was footsteps on the stairs! Then suddenly the door behind the shop counter burst open, and on went the lights!

"It's Mrs. Paterson," Tumtum cried. "*RUN!*"

They all fled to the door, and scrambled out through the broken pane of glass. They could hear Mrs. Paterson shrieking behind them. "Help! Help! I've been robbed!"

"Quick! Get into the car!" the Chief Constable shouted. They all jumped in, the General squeezing between Nutmeg and Mrs. Marchmouse in the back. Then the Chief Constable revved the engine, and off they sped, tearing back toward Rose Cottage.

The passengers all peered anxiously through the back windshield, frightened that Mrs. Paterson might come running after them.

"Do you think she saw us?" Nutmeg asked.

"I don't think so," Tumtum said. "We were very quick—we got out before she came in. And if she had seen us, she'd have come chasing after us by now."

"Doesn't this feel strange?" Mrs. Marchmouse said nervously. "Here we are, running away as though *we* were the criminals!"

"Hmm, isn't it a pity humans can't hear what mice say?" Tumtum said. "Otherwise we could have told Mrs. Paterson the whole story."

As it was, they all felt very glad they had gotten away. For if Mrs. Paterson had seen them, she would have suspected *they* had made all the mess—and they dreaded to think what she would have done if she had caught them.

The Chief Constable did not stop until he reached Rose Cottage. Then he parked beside the garden door, and they crept back inside.

It was late now, and the Mildews had all gone up to bed. The kitchen was very dark. Tumtum led them across the floor with his flashlight and let them into Nutmouse Hall.

They were all very shaken. But they felt better when they had had something to eat. Nutmeg warmed up the cockroach soufflé she had made for the General's birthday feast, and everyone ate hungrily. They washed it down with a strong pot of tea. Nutmeg did not offer them the birthday cake or the strawberry pie or the jellied flies—the mood was not festive enough for party food.

Everyone was very worried about tomorrow. The General was the only one who felt talkative, and the others were too tired to stop him.

"So there we were, speeding off through the bracken yesterday afternoon—and I had no idea what was going on!" he recalled. "Then the scoundrels came clean and told me they were thieves! Well, you should have seen the fuss I made! 'Stop the bus!' I cried, 'I am going to get out and report you to the police!' But believe me, if they didn't tie me up and gag me . . . But I struggled as bravely as I could. . . ."

On and on he boasted, though his memory was a little flawed. ". . . So imagine my relief when you finally arrived in the village shop," he said. "Though I can tell you, I got quite a fright when I saw that toy police—"

"*Oh, no!*" Nutmeg cried, dropping her fork with a clang. "The toy policeman! *WE FORGOT HIM!*"

Everyone groaned. It was true. In their haste to escape, they had left the policeman behind in the shop.

"Oh, well, there's nothing we can do about it now," the Chief Constable said. "And the shop's in such a mess. I'll bet Mrs. Paterson doesn't even notice the policeman tonight. I'll drive over there first thing in the morning and see if I can sneak in and get him back."

"All right," Tumtum said, feeling a little reassured. "Well, I suggest we get some sleep. We've a big day tomorrow, and we shall need to be alert."

Everyone agreed, for they were very tired. Even the General was fading. His eyes had become heavy, and he suddenly hadn't the strength to go on boasting.

Tumtum bolted the front door and locked all the windows. Then Nutmeg showed the guests to their rooms, and they were asleep almost as soon as they had pulled up the covers.

Tumtum fell asleep right away, too. But Nutmeg lay awake worrying. How she wished they hadn't left the toy policeman behind. She felt sure it would lead to trouble. And the next morning her fears were confirmed.

For as the mice breakfasted in Nutmouse Hall, they heard the Mildews' doorbell ringing, so loud it made them jump.

"I wonder who it could be, calling so early on a Sunday morning," Nutmeg said.

They stopped eating and sat with their ears twitching. They heard Mr. Mildew's footsteps coming down the cottage stairs, then a worried voice sounding from the hall.

Nutmeg recognized it at once. "It's Mrs. Paterson!" she said in alarm.

They tried to hear what she was saying. But then Mr. Mildew started talking at the same time, and the conversation became very difficult to follow.

Tumtum and Nutmeg left their guests at the table and crept out under the dresser to try and discover what was going on.

And when they peeked into the kitchen, they saw Mrs. Paterson standing by the table, clearly very upset. She had a wool coat pulled over her dressing gown, and her face was drained of color. In her hand was the toy policeman.

"I was burgled last night, Mr. Mildew. *Burgled!*" she cried. "Oh, you should see the mess! They gobbled my cupcakes and burst open the cornflakes! And they stole *all the sherbet*!"

Arthur and Lucy were standing by the kitchen table, looking very uncomfortable.

"Yes, Mrs. Paterson, I understand that, and I am very sorry," interrupted Mr. Mildew. He was wearing his robe, too, and looked a little bleary. "But I still don't quite see what you think we've got to do with it."

"I told you!" Mrs. Paterson cried, waving the toy policeman. "I found *this* lying on the floor, right next to my pie shells. And look, it's got Arthur's name scratched on it!" She turned the toy over, pointing to where the letters *A. Mildew* had been scratched out on the policeman's plastic back. "Your children must have crept in and dropped this while they were taking my sweets! They'd have had plenty of opportunity—I didn't lock the back door until after nine."

"But we didn't!" Arthur cried.

"Then how do you explain this?" Mrs. Paterson said, turning to wave the toy policeman at him. "And what were you doing snooping around my trash cans this afternoon?"

"Well, do you know anything about this?" Mr. Mildew asked the children. Nutmeg thought his tone was rather sharp.

Arthur and Lucy looked very awkward. They had found Nutmeg's letter earlier that morning, so they knew it was her who had taken their police car to the village shop.

But they didn't know what had happened next. Lucy tried to explain everything — or at least as much as they knew — even though she felt sure the grown-ups wouldn't believe her.

"On Friday night, Arthur's bus disappeared, and we looked everywhere, all over the garden, and all around the house, but we couldn't find it," she began. "Then yesterday afternoon, we thought we saw it driving down the little alley beside the shop....And when we told our friend Nutmeg about it, she said she had discovered that the bus had been stolen by a gang of thieves, led by a wicked villain called Mr. Goldtail....So she borrowed our toy police car and the policeman to go and arrest them, before they could do any more harm....But the bus hasn't come back, and the toy policeman got left behind in your shop.... So Nutmeg's whole mission must have gone wrong!"

"Oh, what nonsense!" Mrs. Paterson cried. "They're making fun of me!"

The children looked at their father anxiously. Surely *he* would believe them. But Mr. Mildew clearly thought it was ridiculous, too.

"This is no time for silly stories. You can see that Mrs. Paterson is upset," he said.

"But it's not a silly story. It's what happened! *It is!*" Arthur cried.

"That's enough," Mr. Mildew said sharply. "I want you both to go up to your room. And you can stay there until you're ready to discuss this sensibly — and tell me how the toy policeman really ended up in Mrs. Paterson's shop."

"Oh, Tumtum, what a terrible mess!" Nutmeg whispered. "The children are in trouble, and it's all our fault!"

"Come on," Tumtum said. "Let's hurry back to Nutmouse Hall and tell the others what's happened. We *must* catch Mr.

Goldtail tonight, otherwise the children will never clear their names!"

The mice hurried back into the broom cupboard. As they let themselves through the front gates, they heard Arthur and Lucy trudging upstairs.

The children felt miserable. It was terrible of their father to take Mrs. Paterson's side.

"Whatever could have happened?" Arthur said, flinging himself down on his bed. "I wonder why Nutmeg left the toy policeman behind."

"Well, she's brought the police car back," Lucy said. "I saw it outside on the doorstep. So at least Nutmeg must have gotten home safely. But I suppose Mr. Goldtail must have gotten away in the bus."

"Well, unless Nutmeg catches him, Mrs. Paterson will *never* believe our story," Arthur said wretchedly. "Oh, dear. Now *everything's* gone wrong!"

Chapter Thirteen

Tumtum and Nutmeg hurried back to Nutmouse Hall to tell the others what had happened.

"Don't you worry, Mrs. Nutmouse," the Chief Constable said confidently. "I'll catch Mr. Goldtail and his good-for-nothing gang, and then the truth will come out!"

He took a pair of handcuffs from his bag and started polishing them with his napkin. It had been several weeks since he made an arrest, and he was very excited.

"I'll teach them what happens to mice who break into village shops!" he said fiercely. "I'll catch them red-handed when they turn up tonight!"

The Chief Constable checked his watch impatiently, wishing the evening would come.

But the General looked grumpy. He wanted to be in charge.

This is my *adventure*, he thought. *And that silly old policemouse has NO RIGHT to tell me what to do! I gave him the tip-off that Mr. Goldtail and his gang were coming to burgle Nutmouse Hall, so I should be allowed to arrest them all by myself!*

The General knew that the arrest of Mr. Goldtail would be reported in the *Mouse Times*, and he couldn't bear to think that the Chief Constable would get all the credit for it. *It should be* my *victory!* he thought.

Then suddenly he remembered something very important.

"You had better leave this arrest to me, Chief Constable," he said grandly. "It will need a general's expertise!"

The Chief Constable looked at him in astonishment. "Leave it to *you*? But you're not a policemouse!" he said indignantly. "This is my job!"

The General smiled smugly. "The problem is, I don't think you can handle this yourself," he said, "seeing as Mr. Goldtail and his gang have sherbet guns——"

"*SHERBET GUNS?*" the Chief Constable spluttered. "Are you sure?"

The General nodded. "They've got one each. They stocked up on powder when they broke into the village shop."

There was a stunned silence. They all knew how dangerous sherbet guns were.

"Oh, help us! What shall we do?" Mrs. Marchmouse cried.

"There is no need to fear," the General said bossily. "The Royal Mouse Army trained all its officers to deal with sherbet attacks. We shall be quite all right, so long as everyone does exactly what I say."

The Chief Constable looked furious. He could see that the General was trying to take command. "You are not telling me what to do!" he snapped.

"Oh, let's not bicker," Nutmeg said. "We must protect ourselves. Chief Constable, do you have any other policemice who might help us?"

The Chief Constable shook his head. "We're understaffed just now," he said. "My lieutenant's away on vacation, and my sergeant's out with the flu. We could summon reinforcements from the police station in town—they've got a dozen officers on duty. But it would take them two days to get here."

"And what about the Royal Mouse Army?" Tumtum asked.

"No good. The troops are down on Apple Farm, seeing off some rats from the hay barn," the General replied briskly. "We'd never be able to fetch them back in time."

"Well, then," Tumtum said anxiously. "It looks as if we're on our own."

"Indeed we are, but we can still win!" the General said with relish. "We must march up to the attic and borrow ammunition from Arthur's toy soldiers—guns, grenades, swords, all the weapons we can carry."

"All right, I'm sure Arthur won't mind," Tumtum said. "But there's to be no marching in Rose Cottage—when we go up to the attic, we shall have to sneak."

The General nodded. "And we shall need a toy tank, too," he said. "Have the children got one?"

"Yes!" Nutmeg cried. "There's one in the living room. It's been there for weeks, tucked behind the sofa. I think Arthur must have forgotten about it."

"Good," the Chief Constable said, trying to take control. "General, Tumtum — you can come with me after dark and drive it back to the kitchen. We can park behind the laundry basket, or somewhere the children won't see us. Then we'll take Mr. Goldtail and his gang by surprise as they sneak in under the garden door! Just think what a scare they'll get when they see an army tank zooming toward them!"

The Chief Constable was thrilled. Guns and tanks sounded much more fun than a baton.

"I'll drive!" the General said grandly. And the Chief Constable was too excited to argue.

So the plan was agreed upon, and the rest of the day was spent preparing for the attack. They intended to catch the thieves before they reached Nutmouse Hall, but even so, they didn't take any chances. Tumtum and the Chief Constable went all around the house, boarding up the doors and windows, and the General set trip wires in the living room and the butler's pantry. Nutmeg and Mrs. Marchmouse stuffed the chimneys with blankets and tapestries, in case they tried to drop sherbet bombs from the roof.

Lunch was a hasty affair, for they were all too nervous to eat much. As soon as the plates had been cleared, Tumtum and the General and the Chief Constable crept out into the kitchen and up to the attic to raid the toy barracks. They reappeared an hour later with toy guns and grenades thrust in their belts.

"Goodness, you do look fierce!" Mrs. Marchmouse said.

"Did you see the children?" Nutmeg asked anxiously.

"Yes, they are in the living room," Tumtum said. "So we won't be able to collect the tank yet — they'd be sure to see us. We shall have to wait until they go upstairs."

"And what about Mr. Mildew?" she asked.

"Oh, he's all right," Tumtum said. "He's shut away in his study. We could hear his typewriter keys clacking."

Nutmeg made a pot of tea, and they sat anxiously in the kitchen. Tumtum and the Chief Constable took turns to creep out of the broom cupboard to see if it was safe to go into the living room. But the children were milling about all afternoon.

Then Mr. Mildew came into the kitchen to make dinner. It

wasn't until nearly eight o'clock that the Chief Constable was able to report that everyone had gone upstairs.

"Come on, we must hurry up and get the tank!" the General said. "Mr. Goldtail might arrive at any minute!"

Tumtum and the General leaped up and followed him to the door.

"Oh, do be careful!" Nutmeg cried.

The mice crept out into the Mildews' kitchen. It was very dark. The lights were out, and they could see stars shining at the window.

Tumtum pulled out his flashlight and they set off toward the living room. But when they were halfway across the kitchen, they heard a sudden roar. And then there was a screech of brakes. The noise was coming from outside. And when they spun around they saw lights flashing under the back door!

"*It's them!*" the General hissed. "Quick! Hide!"

The mice darted behind a pile of telephone books. And when they peeked out, they saw four flashlights being shone around the floor.

"Keep back! Don't let them see us!" Tumtum hissed.

"Ha! Ha!" cried a nasty voice. "Let's go and blast Mr. Nutmouse with sherbet and steal all his valuables! We'll take his chandeliers and his silver candlesticks and his gold tapestries and his fancy soup bowls! Ha! Ha! We'll be as rich as can be!"

"That's Goldtail!" the General said grimly.

"Fetch the tank!" Tumtum whispered. "And HURRY!"

Chapter Fourteen

Mr. Goldtail and his gang had never been to Rose Cottage before. They shone their flashlights around the kitchen, searching for the gates to Nutmouse Hall.

"Where are they?" Mr. Goldtail snarled. He grabbed a piece of paper from his pocket and read the address again.

Nutmouse Hall

The Broom Cupboard

The Kitchen

Rose Cottage

"Well, this is the kitchen. But I don't see a broom cupboard!"

"It must be here somewhere," said Mr. Moody. "Nutmouse Hall is the most famous mouse hole in the whole county."

"We've got to find it. I'm not leaving here empty-handed!" said Mr. Merry.

They continued to hunt, shining their flashlights around the baseboard.

Meanwhile, Tumtum, the General, and the Chief Constable crept out from behind the telephone books and tiptoed into the living room.

Tumtum was glad that Nutmouse Hall was so well hidden. But he knew the thieves would find it in the end.

"We must hurry," he whispered. "Imagine how frightened Nutmeg and Mrs. Marchmouse will be if they break in!"

One of the living room lamps had been left on, and they could just see the tank peeking out from behind the sofa.

They raced over to it. And when the General saw it up close, his eyes glinted. He had driven tanks in the Royal Mouse Army, but never one like this. It was as big as a cake pan and its gun was the size of a rattle!

Tumtum and the Chief Constable had never operated a

tank before, but the General reckoned he knew just what to do. He scrambled up the ladder and wrenched open the hatch. "Quick! Get in!" he shouted. Tumtum and the Chief Constable heaved themselves up after him. Then they all slid into the tank.

It was very dark inside. The General fumbled his way to the driver's seat and started blindly running his paws across the controls. There were dozens of buttons and switches. He felt about, pulling this one and that, until eventually a pale green light came on.

They all looked about them in surprise. They were in a small, gray cabin, and every wall was covered with switches, saying things like FIRE! and FULL SPEED! Everything was mouse-sized.

"Come on, General, let's go!" Tumtum cried, fearing that Nutmeg and Mrs. Marchmouse might be in danger.

The General slammed some more knobs on his dashboard, but he could not get the tank to start.

"Drat!" he said.

Then he saw a big red button labeled GO! He gave it a push. Finally the engine shuddered to life, and the tank ground forward across the carpet.

Vroom!

Grrrrrrr!

Crrrrrrrrrunch!

The tank's huge rubber tracks bulldozed across a game of Snakes and Ladders, and over a discarded candy wrapper.

Tumtum and the Chief Constable crouched behind the General, peering through the bulletproof windshield.

"Straight ahead, left around the coffee-table leg, then hard right into the hall!" Tumtum said.

The General drove faster and faster, until finally they clattered into the kitchen.

"There they are!" the Chief Constable cried.

Mr. Goldtail and his mice were standing at the edge of the dresser, staring at the tank in astonishment.

"By Jove!" Tumtum said. "We got here just in time."

"Let's get them!" the General cried. The Chief Constable swung around the rotating gun until it was pointing toward the dresser.

They could see the thieves in the headlights, frantically loading their sherbet guns.

"Ha! They can't harm us with their silly old sherbet!" The General snorted. "We're in a sealed tank!"

But he spoke too soon. For at that moment the Chief Constable — who was quite senseless with excitement — leaped up and pushed open the hatch.

"Stop! What are you doing?" Tumtum cried. He tried to pull him back, but the Chief Constable was too fast. And before Tumtum could stop him, he poked out his head through the hatch and blew his whistle.

"Paws up! YOU ARE UNDER ARREST!" he cried.

"Get back, you fool!" the General shouted. But it was too late. For suddenly there was a deafening volley of gunfire, and a great blizzard of sherbet tore into the air. There was sherbet everywhere, swirling all around them. Soon the whole windshield was smothered.

The Chief Constable stumbled back inside the tank, gasping for air. He was covered with yellow powder from his ears to his chin, and his throat was burning.

The General leaped up and yanked down the hatch. But before he could shut it properly, a cloud of sherbet floated

down into the tank, turning the air a thick, dusty yellow. The General had his handkerchief clapped to his mouth. But Tumtum was taken unawares and started coughing violently.

"W-w-water!" he spluttered, collapsing beside the Chief Constable on the floor.

The General looked at them in horror. Tumtum's face had turned purple, and the Chief Constable was bright green.

The General knew that if he did not find them water, they would choke. He staggered about the tank, yanking the controls. "Water, water! A tank *must* have water!" he cried. But there was not a drop to be found.

Then, through the haze of sherbet, he saw a lever on the wall, with a sign reading PULL ONLY IN THE EVENT OF AN EXTREME EMERGENCY.

"This is an extreme emergency," he decided. And he grabbed it with both paws and wrenched it down.

Then something wonderful happened.

Three green sprinklers dropped down from the ceiling, and fierce jets of water whooshed out, washing all the sherbet away.

Tumtum and the Chief Constable struggled to their knees, raising their faces to the cool spray. The water stopped their coughing, but it was some moments before they could speak.

"Gracious! I thought we were done for!" Tumtum said weakly.

"Bah! I've survived far worse attacks than that in my Royal Mouse Army days," the General boasted. "Why, you should have seen the Battle of the——"

But he was cut short by another deafening rattle of gunfire, and then the next round of sherbet smacked against the tank.

"I'll show them not to fire their sherbet guns at me!" the General fumed. He flung himself back into the driver's seat and flicked on the windshield wipers. Then he rammed his paw on the accelerator. The tank surged forward, clattering across the tiled floor.

The General could see Mr. Goldtail and his accomplices in the headlights, and he drove straight at them.

The thieves hastily reloaded their guns, hoping to force the tank back with another blast of sherbet.

But while they were fumbling with their ammunition, the General swung around the tank's machine gun until it was pointing straight at them and thumped down his paw on a big red button marked FIRE!

All at once, a ferocious jet of water tore out of the nozzle of the gun and swept the thieves off their feet, rolling them across the floor like tennis balls.

"Ha! Ha! That'll teach you to fire sherbet at us!" the General cried, crashing the tank toward them.

The thieves squealed in terror. There was water blasting all around them, and they could see the hazy outline of the toy tank, grinding closer and closer.

The General whooped with glee. "Fire! Fire!" he cried, hitting the red button.

The jet of water shot out faster and faster from the toy gun, washing Mr. Goldtail and his mice across the floor.

"Stop!" the Chief Constable cried. "Stop firing and let me arrest them!"

But the thieves were being swept closer and closer to the

garden door, until eventually the water sprayed them onto the doormat. They finally managed to scramble upright, clinging to the rough matting.

"Run!" Mr. Goldtail cried. "Run for your lives!" And with the water blasting at their heels, all four mice fled under the door in terror.

"Stop them!" the Chief Constable cried.

He darted for the hatch and pushed it open. "Come on, let's get them!" he shouted.

He was about to fling himself outside, but then suddenly there was a bright glare. The kitchen lights had come on. The Chief Constable ducked back inside. And when the mice peered out through the sherbet-smudged windshield, they got the most terrible fright.

Arthur and Lucy had appeared!

Chapter Fifteen

A rthur stared at the toy tank in astonishment.

"How did that get here?" he said. "I left it in the living room!"

Tumtum and the others looked at each other in horror. If Arthur opened the hatch of the tank, he was sure to find them.

But then suddenly the children heard a strange noise, like an engine revving, and the next moment they saw a light flicker under the garden door.

They ran and opened it, and when they saw outside they both gasped. For there was the toy bus, fleeing down the garden path!

"Catch it!" Arthur cried, stumbling onto the doorstep.

Lucy grabbed the flashlight from the shelf and ran after

him. And as the children rushed into the garden, Tumtum and the others slithered out of the tank and crept under the dresser.

They could hear the toy bus rattling down the path and the children chasing after it.

"Stop!" Arthur shouted—but the bus was going at full speed. Mr. Goldtail was driving it as fast as it would go, speeding toward a gap in the hedge.

"Faster! Faster!" the clowns all shrieked, looking through the back windshield in terror. But the children were catching up with them. And just as the bus was swerving onto the lawn, a huge pink hand appeared in the headlights and clamped over the windshield.

And when the mice looked in the side-view mirror, they saw Arthur unhitching the wagons from behind them. Then the bus was lifted high into the air.

They pushed on the doors, desperately trying to get out. But Arthur had shut the catches on the outside. They were locked in.

"Put us down!" cried Mr. Moody.

"*Let us go!*" roared Mr. Goldtail, thumping the window with rage.

The children shone their flashlight into the bus in amazement.

"It must be Mr. Goldtail and his gang!" Lucy said. "And look, they're even wearing circus clothes!"

Arthur and Lucy had met some very strange mice before. Even so, they were rather astonished.

"Aren't they extraordinary!" Lucy said. "I say, do you think it was *them* who broke into the village shop? Nutmeg said they were very wicked—so breaking into a village shop is just the sort of thing they might do!"

"Let's look in the wagons," Arthur said. "If they did raid the shop, then we'll probably find some of the things they stole."

They carefully carried the bus and the wagons back into the kitchen and set them down on the table.

Tumtum and the others watched excitedly from under the dresser. Then Nutmeg and Mrs. Marchmouse heard the commotion and came running out, too.

"Oh, thank goodness, you're all right," Nutmeg cried, clutching Tumtum's paw. "But you're soaked! Whatever's been going on?"

"Shh!" Tumtum said, pointing into the kitchen. "The children have caught Goldtail and his gang! They all tried to drive off in the toy bus, but Arthur and Lucy chased after them, and now they've locked them in!"

"Oh, how clever of them!" Nutmeg said proudly.

They all crouched very still, watching as the children searched the wagons.

When they opened the door of the first wagon, a beetle bounced out.

"Ugh!" said Lucy, stepping back in fright. The beetle scuttled about the table, then hopped to the floor and disappeared under the sink.

The next wagon was full of tiny cardboard boxes, sealed with brown tape. Arthur took out his pocketknife and carefully slit them open. When the children saw what they contained, they both gasped.

For each box was full of the tiniest, most exquisite

treasures! Gold plates and jugs, and candlesticks, and paintings and tapestries — all mouse-sized.

"These must be all the things they stole," Lucy said, spreading everything out on the table. "We shall have to give them to Nutmeg. I'm sure she can find out which mice they belong to and give them all back!"

The next three wagons were furnished with camp beds and dressing tables, and wardrobes containing tiny suits of clothes. And the last wagon was full of tubes of sherbet!

"Look!" Arthur said. "This must be the sherbet that was stolen from the village shop!" He reached into the wagon and pulled out one of the tubes. It had scuffs at the end, where it had been nibbled.

"Goodness," he said. "Just think how astonished Mrs. Paterson will be when she learns that the robbers were mice!"

Mr. Goldtail and the clowns could not bear to see their loot being touched. They beat the windows so hard that the bus started to shake.

"They're going to get out!" Arthur said anxiously. "Quick, let's get Pa. He'll know what to do with them!"

Lucy ran upstairs. Mr. Mildew was in his study, struggling with his latest invention—a mechanical pencil that could write and spell all by itself. He hardly looked up when Lucy came in. "What is it?" he muttered. "Didn't I say you were meant to stay in your room?"

"Oh, yes, but please come downstairs—we've got something to show you," Lucy said. "Oh, please come and see! Then you'll understand everything!"

Mr. Mildew grudgingly followed her downstairs. But when he saw what was on the kitchen table, his eyes bulged in astonishment.

"They're the ones who robbed the shop," Lucy said. "Look! Here's all the sherbet they took!"

Mr. Mildew hardly knew what to say. "I had better call Mrs. Paterson," he muttered. He picked up the telephone and in a trembling voice told her to come at once.

She hurried to Rose Cottage, wondering what could have happened. And when she saw the mice scrambling about in the toy bus, and the stolen sherbet, she was too astonished to speak.

It took two strong cups of tea to stop her shaking.

"What are we going to do with them?" Mr. Mildew asked, fearing the children might want to keep them as pets.

"Oh, we don't want them at Rose Cottage," Lucy said. "They're too noisy!"

"Perhaps we could find a cage for them, and keep them in the garden shed, where we wouldn't have to look at them all the time," Arthur said.

"*A garden shed!* How dare you!" Mr. Goldtail shouted, not liking the sound of this one bit.

"Oh, stop squealing!" Arthur said, rapping the bus with a spoon.

"They would be a waste in a shed!" Mrs. Paterson said, gathering her wits again. "They are MUCH too good for that. If you don't want them, I shall sell them for you at the village shop! I'll put an advertisement in the local newspaper, and sell them to whoever offers the most money! I'm sure they'll fetch a *fortune*. Four mice dressed in funny clothes, breaking into my shop, and driving off in a toy bus....They must be the cleverest mice in the whole wide world! Why, I wouldn't be surprised if they don't become stars on television!"

"Oh, what a good idea!" Lucy cried.

"And I shall give all the money to you," Mrs. Paterson said kindly. "And you can keep the sherbet, too! It's the least you deserve. I am sorry about this morning, my dears. It was quite wrong of me to accuse you of stealing."

"Oh, don't worry about that," Lucy said. This morning seemed like ages ago — the children were just glad things had turned out so well.

"I'll take them with me now," Mrs. Paterson said, tucking the bus firmly under her arm. "I've a nice big cage back at home I can put them in — it's one my niece used to use for her ger-bils. Then I'll give your bus a good cleaning and bring it back to you in the morning, along with that poor old toy policeman, of course!"

Everyone laughed — and they didn't notice that the pris-oners had become quiet. "The television, just think of that!" Mr. Moody whispered. "We'll be famous!"

"And rich!" Mr. Mirth tittered.

"Sounds better than driving round in a flea-bitten old circus!" Mr. Merry said.

In his mind's eye, Mr. Goldtail suddenly had a delicious image of himself dressed in dark glasses, being chauffeured between television studios in the back of a sleek limousine — and he decided things hadn't turned out so badly after all.

Tumtum and the others all cheered when Mrs. Paterson carried the thieves away.

"I shall telephone the *Mouse Times* first thing in the morning and tell them the burglars have been caught, and that it was all thanks to me — er, us!" the General said boastfully.

"And tomorrow I shall go around to all the local mouse holes, returning the stolen goods to their rightful owners," the Chief Constable said happily. "Just think how pleased they shall all be!"

They hid under the dresser, waiting until Mr. Mildew and the children had gone upstairs. "I shall go up to the attic tonight and write the children a letter, congratulating them on everything they've done," Nutmeg said. "And I shall tell them to keep the wagons. They belong to Arthur's circus now!"

Everyone agreed. "And now let's have a celebration!" Tumtum said, leading them all back to Nutmouse Hall. "Have we anything to eat, dear?" he asked Nutmeg anxiously. "I feel quite hungry!"

"Why, of course!" Nutmeg replied. "We've got the General's birthday feast! The table's still set in the dining room, and the cake hasn't been touched!"

They all cheered. After such a grand adventure, a feast was just what they needed. Tumtum set two extra places for Mrs. Marchmouse and the Chief Constable, and soon they were all sitting down at the big oak table, feasting long into the night.

The birthday party was two days late, but all the better for it. The candles flickered and the cider flowed, and by the time Nutmeg carried in the cake, everyone had quite forgotten how badly the General had behaved.

They all sang "Happy Birthday" while tapping their silver teaspoons on the table.

The General glowed. What a wonderful adventure it had

turned out to be! And just think what a hero he would be when he told the *Mouse Times* how he had chased Mr. Goldtail from Nutmouse Hall in a toy tank!

"*For I'm a jolly good fellow!*" he sang, blowing out his candles with a single puff. And the mood was so merry, even the Chief Constable agreed that a better general there never was!

Arthur and Lucy climbed into bed in the attic, unaware of the festivities going on downstairs. They had left Nutmeg a letter on the chest of drawers, telling her everything that had happened—for they could not have guessed that she had been watching from beneath the dresser.

"Arthur," Lucy said hesitantly, "I was just thinking, and I know it's strange but ... well, we've had so many funny adventures with mice, you don't think *Nutmeg* might be a mouse, do you?"

Nutmeg had told them in a letter that she was a Fairy of Sorts, and the children had always accepted this. But now Lucy was starting to have doubts.

"I suppose it's possible," Arthur said sleepily. "Well, she's

sure to come up here tonight and fetch the letter. Let's take turns to stay awake and watch the chest of drawers!"

"All right," Lucy said. "I'll keep watch first, then I'll wake you up at midnight, and you can take over."

Arthur muttered his agreement and quickly fell asleep. Lucy propped herself up on her pillow, watching the moonlight slip through the curtains. It was very exciting to think she might finally find out who Nutmeg was. But an hour passed, and then another, and her eyes got heavier and heavier. . . . And by the time Nutmeg finally crept upstairs, Lucy was far away, dreaming of mice dressed as clowns, and of fairies with long, nutmeg tails.

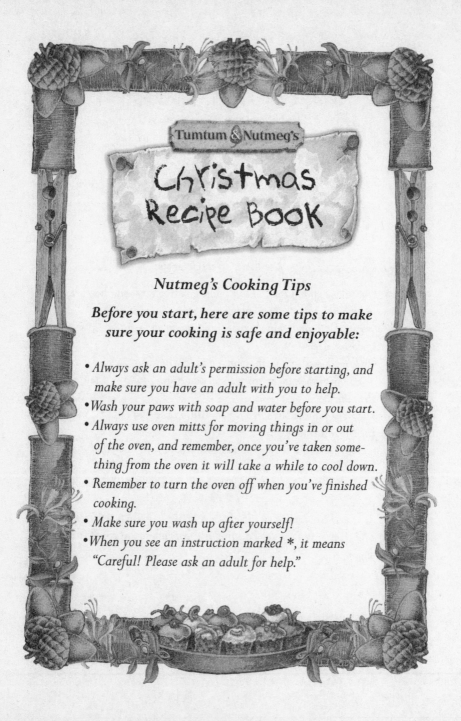

Tumtum & Nutmeg's

Christmas Recipe Book

Nutmeg's Cooking Tips

Before you start, here are some tips to make sure your cooking is safe and enjoyable:

- *Always ask an adult's permission before starting, and make sure you have an adult with you to help.*
- *Wash your paws with soap and water before you start.*
- *Always use oven mitts for moving things in or out of the oven, and remember, once you've taken something from the oven it will take a while to cool down.*
- *Remember to turn the oven off when you've finished cooking.*
- *Make sure you wash up after yourself!*
- *When you see an instruction marked *, it means "Careful! Please ask an adult for help."*

Mini Lemon Pancakes

(Makes roughly 15 pancakes)

Ingredients:

- 1 egg • 1 ⅔ cups of milk • 1 ½ cups of sifted all-purpose flour
- 3 ½ teaspoons of baking powder • 1 tablespoon of sugar
- 3 tablespoons of butter • Juice from a quarter of a lemon
- 2 tablespoons of confectioners' sugar

Cooking Instructions:

1. Place the egg and the milk in a large mixing bowl and beat with a whisk until they form a bubbly mixture.
2. Sift in the flour, baking powder, and sugar, and whisk to form a smooth batter.
3. *Put the butter in a frying pan, and place on the stove on a medium heat. When the butter has melted, add a spoonful of the batter to the pan.
4. Allow the pancake to spread, and wait a couple of minutes until the edges turn golden.
5. Shake the pan to loosen the pancake from the bottom, and then use a spatula to flip it over.
6. Cook another couple of minutes, then turn the pancake out onto a plate.
7. Repeat these steps until all of the pancake mixture is used up, and then serve with lemon juice and a sprinkling of confectioners' sugar.

Bubble and Squeak

Serves 3

Ingredients:

- ½ a cabbage • 1 small onion • 1 cup of leftover chopped vegetables (broccoli, carrots, peas, corn, anything)
- 1 cup of leftover mashed potatoes • 2 tablespoons of butter
- 2 tablespoons of olive oil

Cooking Instructions:

1. *Cut the onion into small pieces and the cabbage into shreds.
2. Heat half of the oil in a small frying pan, then add the cabbage and onion and cook on a medium heat until soft.
3. Mix all the cooked cabbage, leftover vegetables, and mashed potatoes in a bowl, and season with salt and pepper.
4. Add the rest of the olive oil and butter to the pan, and then the vegetable mixture. Pat the mixture down with a spatula to form a large pancake.
5. Cook for about 10 minutes until the underside is golden brown but not burned, then turn over using the spatula.
6. Cook for another 10 minutes until both sides are the same golden brown color.
7. Remove to a plate and cut into slices.

Nutmeg's Tips: The frying pan needs to be small or the bubble and squeak will fall apart when you turn it over. This is also delicious cooked with bacon or ham.

Strawberry Creams
(Makes 24 strawberry creams)

Ingredients:

- ½ cup of confectioners' sugar
- 1 cup of freeze-dried strawberries
(available from most large supermarkets)
- 3 tablespoons of condensed milk
- A few teaspoons of cool water
- Spare confectioners' sugar for dusting

Cooking Instructions:

1. Line a baking tray or large plate with a piece of waxed paper.
2. Put the strawberries in a sealed plastic bag and crush them with a rolling pin until they form a fine red powder, then stir the powder and the confectioners' sugar together in a mixing bowl.
3. Add the condensed milk and mix to form a very firm dough. (If your dough is too dry to come together, add a teaspoon of cool water and continue mixing.)
4. When the dough is ready, dust a board with confectioners' sugar and lay the dough on it. Divide into 2 pieces, and roll each piece into a sausage about 6 inches long.
5. *Cut each sausage into 12 evenly sized pieces, dusting the knife with confectioners' sugar so the dough doesn't stick to it. Roll each slice into a ball and squash the end slightly so that it looks like a strawberry.
6. Place all of the pieces on top of the waxed paper or baking parchment and leave them there to dry for several hours.

Nutmeg's Tip: Freeze-dried strawberries are eaten by astronauts on space flights, and they are the best thing to use for this recipe.

Picnic Pork Pie
Serves 4

Ingredients:

- 1 sheet of frozen or chilled puff pastry
- 16 cooked cocktail sausages
- 1 tablespoon of onion chutney or ketchup
- 1 egg, beaten • Sprinkle of all-purpose flour

Cooking Instructions:

1. *Preheat the oven to 400° F.
2. Thaw the sheet of puff pastry for 30 minutes and then lay it out on a cutting board dusted with flour.
3. Roll the pastry out flat on the board with a rolling pin and cut 4 squares about 5 inches long and 5 inches wide.
4. Spread the onion chutney or ketchup evenly over each square, leaving a ¼ inch gap all around the edge.
5. Stick four cocktail sausages in the middle of each square, and then fold the edges of the square around the sausages and squish the sides together to seal the pie.
6. Brush the beaten egg over the pastry and place on a greased baking sheet. Bake for 20 minutes until golden brown.

Jellied Fruits

Serves 4

Ingredients:

- 1 package of raspberry gelatin
- 1 small basket of blueberries
- 1 small basket of blackberries
- 1 cup of boiling water • ¼ cup of cold water

Cooking Instructions:

1. *Put the gelatin and the boiling water in a heat-proof glass bowl, and stir until all the crystals have disappeared.
2. Let the mixture cool for 30 minutes.
3. Pour the berries into the mixture, then stir in the cold water.
4. Cover the bowl and then return to the refrigerator to set for 24 hours.
5. Serve with cream and mint leaves.

Sugared ants
(Makes a large jarful)

Ingredients:
- 4 ounces of raisins
- ¼ cup of sugar
- ¼ cup of ground almonds
- ⅓ cup of melted butter

Cooking Instructions:

1. Stir the sugar and the ground almonds together in a bowl.
2. *Melt the butter in a pan, and then pour the contents into a separate bowl and stir in the raisins.
3. Use a skimming ladle to transfer the buttery raisins to the sugar and almond mixture, and roll them around until they are fully coated.
4. Spread the sugared ants out spoonful by spoonful on a baking tray lined with foil. Leave them to crystallize overnight, then store in a large jam jar.

Fireside Flapjacks
Makes 12

Ingredients:
- 4 cups of rolled oats
- ¼ cup of sugar
- ⅔ cup of butter
- 4 tablespoons of corn syrup

Cooking Instructions:

1. *Preheat the oven to 375° F and grease a shallow, square baking pan with butter.
2. *Melt the butter, sugar, and syrup in a saucepan on a low heat, then remove from the heat and stir in the oats with a wooden spoon.
3. Pour the mixture into the pan and pat down gently to make sure it is even.
4. Put the pan in the oven for about 20 minutes, until the corners are brown but the middle is soft.
5. Leave to cool for 30 minutes, then cut into squares.